Communications in Computer and Information Science 1970

Rationale

The CCIS series is devoted to the publication of proceedings of computer science conferences. Its aim is to efficiently disseminate original research results in informatics in printed and electronic form. While the focus is on publication of peer-reviewed full papers presenting mature work, inclusion of reviewed short papers reporting on work in progress is welcome, too. Besides globally relevant meetings with internationally representative program committees guaranteeing a strict peer-reviewing and paper selection process, conferences run by societies or of high regional or national relevance are also considered for publication.

Topics

The topical scope of CCIS spans the entire spectrum of informatics ranging from foundational topics in the theory of computing to information and communications science and technology and a broad variety of interdisciplinary application fields.

Information for Volume Editors and Authors

Publication in CCIS is free of charge. No royalties are paid, however, we offer registered conference participants temporary free access to the online version of the conference proceedings on SpringerLink (http://link.springer.com) by means of an http referrer from the conference website and/or a number of complimentary printed copies, as specified in the official acceptance email of the event.

CCIS proceedings can be published in time for distribution at conferences or as post-proceedings, and delivered in the form of printed books and/or electronically as USBs and/or e-content licenses for accessing proceedings at SpringerLink. Furthermore, CCIS proceedings are included in the CCIS electronic book series hosted in the SpringerLink digital library at http://link.springer.com/bookseries/7899. Conferences publishing in CCIS are allowed to use Online Conference Service (OCS) for managing the whole proceedings lifecycle (from submission and reviewing to preparing for publication) free of charge.

Publication process

The language of publication is exclusively English. Authors publishing in CCIS have to sign the Springer CCIS copyright transfer form, however, they are free to use their material published in CCIS for substantially changed, more elaborate subsequent publications elsewhere. For the preparation of the camera-ready papers/files, authors have to strictly adhere to the Springer CCIS Authors' Instructions and are strongly encouraged to use the CCIS LaTeX style files or templates.

Abstracting/Indexing

CCIS is abstracted/indexed in DBLP, Google Scholar, EI-Compendex, Mathematical Reviews, SCImago, Scopus. CCIS volumes are also submitted for the inclusion in ISI Proceedings.

How to start

To start the evaluation of your proposal for inclusion in the CCIS series, please send an e-mail to ccis@springer.com.

Paulraj Dassan · Sethukarasi Thirumaaran ·
Neelakandan Subramani
Editors

Intelligent Computing, Smart Communication and Network Technologies

First International Conference, ICICSCNT 2023
Chennai, India, May 17–18, 2023
Proceedings

 Springer

Editors
Paulraj Dassan
R.M.K. Engineering College
Chennai, Tamil Nadu, India

Sethukarasi Thirumaaran
R.M.K. Engineering College
Chennai, Tamil Nadu, India

Neelakandan Subramani
R.M.K. Engineering College
Chennai, Tamil Nadu, India

ISSN 1865-0929 ISSN 1865-0937 (electronic)
Communications in Computer and Information Science
ISBN 978-3-031-75956-7 ISBN 978-3-031-75957-4 (eBook)
https://doi.org/10.1007/978-3-031-75957-4

Preface

It is with great pride and satisfaction that we present the proceedings of the International Conference on Intelligent Computing, Smart Communication, and Network Technologies (ICICSCNT 2023), successfully conducted by R.M.K. Engineering College on May 17–18, 2023.

This conference brought together scholars, researchers, and practitioners from around the globe to discuss and share their insights on the latest advancements in intelligent computing, smart communication, and network technologies. We received an overwhelming response, with 782 paper submissions covering a wide range of relevant topics. We conducted 3 reviews per paper and after a rigorous double-blinded peer-review process only 36 high-quality papers were selected for publication in these prestigious Springer CCIS proceedings.

We extend our sincere gratitude to all the contributors, reviewers, and participants who made ICICSCNT 2023 a remarkable success. Their dedication and hard work have significantly contributed to the advancement of knowledge in these cutting-edge fields

August 2024

Paulraj Dassan
Sethukarasi Thirumaaran
Neelakandan Subramani

Organization

Chief Patrons

R. S. Munirathinam (Founder & Chairman)
Manjula Munirathinam (Chairperson)

Patrons

R. Jothi Naidu (Director)
R.M. Kishore (Vice-Chairman)
Yalamanchi Pradeep (Secretary)
Durgadevi Pradeep (Vice-Chairperson)
Sowmya Kishore (Management Trustee)

Advisory Committee

M. S. Palanichamy (Advisor)
T. Pitchandi I.A.S (Retd.) (Advisor)
V. Manoharan (Advisor)
Elwin Chandra Monie (Dean – Research)
K. K. Sivagnana Prabhu (Dean – CDC)
S. Pavai Madheshwari (Academic Coordinator)

Convener

K. A. Mohamed Junaid (Principal)

Program Committee Chairs

D. Paulraj   R.M.K. Engineering College, India
T. Sethukarasi   R.M.K. Engineering College, India
S. Neelakandan   R.M.K. Engineering College, India

Organizing Committee

T. Suresh	R.M.K. Engineering College, India
Sandra Johnson	R.M.K. Engineering College, India
S. Thanga Ramya	R.M.K. Engineering College, India
M. Sheerin Banu	R.M.K. Engineering College, India
K. Chidambarathanu	R.M.K. Engineering College, India

Program Committee

A. Thilagavathy	R.M.K. Engineering College, India
B. Jaison	R.M.K. Engineering College, India
C. Geetha	R.M.K. Engineering College, India
K. Saravanan	R.M.K. Engineering College, India
N.R. Gladiss Merlin	R.M.K. Engineering College, India
P. Kavitha	R.M.K. Engineering College, India
P. Umaeswari	R.M.K. Engineering College, India
R.S. Ganesh	R.M.K. Engineering College, India
S. Selvi	R.M.K. Engineering College, India

International Advisory Committee

Ahmad Asari Sulaiman	Universiti Teknologi MARA, Malaysia
Ashish Seth	INHA University, South Korea
Celestine Iwendi	Bangor College China, China
Daniel Chandran	University of Technology Sydney, Australia
Enumi Choi	Kookmin University, South Korea
Hui Xiong	University of New Jersey, USA
Krishnadas Nanath	Middlesex University Dubai, UAE
Md Masoom Rabbani	KU Leuven, Belgium
Mohammad T. Khasawneh	SUNY Binghampton, USA
Naresh Chand	IEEE Photonics Society, USA
San Murugesan	Western Sydney University, Australia
Sanjay Kumar Madria	Missouri University of Science and Technology, USA
Ziming Zhao	University at Buffalo, New York

National Advisory Committee

A. Kannan, Senior	VIT University Vellore, India
D. Sriram Kumar	NIT, Tiruchirappalli, India
Hrishikesh Venkataraman	IIIT, Chittoor, India
J. Klutto Milleth	IIT, Chennai, India
M. D. Selvaraj	IIITDM Kancheepuram, India
N. Venkateswaran	SSN College of Engineering, India
R. Dhanalakshmi	IIIT, Tiruchirappalli, India
Rahul Pandya	IIT, Dharwad, India
S. Raghavan	NIT, Tiruchirappalli, India
Srinivas Talabattula	IISC, India
Surajith Debnath	Women's Polytechnic, India
S. Swamynathan	Anna University, India

Contents

Application of Neural Nets as Computing Network in Augmenting Variables of Abrasive Machining in Polymer Composites

B. Ashwin[1(✉)], K. Gopalakrishnan[1], M. Balasubramanian[2], and S. Madhu[3]

[1] Department of Computer Science and Engineering, Loyola ICAM College of Engineering and Technology, Chennai, India
ashwinblaze111@gmail.com
[2] Department of Mechanical Engineering, R.M.K.College of Engineering and Technology, Thiruvallur, Chennai, India
[3] Department of Autotronics, Saveetha School of Engineering, Chennai, India

Abstract. Advanced industrial sectors such as those for cars, planes, and medical devices have been calling for a metal substitute like fiber-reinforced composites. However, when traditional machining techniques are applied, these composites are severely delaminated. Therefore, it is crucial to research the materials' machining properties when they are subjected to abrasive jet machining. In order to examine the impact of process parameters on the rate of material removal and surface roughness in carbon fibre composite, they were subjected to abrasive jet machining with a new nozzle in this research. The settings were optimised using a neural network. The composites were designed and constructed with internally threaded nozzles. Given that it increases the particle velocities and creates a whirling effect for the abrasive air mixture while cutting any material, the newly created internal threaded nozzle has been demonstrated to give greater possibilities for AJM processing. ANN was successfully used to determine the ideal machining parameters.

Keywords: Machining · roughness · composites · ANN

1 Introduction

The process of machining components made up of brittle materials such as glass, and ceramics, and ductile materials like polymers and metals are essential in optical, biomedical equipment, microfluidic chips, flat panel displays, inertial sensors, aerospace, and auto component manufacturing. An abrasive jet machine is suitable for any secondary machining process like external deburring, edge trimming, and cleaning.

Evaluation of epoxy glass fiber composites was done. Silicon carbide abrasive particles having $60\ \mu$m were blasted through three different diameters (3, 4, 5 mm) tungsten carbide nozzles. The abrasive jet machining process parameters considered were pressure (4, 6, 8 kg/cm^2) and nozzle tip distance (6, 8, 10 mm). Woven laminated glass fiber-reinforced polymer composites with two different thicknesses (8, and 16 mm) were fabricated [1]. Five-factor 2 levels of full factorial design were developed for each

P. Dassan et al. (Eds.): ICICSCNT 2023, CCIS 1970, pp. 1–13, 2024.
https://doi.org/10.1007/978-3-031-75957-4_1

group. The process parameters for the first group were: nominal hole diameter: 6, 8 mm; material thickness: 8, 16 mm; pressure: 150 MPa; SOD: 2 mm. The process parameters for the second group include nominal hole diameter: 6, 8 mm; material thickness: 8, 16 mm; pressure: 150, 200 MPa; SOD: 2, 3 mm [2].

Evaluation of abrasive water jet machining of glass fiber reinforced epoxy laminate composites was done. Bisphenol epoxy resins were used as a matrix while bidirectional glass fabric was used as reinforcement. Experiments were conducted using Taguchi's fractional factorial orthogonal array. The effect of pressure and stand-off distance was investigated. In earlier work, glass fiber-reinforced polymer composites were machined by an abrasive water suspension jet. 2 mm diameter stainless steel nozzle was used for this investigation. The process parameters were: abrasive size: 185,125,105 microns; stand-off distance: 1, 3, 5 mm; abrasive concentration: 2, 3, and 4 (wt. %); feed rate: 100, 125, 150 mm/min [3].

Experiments on CFRP workpieces were conducted. The abrasive particles employed were made of Al2O3. Pressure, SOD, and other process parameters were as follows: 2, 4, 6, and 0.6, 0.8, mm. In this work, a removal rate of 10.417 mm3/min was attained. Abrasive jet drilling was done to assess the variables influencing the rate of metal removal. The workpiece was drilled using Sic abrasive particles [4]. Process parameters included pressure of 4, 6, 8 kg/cm^2, an abrasive flow rate of 3, 4, 5 gm/min, stand-off distance of 6, 8, 10 mm, and nozzle diameter of 2, 3, 4 for this experiment. This work's greatest material removal rate was 0.0657 g/sec [5].

CFRP and Titanium were layered, and abrasive water jet machining was used to investigate. The kerf taper angle is substantially influenced by the water jet pressure and stack arrangements. When the particles struck the top portion with a lot of energy with Titanium on top, the kerf angle was minimal. According to this research, the taper angle lowers as pressure is raised. The jet energy emerging from the nozzle tip will grow with the maximum pressure. Whereas the taper angle of the CFRP is often negative, that of the Ti6Al4V is generally positive. Ti-6Al-4V alloy is inferior to CFRP in terms of machinability index. Thus, the interface region [6] has the highest kerf angle.

1.1 Artificial Neural Network

Neural networks used in artificial intelligence have traditionally been viewed as simplified models of neural processing in the human brain [7, 8]. The human brain, according to most experts, is a form of computer. The first attempts to mimic biological systems' information processing, which frequently relies on parallel processing and implicit instructions based on the recognition of patterns of "sensory" input from outside sources, gave rise to neural networks.

Since a predictive model is crucial for a thorough understanding of the machining process, cost estimation, plant design, and overcoming the problems presented by the trials, recent studies have concentrated on the creation of models for the prediction of machining performance. Various Intelligence (AI) based models were developed to optimize the same. Simple and multiple regression techniques have been mostly used for modeling purposes among many predictive models in several geosciences sectors as well. This is due to the fact that these techniques can be used to determine whether a relationship between a dependent predictive variable and an independent criterion variable is linear or nonlinear when the research problem involves a single dependent variable that is related to two or more independent variables. Nonetheless, the authors are now beginning to pay attention to AI-based models in the fields of both traditional and unconventional machining technologies. Numerous geoscientists have favored the ANN technique over other AI-based models for simulating the behaviour of fiber-reinforced polymer composite materials [9, 10].

The ANN's capacity for learning and generalizing interactions among numerous variables is the primary factor in its performance. In terms of the ability to learn from instances when there is a nonlinear relationship between the dependent and independent variables as well as a multidisciplinary nature that is well-liked by researchers, planners, and designers, it is extremely successful when compared to conventional approaches. It performs admirably in terms of non-linear multivariable problem-solving. Continuous retraining of new data meaning is possible to support the traditional application. The degree of nonlinearity among the variables is not assumed by the artificial neural network.

2 Materials and Methods

Fabricated specimens were cleaned for any unwanted particles from the surface, Fig. 1. Trial experiments were conducted in the laboratory, and from previous literature, the process parameter and its level were agreed upon for investigation [11–15]. Holes were made on the laminates with parameters presented in Table 1. Experiments were conducted as follows.

(i) CFRP is being machined with a new nozzle that has internal threading.
(ii) CFRP sample machining with a new nozzle that has no internal threads.

The abrasive particle with variable grits (50, 70, 90,110,130 microns) was blasted through the nozzles at varied blasting pressure. The investigation to understand the machining characteristics of the polymeric composites were carried out as per the flow chart depicted in Fig. 2.

Fig. 1. Sample for Investigation

An artificial neural network (ANN) was developed for predicting the optimum conditions at which the material removal rate is high and surface roughness (Ra) is low in the abrasive jet process using threaded (TN) and unthreaded nozzle (UTN) in the machining of carbon fiber and glass fiber reinforced polymer composites. In this development, the machining parameters of abrasive jet machining were optimized using a feed-forward neural network.

To investigate artificial neural nets (ANN) technology in the novel application field of design optimization, a methodology utilising back-propagation neural network models was created [16]. In this method, the boundaries between the feasible and the infeasible design regions are determined using pattern classification with a back-propagation network, the neural network application with the highest shown power. Up until convergence requirements are met, iteration is continued. This study demonstrates how artificial neural networks can provide solutions that are both good and close to ideal from a broad view of the entire design space. In fact, ANN has the potential to be a tool for improving the design. Optimization is accomplished using backpropagation. Figure 3 shows the artificial neural network controller used for this investigation.

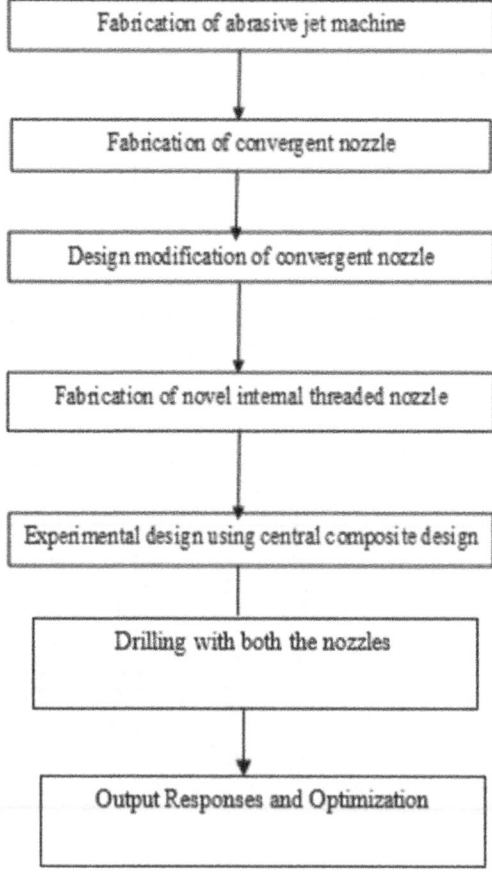

Fig. 2. Methodology

To investigate artificial neural nets (ANN) technology in the novel application field of design optimization, a methodology utilising back-propagation neural network models was created [16]. In this method, the boundaries between the feasible and the infeasible design regions are determined using pattern classification with a back-propagation network, the neural network application with the highest shown power. Up until convergence requirements are met, iteration is continued. This study demonstrates how artificial neural networks can provide solutions that are both good and close to ideal from a broad view of the entire design space. In fact, ANN has the potential to be a tool for improving the design. Optimization is accomplished using backpropagation. Figure 3 shows the artificial neural network controller used for this investigation.

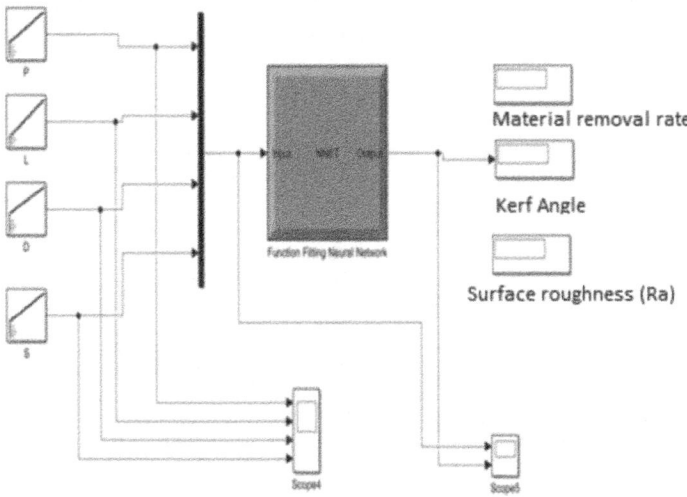

Fig. 3. ANN Controller for optimizing the output parameters

Table 1. Levels of Parameters

S. No	Factor	Notation	Unit	Levels				
				(−2)	(−1)	(0)	(+1)	(+2)
1	Abrasive Pressure	P	MPa	0.20	0.30	0.40	0.50	0.60
2	SOD	L	mm	0.50	1.0	1.50	2.0	2.50
3	Diameter of Nozzle	D	mm	1.50	2.0	2.50	3.0	3.50
4	Size of abrasive	S	microns	50	70	90	110	130

31 tests were carried out with the process parameters presented in Table 1. The metal removal rate ranged between 0.019 and 1.422 g/s obtained for the threaded nozzle and the unthreaded nozzle it was between 0.003 and 0.1 g/s. Surface roughness ranged between 0.663 and 1.282 μm for the unthreaded nozzle and 0.315 and 1.028 μm for the threaded nozzle.

3   Results

3.1   Effect of Process Parameters

3.1.1   Pressure and SOD on MRR

According to the investigation, a UTN operating at a pressure of 0.2 MPa, which is minimum, achieved a material removal rate of 0.006 g/sec. The MRR measured at 600 kPa was 0.1 g/sec. The threaded nozzle was 0.113 g/sec at 200 kPa and 1.422 g/sec at 600 kPa. For unthreaded threaded nozzle in the current experiment was 0.066 g/sec at a minimum stand-off distance of 0.5 mm. 0.105 g/sec was removed when the SOD was high (2.5mm) was 0.105 g/sec. 0.016 g/sec was obtained at 0.5 mm and 0.026 g/sec at 2.5 mm for the nozzle without thread, Fig. 4.

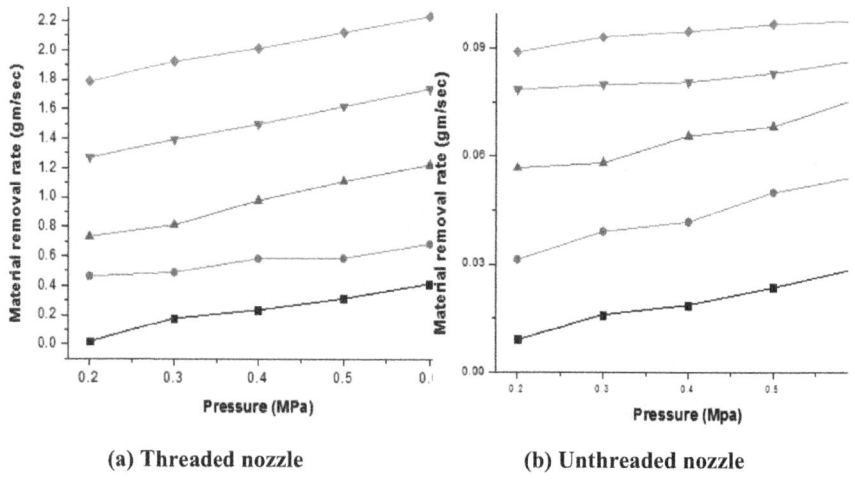

(a) Threaded nozzle   (b) Unthreaded nozzle

Fig. 4. P and SOD on MRR

3.1.2   Pressure vs Nozzle Diameter on MRR

The volume of material removed using the UTN of size 3.5 mm was 0.027g/sec and for the TN, 0.073 g/sec. Using the UTN of size 1.5 mm, 0.009g/sec was obtained and for the TN, it was 0.044 g/sec. Figure 5 illustrates the effect of P and nozzle size on the volume of material removal using a TN.

The swirling abrasive particle coming out of the threaded nozzle increases the MRR as compared to the unthreaded nozzle. The volume of material removal was found to be high in both the nozzles of higher size.

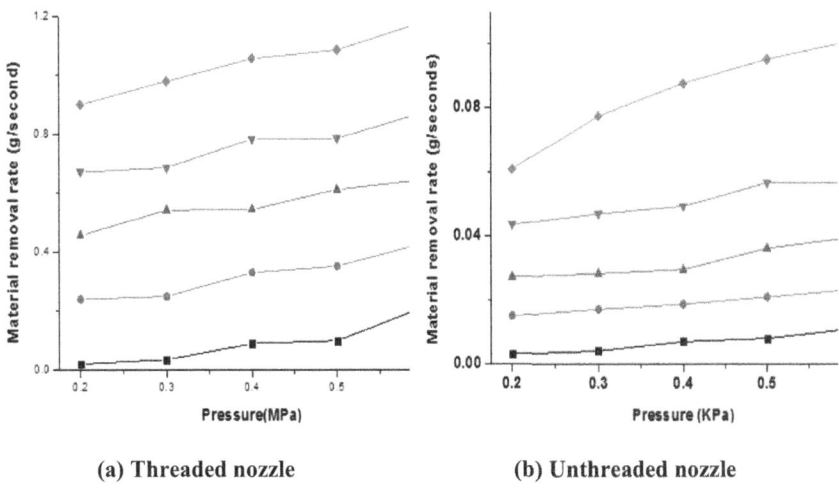

(a) Threaded nozzle   (b) Unthreaded nozzle

Fig. 5. Pressure Vs Nozzle diameter on MRR

3.1.3 Result of Abrasive Proportions on MRR

The MRR for a 50 mm particle supplied via a UTN was 0.01 g/sec, whereas the MRR for a threaded nozzle was 0.080 g/sec. The MRR obtained with the 130 micron abrasive particle in the unthreaded nozzle was 0.024 g/sec, in the threaded one it was 0.466 g/sec. The volume of material removal increased with maximum abrasive particle size. Yet with a newly created nozzle with a thread, the rate of removal was high. Threading amplified the depth of craters on the workpiece's surface as well as the push of silicon carbide abrasive particles onto the workpiece. As a result, CFRP composites achieved greater degradation and MRR. The MRR of 1.422 g/sec could be attained in the current experiment employing a unique internal threaded nozzle, which is much better than what was observed in a few prior studies.

3.1.4 Pressure and SOD on Surface Roughness

The internally threaded nozzle's roughness measured 0.315 m at the pressure's maximum (0.6 MPa). Yet, the interior thread-free nozzle generated a roughness value of 0.663 m. Surface roughness measured with an internally threaded nozzle measured 0.669 m and without an internally threaded nozzle measured 1.282 m at the lowest abrasive jet pressure (0.2 MPa). The surface roughness measured was 0.469 m when the internally threaded nozzle tip was closest to the workpiece. When compared to the 0.703 m roughness measurement from the nozzle without a thread, this value was low. With a threaded nozzle and a higher SOD, 0.618 m of roughness was measured and 0.732 μm without thread.

The impact of P and L during machining on surface roughness is described in Fig. 6(a & b). The graphic illustrates maximum pressure and minimal SOD lowering the surface roughness in both nozzles. The abrasive particles were reduced in size as the pressure of the abrasive jet grew. Also, the particles spin around, and as their

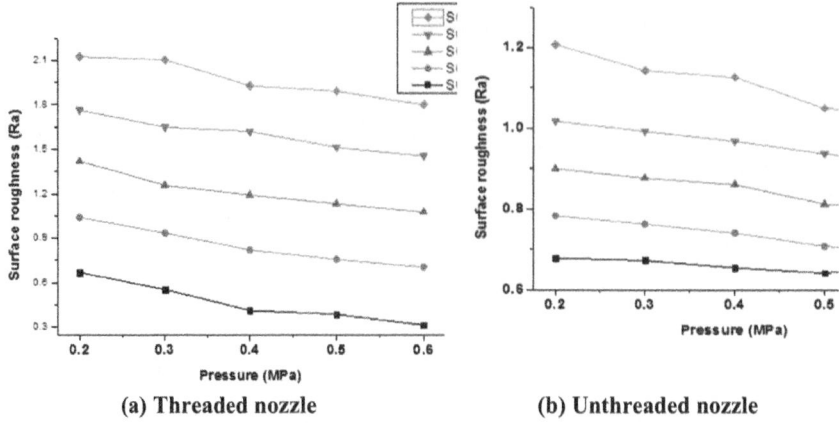

(a) Threaded nozzle   (b) Unthreaded nozzle

Fig. 6. Pressure and SOD on Roughness

kinetic energy rose, the smoothness of the workpiece surface increases. Minimum SOD decreased the roughness.

3.1.5 Pressure vs Nozzle Diameter on Surface Roughness

Surface roughness measurements from internally threaded nozzles of 1.5 mm and 3.5 mm diameter were 0.622 m and 0.722 m, respectively. Surface roughness measured with a UTN using a minimum diameter of 1.5 mm and a maximum diameter of 3.5 mm was 0.727 m and 0.746 m, respectively (Fig. 7). (a & b).

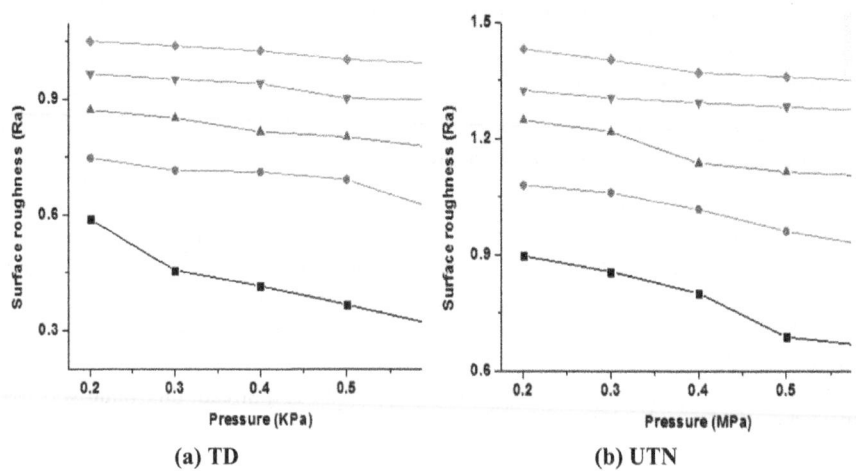

(a) TD   (b) UTN

Fig. 7. P and D on surface roughness (Ra)

3.1.6 P and Abrasive Size on Roughness

The surface roughness of the 130 micron particle blasted from the internally threaded nozzle was 0.509 μm. In comparison to an unthreaded nozzle, which produces 0.726 microns, an abrasive particle of 50 microns produced a surface roughness of 0.372 m. The 130 micron-sized particles flowing through the unthreaded nozzle yielded 0.852 microns (Fig. 8). (a & b).

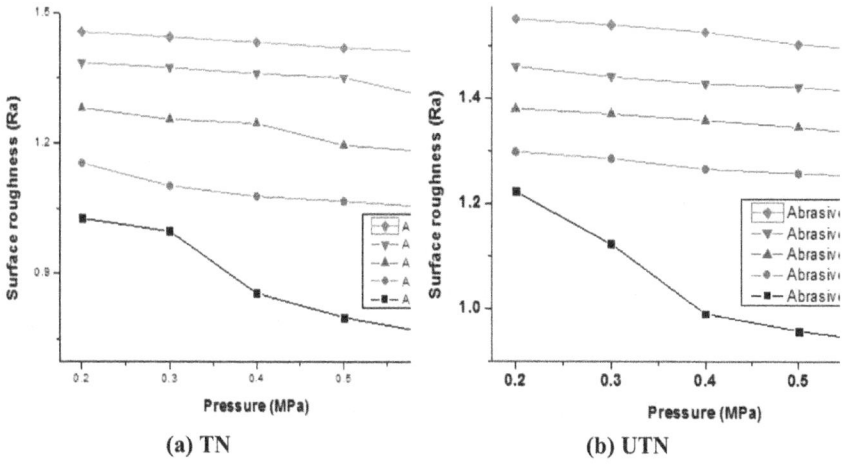

(a) TN   (b) UTN

Fig. 8. P and S on Roughness (Ra)

4 Discussions

The rate of material removal galloped with a rise in pressure for both threaded and unthreaded nozzles. The MRR decreased as the pressure decreased. Comparing an unthreaded nozzle, the volume of material removed in a threaded nozzle was the highest. The particle's kinetic energy rose as the pressure increased, speeding up the removal of material. The particles may lose their kinetic energy when air and abrasive particles flow through a nozzle with a greater diameter. Particles flow in a swirling motion because the nozzle has an internal thread. It aids in the particles' high kinetic energy acquisition. The flow rate is increased by further decreasing the nozzle diameter. Thus smaller diameter TD offers a superior surface finish. The large diameter nozzle's internal thread created a whirling motion in the particle flow, which decreased the surface roughness. In comparison to the UTN, the roughness value attained in the threaded nozzle was the lowest.

The flow of the mass rose with a rise in nozzle diameter, which also increased the rate of material removal. TN resulted in a higher mass flow rate combined with a swirling motion that offers a reasonably high material removal rate (MRR).

The acceleration of particles towards the surface and causing deeper craters may be the cause of the rise in MRR with the use of a unique TN. This causes the work product to erode more quickly, which results in more material being removed. The diameter of the jet also rose as the stand-off distance did. As a result, the jet's kinetic energy was reduced before it reached the work surface. As a result, surface roughness increased as the stand-off distance rose.

Surface roughness in both nozzles was minimised by the abrasive particle with the largest possible mesh size. The threaded nozzle's internal rotation of the abrasive particles reduced the surface roughness. The particle swirl velocity was also accelerated by the internally threaded nozzle. Glass fibre composites have a superior surface quality as a result. As a result, when compared to an unthreaded nozzle, the threaded nozzle's surface roughness was reduced by the abrasive particle size with a big mesh.

5   Optimization of Process Parameters

5.1   Prediction of Optimum Process Parameters to Maximize MRR

Figure 9 indicates the validation of performed experiments and values obtained from ANN. The experimental result showed that 1.422 g/sec was obtained from the internal thread nozzle. Hence at 2.025 mm D, 1.625 mm L, 81 microns S, and 0.505 MPa P, a maximum MRR of 1.532 g/sec could be obtained. In this MRR prediction model, a 0.065 error was obtained. Confirmation experiments were performed with a nozzle diameter of 2 mm SOD of 1.6 mm, abrasive size of 80 microns, and pressure of 5 MPa. The MRR obtained was closer to the value predicted by ANN.

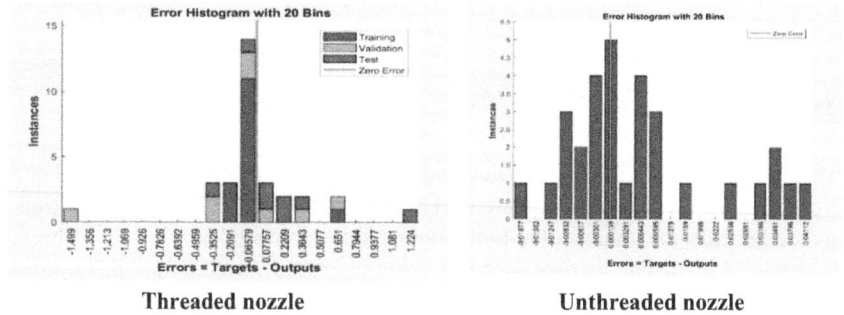

Threaded nozzle   Unthreaded nozzle

Fig. 9. Validation of experimental and predicted MRR

The experimental result showed that the machining maximum from the unthread nozzle was 0.1 g/sec. The ANN results showed that the predicted material removal rate was 0.098 g/sec at a P of 0.565 MPa, L: 0.74 mm, D: 1.63 mm, and S: 63 microns. The error value obtained is 0.0001. Figure 5 indicates the validation of experimental and predicted values of material removal rate in the unthreaded nozzle. The predicted result was closer to the experimental results. Table 2 explains the predicted and experimental values of the material removal rate.

Table 2. Prediction of optimum parameter for maximum MRR

	Threaded		Unthreaded	
	Predicted	Experimental	Predicted	Experimental
P	0.505	0.6	0.565	0.6
L	1.625	1.5	0.74	0.5
D	2.025	2.5	1.63	1.5
S	81	90	63	50
Maximum MRR (g/sec)	1.532	1.422	0.098	0.1

5.2 Prediction of Optimum Process Parameters to Minimize Surface Roughness

The experimental result showed that the minimum surface roughness obtained from the internal thread nozzle was 0.315 μm. The ANN predicted surface roughness was 0.339 μm for the optimized process parameters of P: 0.37 MPa, L: 1.35 mm, D: 2.35 mm, S: 84 microns. Figure 10 indicates the error of surface roughness in both threaded and unthreaded nozzles with a surface roughness prediction model of -0.065 error.

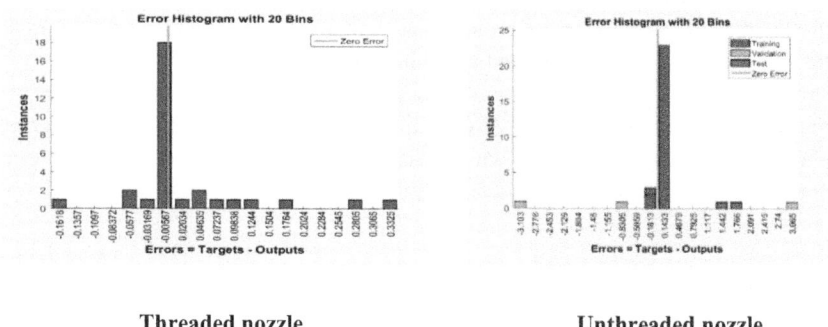

Threaded nozzle   **Unthreaded nozzle**

Fig. 10. Validation of experimental and predicted Surface Roughness

The minimum surface roughness obtained from the unthreaded nozzle was 0.663 μm. The ANN predicted surface roughness was 0.489 μm for the optimized process parameters of P: 0.445 MPa, L: 1.725 mm, D: 2.725 mm, S: 99 microns. The error value obtained is 0.1433. Table 3 explains the predicted and experimental values of surface roughness in the machining carbon fiber-reinforced polymer composites using threaded and unthreaded nozzles.

Table 3. Prediction of optimum parameter for minimizing the surface roughness

	Threaded		Unthreaded	
	Predicted	Experimental	Predicted	Experimental
P	0.37	0.4	0.445	0.4
L	1.35	1.5	1.725	1.5
D	2.35	2.5	2.725	2.5
S	54	50	99	130
Minimum surface Roughness (μm)	0.3339	0.315	0.6897	0.663

6 Conclusions

Internally threaded nozzles were utilized in machining the CFRP composites using abrasive jet machining. Neural networks were utilized in identifying the critical level of process parameters. Based on the observations made the following conclusions are drawn:

a) Highest machining is identified as the highest (85.3 g/min) at a pressure of 4 bar, stand-off distance of 1.5 mm, nozzle diameter of 2.5 mm, and abrasive particle size of 90 with the use of a novel internally threaded nozzle, which is 10 times more than the material removal rate obtained from the conventional nozzle.
b) With 0.505 MPa, SOD of 1.625 mm, nozzle size of 2.025 mm, and particle size of 81 microns with a threaded nozzle, ANN estimated a maximum MRR of 1.535 g/sec.
c) With a threaded nozzle at 0.4 MPa pressure, 1.5 mm gap, 2.5 mm nozzle size, and 50 micron abrasive size, the surface roughness value of 0.339 m was attained.

References

1. Nageshwar, K.R., Srikanth, D.V., Vijayasree, K.: Optimization of process parameters of abrasive jet machining on epoxy glass fibre composite. Int. J. Sci. Res. Educ. 3(9), 4577–4587 (2015)
2. Hussein, M.A.I., Asif, I., Majid, H.: Numerical optimization of hole making in GFRP composite using abrasive water jet machining process. J. Chin. Inst. Eng. 38(1), 66–76 (2015)
3. Deepak, D., Basavanna, S., Devineni, A., Yagnesh, S.N.: An investigation of abrasive water jet machining on Graphite/Glass/Epoxy composite. Int. J. Manuf. Eng. 627218 (2015)
4. Satyendra, S., Shrivas, S.P., Sailesh, D.: Analysis the machining effect of CFRP material using AJM. J. Harmonized Res. 3(4), 151–155 (2015)
5. Srikanth, D.V., Sreenivasa Rao, M.: Machining of FRP composites by abrasive jet machining optimization using Taguchi. Int. J. Mech. Aerosp. Ind. Mechatron. Manuf. Eng. 8(3), 632–636 (2014)
6. Alberdi, A., Artaza, T., Suarez, A., Rivero, A., Girot, F.: An experimental study on abrasive waterjet cutting of CFRP/Ti6Al4V stacks for drilling operations. Int. J. Adv. Manuf. Technol. 86(1), 691–704 (2016)

7. Ulas, C., Ahmet, H.: A study on surface roughness in abrasive waterjet machining process using artificial neural networks and regression analysis method. J. Mater. Process. Technol. **2**(2), 574–582 (2008)

8. Surendra, K.S.: Response parameters modeling of abrasive jet machined composite using artificial neural network. Mater. Today: Proc. **62**(6), 3860–3864 (2022)

9. Venkatesh, C., et al.: An experimental and empirical assessment of machining damage of hybrid glass-carbon FRP composite during abrasive water jet machining. J. Market. Res. **19**, 1148–1161 (2022)

10. Vikas, G., Aniket, N.: Experimental investigations of abrasive water jet machining on hybrid composites. Mater. Today: Proc. (2022). https://doi.org/10.1016/j.matpr.2022.05.372

11. Sathishkumar, N., Selvam, R., Kumar, K.M., Abishini, A.H., Khaleelur Rahman, T., Mohanaranga, S.: Influence of garnet abrasive in drilling of Basalt–Kevlar–Glass fiber reinforced polymer cross ply laminate by Abrasive Water Jet Machining process. Mater. Today: Proc. **62**(2), 1361–1368 (2022)

12. Dipak, K.J., Ramesh, K.N.: Sensitivity analysis of abrasive air-jet machining parameters on machinability of carbon and glass fiber reinforced hybrid composites. Mater. Today Commun. **25**(101624) (2020)

13. Jesthi, D.K., Nayak, R.K., Nanda, B.K., Diptikanta, D.: Assessment of abrasive jet machining of carbon and glass fiber reinforced polymer hybrid composites. Mater. Today: Proc. **18**(7), 3116–3121 (2019)

14. Madhu, S., Balasubramanian, M.: Evaluation of delamination damage in carbon epoxy composites under swirling abrasives made by modified internal threaded nozzle. J. Compos. Mater. **53**(6), 819–833 (2018)

15. Madhu, S., Balasubramanian, M.: Effect of the threaded nozzle on delamination and surface texture of PEEK CF30 composite machined by abrasive jet. World J. Eng. Vol. ahead-of-print No. ahead-of-print (2021). https://doi.org/10.1108/WJE-03-2021-0186

16. Shuo-Jen, L., Henzer, C.: Design optimization with back- propagation neural networks. J. Intell. Manuf. **2**, 293–303 (1991)

Design of Metaheuristically Supervised Linear ADRC for a Magnetic Levitation System Control

Dephney Blossom and Vidya S. Rao$^{(\boxtimes)}$ (iD)

Manipal Institute of Technology, MAHE, Manipal 576104, India
dephney.blossom@learner.manipal.edu, rao.vidya@manipal.edu

Abstract. A magnetic levitation system exhibits features of frictionless motion, low maintenance, and environmental isolation along with characteristics like chaos, uncertainty, multiple disturbances, and is highly nonlinear. These characteristics require a controller that doesn't depend on a precise mathematical model, improves stability, dynamic response, and is robust against disturbances. Hence a linear active disturbance rejection controller (LADRC) is introduced into the system. A metaheuristic algorithm called cuckoo search has been used to fine-tune the parameters of LADRC, offering optimal solution with strong global search capabilities. To analyze the effectiveness of the proposed intelligent controller LADRC using cuckoo search is compared with a traditional PID controller based on their dynamic response and tracking performance. The analysis showed that LADRC with cuckoo search algorithm had a better tracking performance and dynamic response.

Keywords: Linear active disturbance rejection control · cuckoo search · magnetic levitation system

1 Introduction

Magnetic levitation technology has developed rapidly in recent years and it's widely used in many industrial and research fields, such as magnetic trains, magnetic pumps, and magnetic bearings. It is highly nonlinear and exhibits characteristics such as hysteresis, chaos and it is unstable even in it's open loop form. Though intelligent controllers have made progress recently, PID remains one of the widely used controllers in magnetic levitation technology solely because it has fewer tuning parameters compared to advanced controllers and also it's easy for a hardware implementation [1].However, PID controller may not achieve desired performance for highly nonlinear systems. PID has an error feedback control strategy and the derivative action provides one step error prediction. Hence the uncertainties and disturbances of a nonlinear system may not be compensated fast enough. PID also need a precise mathematical model of the plant.

Advanced controller like active disturbance rejection controller (ADRC) addresses the problems such as absence of precise mathematical model, robustness towards process parameter variations. ADRC considers the total disturbance that can be estimated from the output of the system. By estimating the uncertainty, extended state observer rejects

P. Dassan et al. (Eds.): ICICSCNT 2023, CCIS 1970, pp. 14–26, 2024.
https://doi.org/10.1007/978-3-031-75957-4_2

it in the feedback loop. It provides better robustness and stronger adaptive abilities. However ADRC has complex structure with many tuning parameters. To overcome this Gao proposed a linear ADRC and it's tuning method and design in his work [2]. LADRC has two tuning parameters namely observer bandwidth and controller bandwidth. Gernot discussed about the modifications needed in the design of discrete linear ADRC for a faster real time implementation in high dynamic applications [3].

Manually tuning these parameters is tedious and the tuned values may not be optimal. Hence the application of some sort of intelligence may make the tuning process easier. Bingwei Gao proposed an application of gray wolf optimization technique to tune the ADRC parameters optimally in an Electro-Hydraulic servo unit [4]. Amjad in his paper presented a design and implementation of nonlinear extended state observer (NESO) and linear extended state observer (LESO). He used Particle swarm optimization (PSO), a metaheuristic algorithm to obtain the necessary parameters.His study shows that LESO estimates better than NESO for a magnetic levitation system [5]. Amjad further continued in another work about implementing PSO for the tuning parameters of ADRC. He then implemented this on magnetic levitation system [6]. Recent years have seen the implementation of intelligent algorithms to many complex problems. The aim of intelligent control is to build upon and improve conventional control methodologies in order to solve novel, complex control issues.

Hence parameter tuning through a metaheuristic approach can be an appropriate method. Libing discussed about implementing cuckoo search (CS) with levy flights as a searching algorithm in LADRC. CS tuned two parameters, controller bandwidth, and observer bandwidth. And this novel combination of LADRC with searching algorithm was designed for controlling magnetic levitation system. He analysed CS, PSO and adaptive PSO (APSO) and stated that CS with LADRC has better dynamic response [7]. Xin She Yang and Suash Deb came up with the novel metaheuristic algorithm, Cuckoo search via Levy flights [8].He further continued his work with Xing Shi and discussed about CS algorithm and Bat algorithm in their book and provided a pseudo code and a working code in MATLAB [9].

Cuckoo search has better global search capabilities when compared with prominent metaheuristic algorithms like PSO, APSO [7]. This paper aims at building LADRC for a highly nonlinear magnetic levitation system. The parameters of LADRC will be tuned with CS algorithm for further improvement of the controller. The LADRC is modified for a faster real time implementation [3]. The effectiveness of this intelligent controller is compared with traditional controller, PID in Matlab environment for different test cases and evaluated on the basis of performance indicators.

2 Mathematical Modelling

In this study, feedback instruments 33-210 magnetic levitation system have been used. It consists of electromagnetic coil, infrared light sensors, metal ball, analogue and digital interface and controller from computer [10]. When current passes through the electromagnetic coil it produces a necessary magnetic field forcing the metallic ball to experience the lifting force. When an object levitates, some portion of the IR light is blocked. Generation of voltage depends on the amount of light falling on the IR receiver which reveals the position of the object. IR Sensor output will be passed through an analog to digital converter pin. The digital to analogue converter pin receives the calculated control law from the controller. The current in the coil decreases as the object gets closer to the magnet, and vice versa.

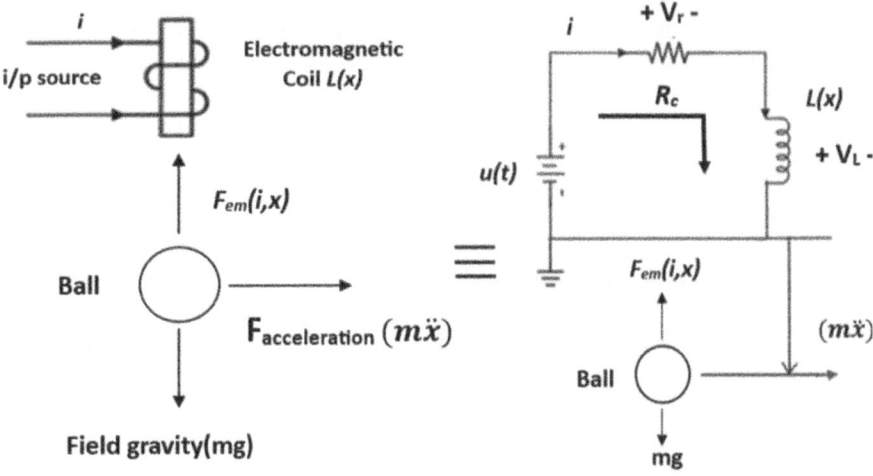

Fig. 1. Net forces acting on the ball.

Figure 1 shows the net forces acting on the ball. In an electromagnetic coil, the relationship between control voltage and current can be described by Kirchoff's law and Faraday's law as follows [11].

$$U_{in} = V_r + V_L = iR + L\frac{di}{dt} \tag{1}$$

R denotes coil's resistance, U_{in} denotes input voltage applied, i denotes current present in the coil, L indicates the self inductance of the electromagnetic coil.

Applying Newton's third law of motion for suspended object neglecting damping force:

$$m\ddot{x} = mg - F_{em}(i, x) \tag{2}$$

$F_{em}(i, x)$ denotes electromagnetic force, g denotes gravitational constant, m denotes mass of the ball, x indicates ball position, $\ddot{x}$ denotes the acceleration of the ball at position x. The ball's acceleration is zero when it is in the equilibrium position and it is expressed as

$$F_{em}(i_0, x_0) + mg = 0 \tag{3}$$

where $F_{em}(i_0, x_0)$ denotes the electromagnetic force at equilibrium point (i_0, x_0).

Force experienced by the levitated object $F_{em}(i,x)$ is a function of ball position, x and the current supplied to the magnet, i. This is given by Ampere's and Faraday's laws.

$$F_{em}(i, x) = K \frac{i^2}{x^2} \qquad (4)$$

Linearize the modelling of maglev by applying Taylor's series expansion at the equilibrium point (i_0, x_0). The $F_{em}(i_0, x_0)$ after the application of Taylor's series

$$F_{em}(i, x) = F_{em}(i_0, x_0) + F_x(i_0, x_0)(x - x_0) + F_i(i_0, x_0)(i - i_0) + F_{hi}(i, x) \qquad (5)$$

where $F_{hi}(i,x)$ denotes all the remaining higher order terms. Define K_i and K_x

$$K_i \triangleq F_i(i_0, x_0)$$
$$= \frac{2Ki_0}{x_0^2} \qquad (6)$$

$$K_x \triangleq F_x(i_0, x_0)$$
$$= -\frac{2Ki_0^2}{x_0^3} \qquad (7)$$

Neglecting the higher order term $F_{hi}(i,x)$ and applying Eq. (3) into Eq. (5), it can be modified as

$$m\ddot{x} = 0 + K_x(x - x_0) + K_i(i - i_0) \qquad (8)$$

Substituting Eq. (6) and (7) into (8)

$$\ddot{x} = \left(\frac{2Ki_0}{mx_0^2} \right) i - \left(\frac{2Ki_0^2}{mx_0^3} \right) x \qquad (9)$$

Solving mass(m) from Eq. (3) and substituting it in Eq. (9) gives

$$\ddot{x} = \left(\frac{-2g}{i_0} \right) i + \left(\frac{2g}{x_0} \right) x \qquad (10)$$

Modern control theory deals systems like

$$\ddot{x}(t) = f(t, x(t), \dot{x}, w(t)) + b(t)u(t) \qquad (11)$$

$$y(t) = C_1 x(t) + sn(t) \qquad (12)$$

where $x(t)$ represents the position output, $b(t)$ denotes the time varying coefficient, $w(t)$ denotes unknown disturbance such as friction, torque disturbances, vibrations and sn represents sensor noise.

Consider the Eq. (9) which is similar to (10). Now let us take $x_1 = U_{out} = position(x)$ and $x_2 = U_{in} = current$ then state space equations are given as

$$\dot{x}_1 = x_2 \tag{13}$$

$$\ddot{x} = \left(\frac{-2g}{i_0}\right) U_{in} + \left(\frac{2g}{x_0}\right) x_1 \tag{14}$$

$$y = x_1 = position \tag{15}$$

Equation (13) can be rewritten as

$$\dot{x}_2 = f(i, x_1) + b_0 U_{in} \tag{16}$$

Hence

$$b_0 = \frac{-2g}{i_0} \tag{17}$$

$$f = \left(\frac{2g}{x_0}\right) x_1 \tag{18}$$

Table 1. Feedback Instruments Maglev 33-210 parameters [10].

Parameter	Symbol	Value
Mass of the object (ball)	m	0.068 kg
Equilibrium value of current	i_0	0.8A
Equilibrium value of position	x_0	−1.5V, 0.009 m
Acceleration due to gravity	g	9.81m/s^2
Sensor gain	k_2	143.48 V/m
Control voltage input level	u	±5 V
Control voltage to coil current gain	k_1	1.05 A/V
offset	η	−2.8 V
Sensor output level	x_v	+1.25 V to −3.75 V
Internal Inuctance	L_0	0.018761 H

Table 1 gives the value of the physical parameters used for the setup. If one converts the Eqs. (13) (14) into a transfer function, the open loop poles are found at the right half side of the s-plane. Hence magnetic levitation is an unstable open loop system which requires a controller to compensate it's instability.

3   Linear Active Disturbance Rejection Controller

The magnetic levitation plant designed is a second order system, hence LADRC uses a second order LESO and linear proportional PD feedback controller. Figure 2 represents the block diagram of LADRC.

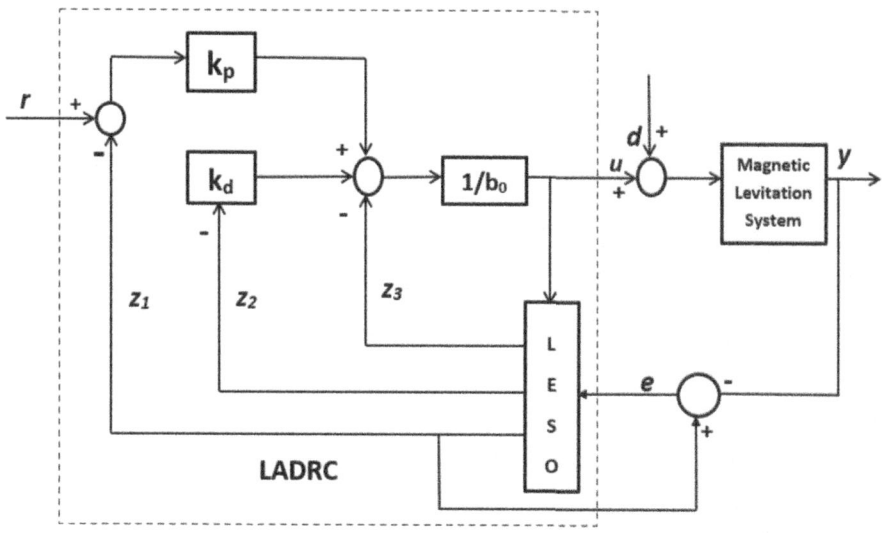

Fig. 2. Block Diagram of LADRC

As the name suggests extended, in LESO the states are indeed extended. Magnetic levitation system designed has two states (13) (14), but LESO states are extended to three x_1, x_2 and x_3 [2]. Here x_3 state represents all the unknowns or disturbances acting on the system which is given by Eq. (18). The design scheme of LESO [12] for easy implementation contains β_1, β_2 and β_3 as the observer gains. The output of LESO should be $z_1 \approx x_1$, $z_2 \approx x_2$ and $z_3 \approx f(t)$.

Solve Eq. (16) for control law, u, obtained from linear PD feedback controller

$$u(t) = \frac{K_p(r - z_1) - K_d z_2 - z_3}{b_0} \tag{19}$$

where K_p, K_d are adjustable parameters. While formulating the PD gains [3] one can tune the closed loop to a desired 2% settling time and critically damped behavior. So closed loop controller bandwidth is the only tuning parameter for linear PD feedback controller. If the observer gains are formulated [3] on the basis of practical approach then observer bandwidth w_o is the only parameter to tune in LESO. So both the tuning parameters of LADRC, w_c and w_o are tuned using cuckoo search algorithm.

4 Cuckoo Search Algorithm

When certain species of cuckoos following brood parasitism lay their eggs in the host bird's nest, the latter can engage in a direct conflict with the intruders or discard the intruder's eggs or simply depart its nest if it realizes the eggs are not their own [8, 9]. Current research shows that CS is considerably more reliable than genetic algorithm and PSO [11].

This algorithm implements L´evy flights obeying the L´evy distribution which acts as a searching pattern followed by the cuckoos to forage for the host nests. The updating of the position or solution in L´evy flights should be in such a way that it converges to the final algorithm. It has a good global search ability because it searches short distances and occasionally walks long distances preventing it from falling into a local optimum. Thus L´evy flight fits cuckoo search [8, 9, 14].

An assumption is made that a population of m cuckoos and m nests represents the cuckoo-host system. An egg's location serves as a solution vector. While designing the CS three idealized rules [8] should be followed. Compare the random number in the range [0, 1] with probability of finding intruder's eggs(P_a). [7] If $r > P_a$, then the host bird's nest position gets randomly modified once else the position remains unchanged.

Performance indicators namely RMSE is used by CS algorithms as a fitness function in this work which should be minimized for the convergence of the algorithm. Once it reaches the convergence, w_c and w_o values are obtained, which are later used in tuning LADRC.

5 Simulation

The simulation of metaheuristically supervised LADRC is carried out in MATLAB environment. This simulation introduces PID and LADRC using CS for comparison and analysis. The controllers are analysed for different tracking signals and a impulse disturbance interference signal.

LADRC needs two tuning parameters w_c and w_o respectively. These optimal value of tuning parameters are obtained from CS which are then used to calculate K_p, K_d, β_1, β_2 and β_3 from Eq. (25) and (26). The sampling time considered here is 0.001 s. The fitness function taken to solve CS algorithm in the design is RMSE. The goal of CS is to minimize RMSE. RMSE is given as

$$RMSE = \sqrt{\frac{1}{T}\sum_{t=0}^{T}(r(t) - y(t))^2} \qquad (20)$$

The population in general will be 25 [9]. Probability of host finding the cuckoo eggs is 0.25 [8, 9]. An initial fitness function value generally considered for any functions are 10^{10}. The scaling parameter λ will be considered as 3/2 [13].

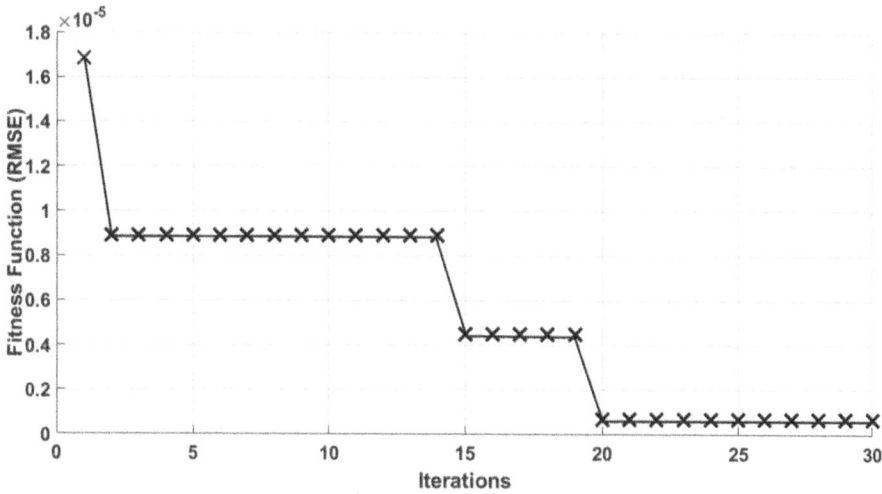

Fig. 3. Individual optimal fitness values obtained from CS

Figure 3 shows the fitness value (RMSE) given for each iterations. It converges to zero quite fast and has higher ability to find the optimal tuning parameters without falling into local minima.

Table 2. Values obtained from CS

Iteration	Fitness function	w_c	w_o
25	0.000017	$-1.225113e + 02$	-236.883086

Table 2 shows the controller and observer bandwidth values obtained from CS. Using these values the tuning parameters of LADRC are calculated according to Eq. (25) and (26) and are tabulated in Table 3.

Table 3. Metaheuristically obtained tuning parameters of LADRC

K_p	K_d	β_1	β_2	β_3
$1.5009e + 04$	245.0226	710.6493	$1.6834e + 05$	$1.3292e + 07$

Figure 4 shows step tracking signal and it can be seen that output response obtained from the LADRC is tracking the given reference with no overshoot effectively. It can also be seen that the impulse disturbance added to the system, $d = 10$ is effectively rejected by LADRC using cuckoo search. But PID controller response and disturbance rejection ability is not good as LADRC. PID exhibits an overshoot of 0.5% whereas LADRC using CS has an overshoot of 0.05%.

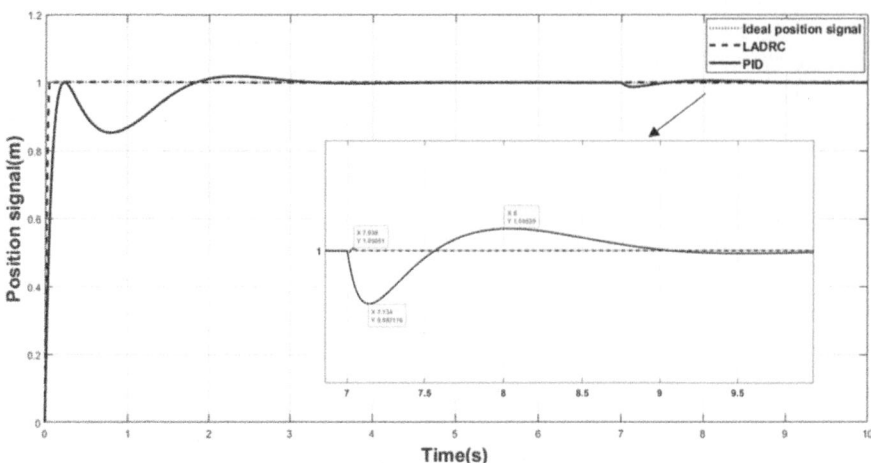

Fig. 4. Step Position tracking

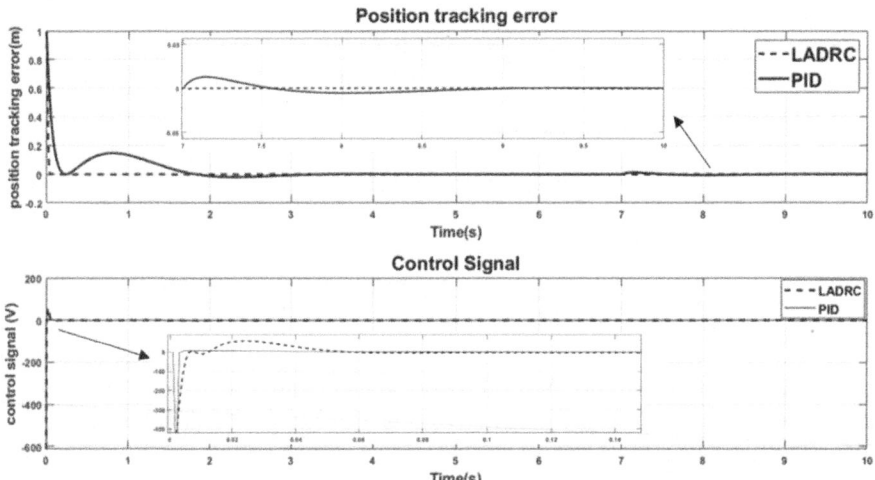

Fig. 5. Step signal tracking error and Control Law

Figure 5 shows the magnitude of position tracking error and the control signal, used to control the magnetic levitation system. It's evident from the Fig. 5 that position tracking error is less in LADRC than PID implying LADRC tracks the position better. The control signal of LADRC is however slightly more than PID due to the intelligence added to the system. Figure 6 compares the output obtained from the LESO that are the estimated states z_1, z_2 and z_3 with the actual state x_1, x_2 and uncertainty term f. As can be seen that LESO estimates the states accurately with very less deviation.

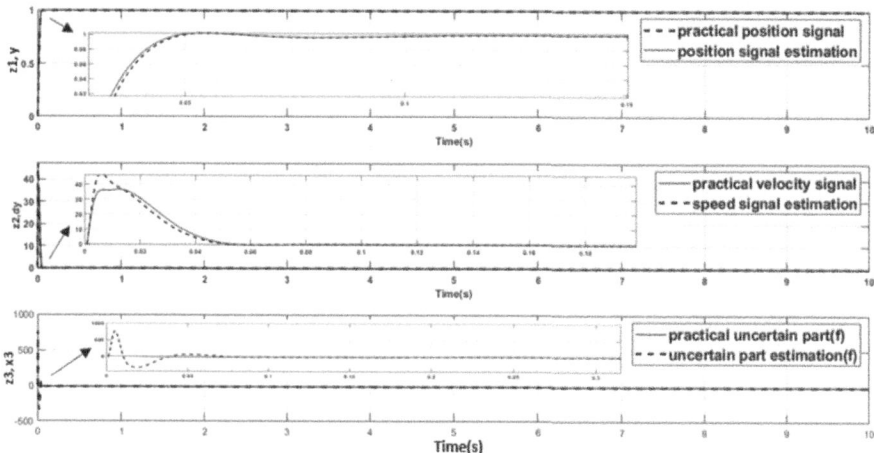

Fig. 6. LESO for Step reference

A sine signal is also passed to evaluate the efficiency and performance of the proposed controller, and the same conclusions are drawn as with a step signal. The responses can be seen in Fig. 7, Fig. 8 and Fig. 9.

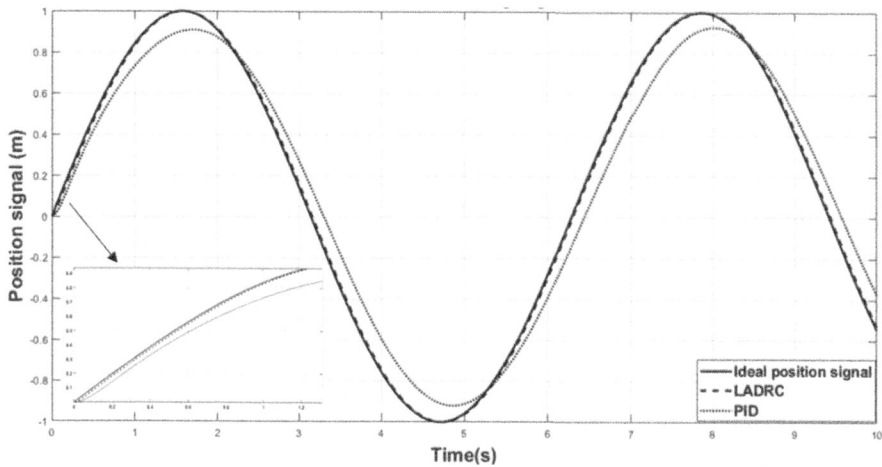

Fig. 7. Sine Position tracking

Prior to comparing the controllers, performance indicators such as RMSE, IAE, ITAE and IAE should be calculated for the controllers under comparison. This is tabulated in the Table 4. Based on the percentage change of improvements mentioned in the Table 4, it can be said that the performance of LADRC using cuckoo search is relatively good than traditional controller PID. It improves the stability of the controller.

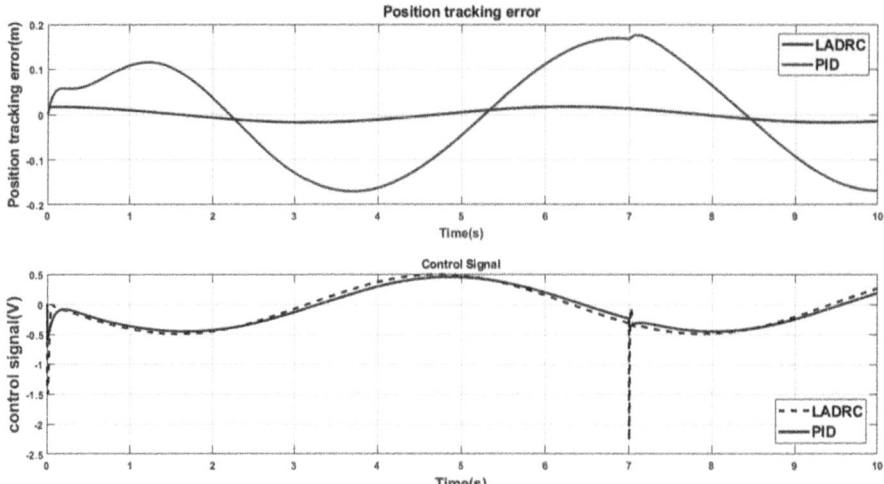

Fig. 8. Sine signal tracking error and Control Law

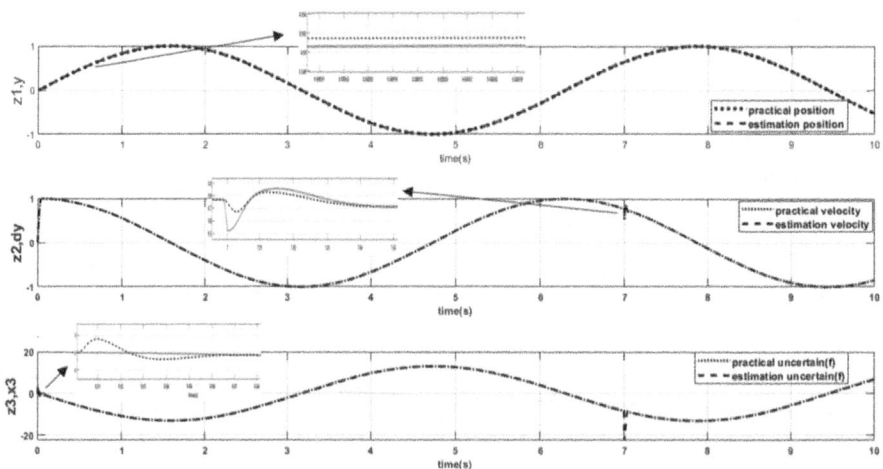

Fig. 9. LESO for Step reference

Table 4. Comparison of the Performance Indicators.

Signal	Controller	ISE	RMSE	ITAE	IAE
Step	PID	0.047547	0.002181	0.245058	0.216239
	LADRC-CS	0.011440	0.001070	0.000421	0.017426
	Improvement	75.9395%	50.9399%	99.8282%	91.9413%
Sine	PID	0.128434	0.003584	5.368337	1.013220
	LADRC-CS	0.001569	0.000396	0.581749	0.113353
	Improvement	98.7783%	88.9508%	89.1633%	88.8125%

6 Conclusion

This paper proposes a metaheuristically supervised LADRC to control the magnetic levitation system. The implementation of metaheuristic algorithm that is CS, eased the tuning process of LADRC and obtained the controller and observer bandwidth. This paper further analyses the tracking performances, dynamic response and disturbance rejection ability of PID and LADRC using CS. Disturbance rejection ability of intelligent controller LADRC using CS is improved 90% more than traditional controller PID and it settles 98% faster than PID. The tests indicates that LADRC has better stability and good performance indicators compared to PID. Hence the proposed technique solves the problem of optimally tuning the parameters of LADRC and improves the efficiency and performance of the magnetic levitation system.

References

1. Tan, W., Fu, C.: Linear active disturbance-rejection control: analysis andtuning via IMC. IEEE Trans. Ind. Electron. **63**(4) (2016). https://doi.org/10.1109/TIE.2015.2505668
2. Yoo, D., Yau, S.S.-T., Gao, Z.: Optimal fast tracking observer bandwidth of linear extended state observer. Int. J. Control **80**(1) (2007). https://doi.org/10.1080/00207170600936555
3. Herbst, G.: A simulative study on active disturbance rejection control (ADRC) as a control tool for practitioners. MDPI Electron. **2**, 246–279 (2013)
4. Gao, B., Guan, H., Shen, W., Ye, Y.: Application of the GrayWolf optimization algorithm in active disturbance rejection control parameter tuning of an electro-hydraulic servo unit. Machines **2022**, 10, 599(2022). https://doi.org/10.3390/machines10080599
5. Humaidi, A.J.: Experimental design and verification of extended state observers for magnetic levitation system based on PSO. Open Electr. Electron. Eng. J. **12** (2018). https://doi.org/10. 2174/1874129001812010110
6. Humaidi, A.J., Badr, H.M., Hameed, A.H.: PSO based active disturbance rejection control for position control of magnetic levitation system. In: International Conference on Control, Decision and Information Technologies (2018). https://doi.org/10.1109/CoDIT.2018. 8394955
7. Wei, L., Fan, K., Zhou, X., Hu, L., Tan, W.: Linear active disturbance rejection control of magnetic levitation system based on cuckoo search. In: IEEE 11th Data Driven Control and Learning Systems Conference (2022). https://doi.org/10.1109/DDCLS55054.2022.9858490

8. Yang, X.-S., Deb, S.: Cuckoo Search via Levy Flights, World Congress on Nature Biologically Inspired Computing (2009). https://doi.org/10.1109/DDCLS55054.2022.9858490

9. Yang, X.-S., He, X.-S.: Bat Algorithm and Cuckoo Search Algorithm, Chapter 2, Nature-Inspired Computation and Swarm Intelligence (2020). https://doi.org/10.1016/B978-0-12-819714-1.00011-7

10. Magnetic Levitation Control Experiments 33-942S (For use with MATLAB R2006bversion 7.3), UK

11. Ahmad, I., Javaid, M.A.: Nonlinear model and control design for magnetic levitation system. In: 9th WSEAS International Conference on SIGNAL PROCESSING, ROBOTICS and AUTOMATION (2010)

12. Guo, W., Zhao, Y., Li, R., Ding, H., Zhang, J.: ActiveDisturbance Rejection Control of Valve-Controlled Cylinder Servo Systems Based on MATLAB-AMESim Cosimulation, Hindawi Complexity Volume (2020), Article ID 9163675. https://doi.org/10.1155/2020/9163675

13. MathWorks Homepage. https://in.mathworks.com/matlabcentral/fileexchange/29809c uckoo-search-cs-algorithm. Accessed 17 Mar 2023

14. Yang, X.-S.: Cuckoo Search, Chapter 9, Nature-Inspired Optimization Algorithms (2014). https://doi.org/10.1016/B978-0-12-416743-8.00009-9

Heart Disease Prediction Using Machine Learning Techniques

Segu Parameswara Reddy$^{(\boxtimes)}$, Chetipattu Vinesh Kumar Reddy, M. Sambath, and J. Thangakumar

Hindustan Institute of Technology and Science, Chennai 603103, India
{19113120,19113108}@student.hindustanuniv.ac.in

Abstract. For the purpose of making wise decisions, the health care sectors gather enormous amounts of data that may contain some hidden information. Some sophisticated data mining techniques are employed for producing acceptable findings and making sensible judgement based on data. In this paper, a Heart Disease Prediction System (HDPS) is created to predict the risk level of heart disease utilizing the Logistic regression and Decision Tree algorithms. The algorithm makes predictions using 15 medical characteristics, including age, sex, blood pressure, cholesterol, and obesity. The HDPS forecasts the probability that people may develop heart disease. It makes possible important knowledge. For instance, it is necessary to identify relationships between patterns and medical factors associated to heart disease. The training approach we used was a by different algorithms in machine learning like Random Forest, KNN, Support Vector Machine, Logistic Regression Decision Tree. The outcomes show that the developed diagnostic system can accurately identify the risk level of cardiac illnesses.

Keywords: HDPS · Support Vector Machine (SVM) · Decision Tree · Logistic Regression · Random Forest · KNN

1 Introduction

Predicting the risk of heart disease is essential for early diagnosis and preventative interventions since heart disease is a major public health concern. The ability of machine learning algorithms to predict cardiac disease has been demonstrated. These methods train prediction models that can recognize people at high risk of developing heart disease. The data used in these methods includes patient demographics, medical history, and lifestyle factors. In comparison to conventional methods, the use of machine learning has been proven to increase the speed and accuracy of cardiac disease prediction. One of the biggest causes of death worldwide is cardiovascular disease.

Heart disease can be predicted and diagnosed early to significantly improve patient outcomes and cut mortality rates. The accurate prediction of cardiac disease utilizing patient data including demographics, medical history, and test results has shown significant potential thanks to machine learning algorithms. In order to estimate a patient's risk of developing heart disease, the approaches can examine patterns and relationships

P. Dassan et al. (Eds.): ICICSCNT 2023, CCIS 1970, pp. 27–35, 2024.
https://doi.org/10.1007/978-3-031-75957-4_3

in this data. In this manner, machine learning has the potential to be a useful tool for medical practitioners in the diagnosis and treatment of cardiac disease.

Given that heart disease is one of the major causes of death in the world, its early detection and prevention are essential for minimizing its effects. By examining a variety of variables, including demographics, lifestyle, medical history, and test results, machine learning is a potent tool that can assist in predicting a person's chance of getting heart disease. Healthcare practitioners can decide on preventative measures and treatment approaches with greater knowledge by employing machine learning techniques.

The goal of this research is to create a machine learning model for heart disease prediction. Various techniques, including decision trees, random forests, support vector machines, and neural networks, will be used to train the model using a sizable datasets of patient data. Then, after optimizing the best-performing model, we will deploy it in a real-world context. We will assess the model's performance and compare it to other models. Our objective is to show how machine learning can effectively diagnose cardiac disease early and enhance patient outcomes.

2 Related Work

Srinivas Polipireddy et al.[1] are investigated goal in this study is to give a thorough overview of the most recent advancements made in the field of machine learning-based cardiac disease prediction. We will look at the numerous methods that have been used to this problem, such as decision trees, random forests, support vector machines, and neural networks. A review of the most current and influential studies in this field will be done, and the advantages and disadvantages of various strategies will be examined. An up-to-date assessment of the state of the field is provided in this survey, along with potential areas for further investigation. We intend to spur additional investigation into the creation of powerful and effective machine learning models for heart disease prediction by comprehending the most current developments in the field. These models could enhance patient outcomes and lessen the effect of cardiovascular disease globally.

Aditi Gavhane et al. [2] To recognize the signs of a heart attack early on and avoid it, a system must be put in place. A method that is practical and trustworthy in determining the likelihood of heart disease must be in place because it is unrealistic for the average person to routinely undertake expensive tests like the ECG. Prediction of Heart Disease Using Machine Learning Determining the vulnerability of a cardiac illness based on basic symptoms like age, sex, pulse rate, etc., is what we propose to do. The suggested system uses the machine learning algorithm neural networks because it has been shown to be the most accurate and dependable algorithm.

Y. Hasija et al. [3] investigated in fatal diseases like some diseases and when it comes to predicting the likelihood of a lethal condition like melanoma being cured, early diagnosis is crucial. We think the use of automated approaches will aid in early diagnosis, particularly when a batch of phot os has a variety of diagnoses. As a result, we describe in this article a fully automated approach for identifying dermatological diseases from lesion photos. Our machine intervention differs from traditional medical personnel-based identification. Automated detection of dermatological disorders through image-processing and machine learning Our model is designed into three phases compromising of data collection and augmentation, designing model and finally prediction.

S. Uddin et al. [4] introduced the heart disease background In the realm of data mining, supervised machine learning methods have dominated. Recent research has indicated a potential application area for these techniques: disease prediction using health data. This study aims to discover the important trends in the performance and use of several supervised machine learning algorithms for illness risk prediction. Methods In this analysis, a great deal of work was put into finding papers that used more than one supervised machine learning algorithm to predict a particular disease.

R. Katarya et al. [10] implemented to treat cardiac disorders, hospitals and other institutions provide pricey therapies and procedures. So, being able to detect cardiac disease in its early stages will help individuals all over the world take the required precautions before it becomes severe. Predicting Heart Disease at Early Stages using Machine Learning. The consumption of alcohol, tobacco, and a lack of exercise are the main causes of heart disease, which has become a serious issue in recent years. Machine learning has proven successful over time at making judgement and predictions from the vast amount of data generated by the healthcare sector. To treat cardiac disorders, hospitals and other institutions provide pricey therapies and procedures. So, being able to detect cardiac disease in its early stages will help individuals all over the world take the required precautions before it becomes severe.

It is imperative to create a method to precisely and effectively anticipate heart diseases given the rise in fatalities from cardiovascular causes. The goal of the study was to identify the most effective ML algorithm for heart disease identification. Using data from the UCI machine learning repository, this study evaluates the accuracy of the Decision Tree, Logistic Regression, Random Forest, support vector machine, Random Forest and KNN algorithms in predicting heart disease. Collect relevant data from various sources such as medical records, clinical trials, or public datasets. The data should include various patient attributes such as age, sex, cholesterol levels, blood pressure, family history, etc.

3 Methodology

Data processing, which includes feature extraction and data cleansing, is the initial part. To do this, the data must be cleaned up and apply the algorithms machine learning, entails creating a model for predicting cardiac disease using machine learning methods. On the previously cleaned up data, the model is trained. The third element is model evaluation, which entails gauging the effectiveness of the model using several measures like recall, accuracy, and precision. The final step is heart disease prediction, which entails determining a patient's risk of developing heart disease using the trained model elements that are crucial for forecasting heart disease must be chosen. The following element. This architecture diagram summaries the steps involved in developing and deploying a machine learning-based system for heart disease prediction (Fig. 1).

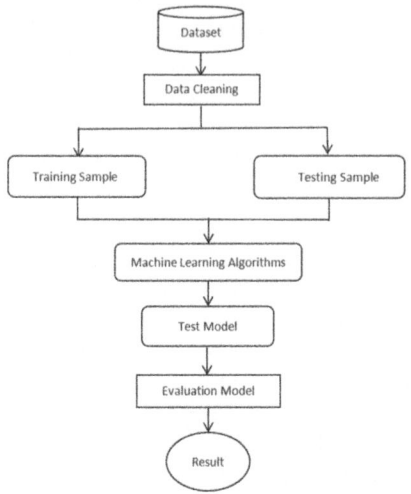

Fig.1. Application Architecture

4   Algorithm

Processing of heart disease prediction using machine learning.To classify the data set and apply different machine learning algorithms to the data simply the train model and projecting the good accuracy Following are the steps:

Step 1: Import the necessary modules and libraries, including Numpy, and Pandas,TensorFlow and Seaborn.

Step 2: Load the data set into a pandas Data Frame with the "Text" and "Label" columns selected.

Step 3: By categorical feature If the 'sex' column contains the values 'male' and 'female', the Label Encoder() instance will map 'male' to 0 and 'female' to 1.

Step 4: Similarly, if the 'cp' column contains the values 'typical angina', 'atypical angina', 'non-angina pain', and 'asymptomatic', the Label Encoder() instance will map these values to 0, 1, 2, and 3, respectively.

Step 5: By Numerical Feature then loops over each numerical feature (except the last one) to create a histogram of its distribution using the dist-plot() function from Seaborn.

Step 6: These visualizations provide an overview of the distribution of heart disease cases in the data set, which can help guide further analysis and modeling.

Step 7: In these step the visualizations allow us to see if there is a clear relationship between each numerical feature and the presence or absence of heart disease.

Step 8: Min Max scales the features to a range between 0 and 1, while Standard Scale standardizes the features to have a mean of 0 and a standard deviation of 1.

Step 9: Positive correlations are shown in shades of red, while negative correlations are shown in shades of blue.

Step 10: Now apply algorithms and Divide the data set into training and testing data.

Step 11: This sets the random seed for the random number generator used by the logistic regression algorithm.

Step 12: The SVC class is then used to train a linear SVM classifier model on the training set. The fit() method is used to train the model on the training set.

Step 13: The DecisionTree Classifier class from the sklearn.tree library is used to train a decision tree classifier model on some training data.

Step 14: The class takes in several hyper parameters that can be set during initialization to configure the model. In the example you provided, max_depth and random_state are set.

Step 15: Here the leaf_size is an integer parameter that controls the size of the leaf nodes in the KD tree used by the KNN algorithm.

Step 16: Then, the accuracy_score, precision_score, recall_score, and f1_score functions from the sklearn.metrics library are used to calculate the evaluation metrics.

Step 17: As of now the result for the above data set the Support Vector Machine provides the good accuracy for heart disease prediction using machine learning algorithms.

5   Results and Discussion

Comparison Of Algorithms.

Five machine learning algorithms techniques were used. Confusion matrix in the two class classification assignments, there are five machine learning algorithms (A) Logistic Regression (B) Support Vector Machine (C) Decision Tree (D) Random Forest (E) K Nearest Neighbor.

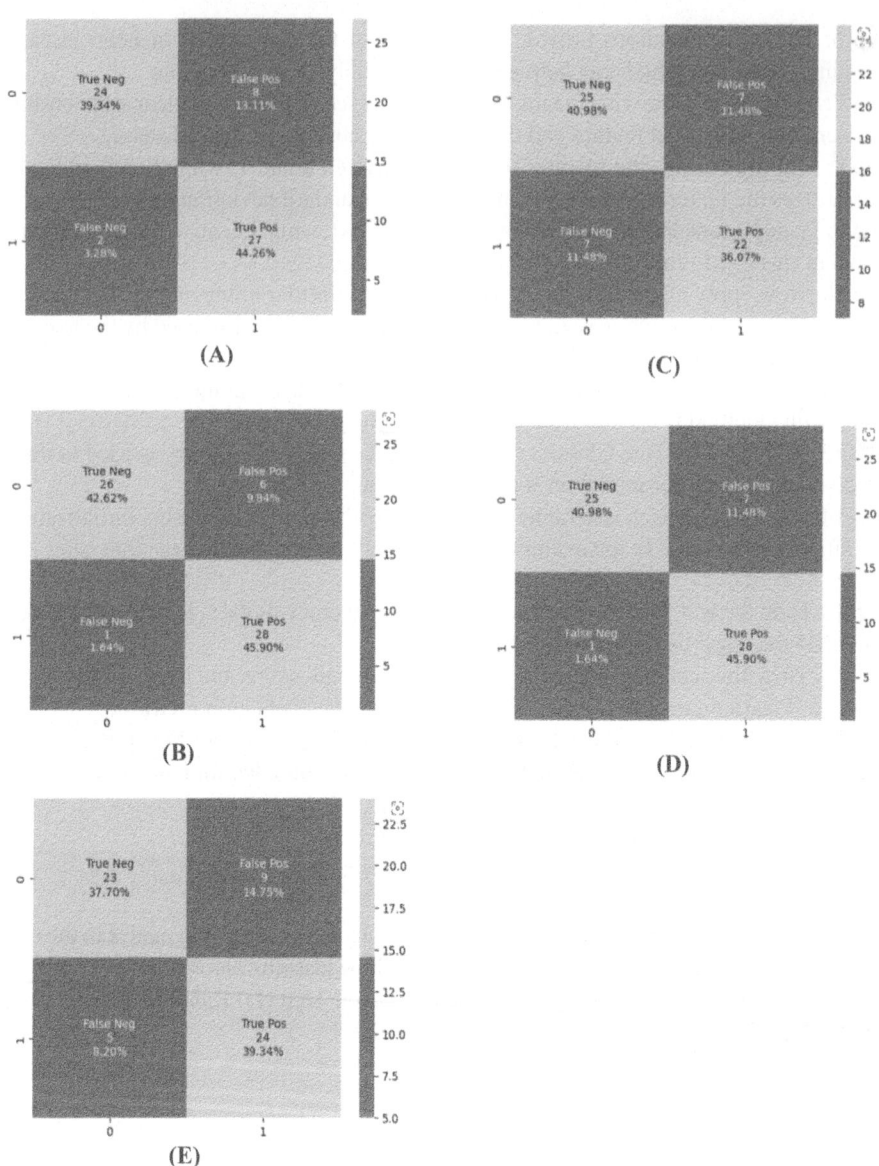

(A)

(B)

(C)

(D)

(E)

A confusion matrix is a table that is used to evaluate the performance of a machine learning algorithm by comparing the predicted and actual values of the target variable. A typical confusion matrix includes four cells: true positive (TP), false positive (FP), true negative (TN), and false negative(FN). True positives (TP): the number of cases where the model predicted the positive class correctly. False positives (FP): the number of cases where the model predicted the positive class incorrectly. True negatives (TN): the number of cases where the model predicted the negative class correctly. False negatives (FN): the number of cases where the model predicted the negative class incorrectly.

The elements of the matrix can be used to calculate various evaluation metrics such as accuracy, precision, recall, and F1 score.

ROC_AUC curve for the five-machine learning algorithm and ensemble method. (a) Logistic regression. (b) Support Vector Machine. (c) Decision Tree. (d) Random Forest and (e) K Nearest Neighbor. The Y-axis displays the true-positive rate, and the X-axis displays the false-positive rate.

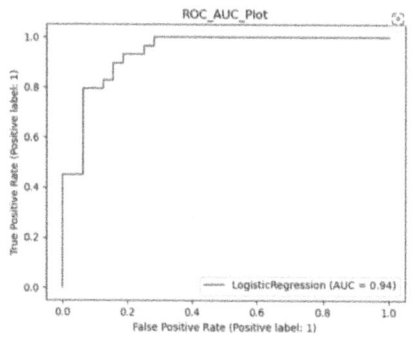

Fig. 2. ROC_AUC_Plot Logistic Regression Graph.

Fig. 3. ROC_AUC_Plot SVM Graph.

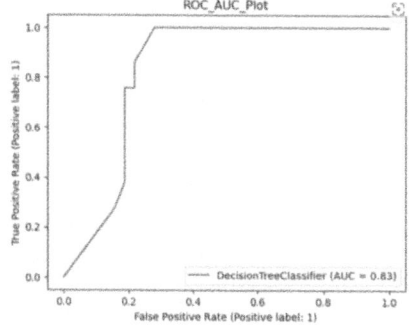

Fig. 4. ROC_AUC_Plot Decision Tree Graph

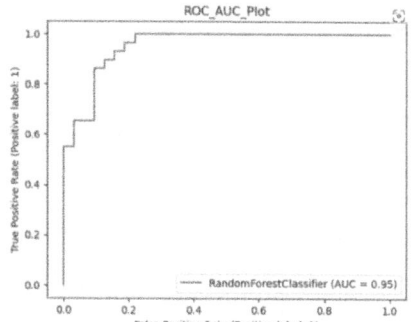

Fig. 5. ROC_AUC_Plot Random Forest Graph.

The Receiver Operating Characteristic (ROC) curve is a graphical representation of the performance of a binary classification model. The curve displays the trade-off between sensitivity and specificity for different classification thresholds. The area under the curve (AUC) is a commonly used metric that summarizes the overall performance of the model. A perfect model would have an AUC of 100%, while a model with no predictive power would have an AUC of 0.5. The Fig. 2 shows an ROC curve for a logistic regression model predicting. The AUC for this model is 0.94. The Fig. 3 shows an ROC curve for a SVM model and AUC for this model is 0.95. The Fig. 4 shows an ROC curve for a Decision Tree model and AUC for this model is 0.83. The Fig. 5 shows an ROC curve for a Random Forest classifier and AUC for this model is 0.95. The Fig. 6 shows an ROC curve for a KNN and AUC for this model is 0.82.

The performance of a support vector machine (SVM) model in heart disease prediction depends on various factors, such as the quality of the data, the choice of features,

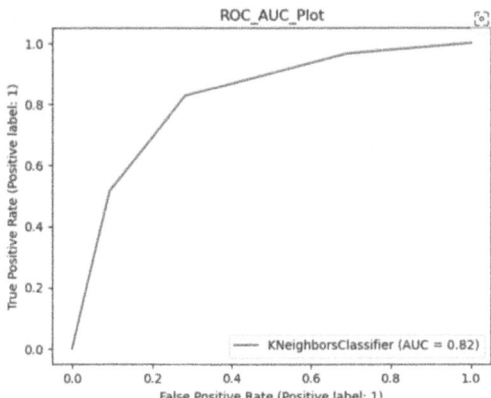

Fig. 6. ROC_AUC_Plot KNN Graph.

Sl.no	Algorithm	Cross Validation score	ROC_AUC Score	Accuracy
1.	LG	88.60%	84.05%	83.61%
2.	SVM	87.86%	88.90%	88.52%
3.	Decision tree	79.47%	76.99%	77.05%
4.	Random forest	89.60%	87.34%	86.89%
5.	KNN	76.62%	77.32%	77.05%

Fig. 7. Algorithm Result Table

the selection of hyper parameters, and the evaluation metrics used. Some common evaluation metrics used for classification tasks like heart disease prediction are accuracy, precision, recall, F1 score, and area under the receiver operating characteristic curve (AUC-ROC). The exact performance of an SVM model in heart disease prediction will depend on the specific data and evaluation metrics used.

SVM have been found to perform well on heart disease prediction tasks, achieving accuracy in the range of 75–90% on various datasets. However, the performance can vary depending on the specific data-set.

Figure 7 The SVM classifier did the best and achieved a (88.61%) accuracy in the two-class classification task, followed by the other classifiers that are the highest precision was obtained by Random Forest (86.89%), followed by Logistic Regression (83.61%), Decision Tree (77.05%), and KNN (77.05%).

References

1. Golande, A., Pavan Kumar, T.: Heart disease prediction using effective machine learning techniques. Int. J. Recent Technol. Eng. **8**(1), 944–950 (2019)

2. Nagamani, T., Logeswari, S., Gomathy, B.: Heart disease prediction using data mining with mapreduce algorithm. Int. J. Innov. Technol. Exploring Eng. (IJITEE) ISSN, 2278–3075 (2019)
3. Alotaibi, F.S.: Implementation of machine learning model to predict heart failure disease. Int. J. Adv. Comput. Sci. Appl. **10**(6) (2019)
4. Repaka, A.N., Ravikanti, S.D., Franklin, R.G.: Design and implementation heart disease prediction using natives Bayesian. In: International Conference on Trends in Electronics and Information (2019)
5. Thomas, J., Princy, R.T.: Human heart disease prediction system using data mining techniques. In: 2016 International Conference on Circuit, Power and Computing Technologies (ICCPCT), pp. 1–5. IEEE (2016, March)
6. Gavhane, A., Kokkula, G., Pandya, I., Devadkar, K.: Prediction of heart disease using machine learning. In: 2018 Second International Conference on Electronics, Communication and Aerospace Technology (ICECA), pp. 1275–1278. IEEE (2018, March)
7. Hasija, Y., Garg, N., Sourav, S.: Automated detection of dermatological disorders through image-processing and machine learning. In: 2017 International Conference on Intelligent Sustainable Systems (ICISS), pp. 1047–1051. IEEE (2017, December)
8. Oyeleye, M., Chen, T., Titarenko, S., Antoniou, G.: A predictive analysis of heart rates using machine learning techniques. Int. J. Environ. Res. Public Health **19**(4), 2417 (2022)
9. Uddin, S., Khan, A., Hossain, M.E., Moni, M.A.: Comparing different supervised machine learning algorithms for disease prediction. BMC Med. Inform. Decis. Mak. **19**(1), 1–16 (2019)
10. Katarya, R., Srinivas, P.: Predicting heart disease at early stages using machine learning: a survey. In: 2020 International Conference on Electronics and Sustainable Communication Systems (ICESC), pp. 302–305. IEEE (2020, July)
11. Kohli, P.S., Arora, S.: Application of machine learning in disease prediction. In 2018 4th International Conference on Computing Communication and Automation (ICCCA), pp. 1–4. IEEE (2018, December)
12. Chapman, B., DeVore, A.D., Mentz, R.J., Metra, M.: Clinical profiles in acute heart failure: an urgent need for a new approach. ESC Heart Failure **6**(3), 464–474 (2019)
13. Haq, A.U., et al.: Feature selection based on L1-norm support vector machine and effective recognition system for Parkinson's disease using voice recordings. IEEE Access **7**, 37718–37734 (2019)
14. Mohammadi, A.G., Mehralian, P., Naseri, A., Sajedi, H.: Parkinson's disease diagnosis: the effect of autoencoders on extracting features from vocal characteristics. Array **11**, 100079 (2021)

Mental Health Classifier Using Support Vector Machines

Pavitra Golchha, Payel Paul, and P. Saranya[✉]

Department of Computing Technologies, SRM Institute of Science and Technology,
Kattankulathur, TamilNadu 603203, India
saranyap@srmist.edu.in

Abstract. An individual's quality of life may be significantly impacted by mental health conditions like anxiety, depression, and substance misuse. Yet, the stigma associated with mental illness frequently discourages people from seeking help, and the mental health care system struggles to provide effective care due to several technical issues. To address these challenges, this study proposes a mental health analysis system that utilizes a Support Vector Machine (SVM) model to detect whether a user needs a therapist or not. The system was trained and tested on a dataset consisting of mental health records, and the SVM model outperformed other machine learning models with an accuracy rate of 83%. This system can be used as a screening tool for mental health professionals to identify individuals who may require professional help, enabling early intervention and improving the overall mental health of individuals. Addressing the challenges in the mental health care system requires the integration of innovative technological solutions, increasing access and affordability, and reducing the stigma surrounding mental illness.

Keywords: Mental health · supervised learning · classification model · support vector machines

1 Introduction

Mental health conditions are a growing concern worldwide, with millions of people affected by various types of mental health disorders. These conditions can range from mild to severe and can significantly impact an individual's life, affecting their relationships, work, and overall well-being [1]. The prevalence of mental health disorders has prompted researchers to explore new ways of addressing and managing these conditions [2]. One of the most promising methods is the use of machine learning models to analyze mental health data and predict the likelihood of an individual requiring therapy [3].

In this paper, we propose the development of a mental health analysis system that utilizes a Support Vector Machine (SVM) model to detect whether a user needs a therapist or not. The proposed system can be used as a screening tool for mental health professionals to identify individuals who may require professional help. Early intervention can significantly improve an individual's mental health, preventing their condition from worsening and reducing the risk of severe mental health issues.

P. Dassan et al. (Eds.): ICICSCNT 2023, CCIS 1970, pp. 36–45, 2024.
https://doi.org/10.1007/978-3-031-75957-4_4

The study focuses on analyzing a dataset consisting of mental health records from various individuals. The dataset was preprocessed, and feature selection was performed to select the most relevant features for the analysis. The SVM model was then trained using the selected features to classify the mental health status of the users as either needing a therapist or not. Several machine learning models, including Logistic Regression, Gradient Boosting Machine, and Stochastic Gradient Descent, were used to compare the accuracy of the SVM model. The study aims to address the growing concern of mental health conditions and provide an efficient method for mental health professionals to identify individuals who may require professional help. The proposed system can be used as a tool to reduce the burden on mental health professionals, allowing them to focus on individuals who require immediate attention.

The goal of the study is to add to the expanding body of knowledge on the application of machine learning to mental health studies. Machine learning models can provide accurate predictions, reducing the chances of misdiagnosis and improving the overall mental health of individuals. The proposed system has several potential benefits, such as reduced waiting times for mental health professionals, early intervention for individuals, and improved mental health out- comes. A person's mental health is an important part of their overall well-being, and the suggested method can greatly enhance the quality of life for those who are struggling with mental health issues.

The paper is structured in such a manner: Section 1 contains the introduction of the model. Section 2 contains the literature review of the related work for mental health prediction models for several types of research. Section 3 proposes and details about the implementation of the SVM based classifier model. Section 4 discusses the dataset; results obtained and compares the models' performance. Section 5 concludes all the work.

2 Literature Review

Konda Vaishnavi et al. did research on the application of machine learning algorithms to diagnose mental disorders [4]. The study aimed to identify the most accurate machine learning technique for determining mental health issues using various accuracy conditions. It's used five different machine learning techniques including Logistic Regression, KNN Classifier, Decision Tree Classifier, Random Forest, and Stacking. The results showed that the Stacking technique had the highest prediction accuracy of 81.75%. The study concluded that machine learning algorithms have great potential in the early detection and treatment of mental health issues.

M Srividya et al. presented a research work on behavioural modelling for mental health using machine learning algorithms [5]. Mental health is a crucial aspect of an individual's well-being that affects their thoughts, feelings, and actions. In order to determine the condition of mental health in a target group, this study suggested using machine learning methods including support vector machines, decision trees, naive Bayes classifiers, K-nearest neighbour classifiers, and logistic regression. The responses from the target group were used to train an unsupervised learning algorithm, and the labels generated by clustering were validated by computing the Mean Opinion Score. The study analyzed the application of these algorithms on the target groups and suggested directions for future work. The results of this study suggest that machine learning algorithms can be a valuable tool for predicting the onset of mental illness and serving as a monitoring tool for individuals with deviant behavior.

S Monlya and T R Pragathi proposed a paper that aims to create an AI-ML based conversational agent that acts as a real-time therapist, analyzing the user's emotions and providing appropriate responses and feedback [6]. The paper highlights the potential benefits of AI chatbots in the field of mental health, such as their ability to destigmatize seeking help and provide more accessible support at any time. Furthermore, the chatbot is designed to be used on the go and includes a self-healing kit with various mental and physical exercises to improve the user's health. The study provides insights for the evaluation and design of emerging chatbot agents.

Sreevidya Iyer et al. presents a virtual mental health assistant with chat, psychological assessment, emotion detection, and a recommendation system to improve the user's mood [7]. The study used Naive Bayes and neural networks for sentiment analysis and found that Naive Bayesian had a higher accuracy. The assistant aims to provide continuous and convenient support for individuals to manage their mental health.

Genta Indra et al. proposes a virtual coaching platform designed to help individuals cope with the negative emotions and loneliness brought on by the COVID-19 pandemic and social distancing measures [8]. The platform uses natural language understanding, records the user's emotions, sentiment, and stress, and recommends exercises for a healthy daily routine. It also includes a social community feature for users to connect and share experiences. The platform aims to provide continuous support for individuals to improve their well-being.

Ava Podrazhansky et al. proposed a mobile application for mental health uses machine learning algorithms to determine if an individual has a mental illness and suggest prevention methods [9]. The application improves upon existing counselling apps by incorporating natural language processing and continuous emotion dialogue analysis for more natural and intelligent interactions. The study highlights the importance of NLP, continuous emotion dialogue analysis, and audio/sentence generation in the design of mental health chatbots. The application aims to provide a more efficient and accessible way for individuals to manage their mental health.

The previous studies on mental health and machine learning algorithms have shown promising results in identifying mental health issues. However, this paper proposes a more focused approach by developing an SVM based classification model for predicting the onset of mental illness in individuals. The proposed model aims to improve the accuracy of predictions by analyzing various features and characteristics of the individual's behavior and mental state. The aim of this paper is to address the challenges in this field, such as the lack of large-scale data sets and the need for personalized and culturally sensitive interventions. The proposed model can be integrated into existing mental health services to provide more accurate and effective early detection and treatment of mental health issues.

3 Implementation

A labelled dataset is used to train the algorithm in supervised learning, a type of machine learning. The input data and corresponding output labels comprise the labelled dataset. To make predictions on new, unlabeled data, the algorithm learns from this data. In the case of classification models, the output labels are categorical variables that divide the data into distinct classes. The main objective of a classification model is to accurately predict the set of a new observation based on its input variables.

One commonly used algorithm for classification tasks is Support Vector Machines (SVMs). SVMs [10] are powerful and flexible models that can handle both linear and non-linear datasets. In order to effectively distinguish the two classes of data, SVMs build a hyperplane in a high-dimensional environment. This hyperplane is determined by selecting a subset of the input variables that provide the most discriminating power between the classes. SVMs try to maximize the margin, which is the separation between the hyperplane and the nearest data points on either side.

In addition to their effectiveness in classifying data, SVMs have several advantages over other classification models. One advantage is that they are less prone to overfitting than other models, meaning they can generalize well to new, unseen data. Another advantage is that they can handle datasets with a large number of variables, as long as the number of observations is sufficient. However, training SVMs can be computationally expensive, especially for large datasets, and the model's performance can be significantly impacted by the kernel function selected for data transformation [11]. Overall, SVMs are a valuable tool in the field of supervised learning and have numerous applications in areas such as image classification, natural language processing, and bio-informatics.

The architecture of this machine learning model consists of five key steps: dataset preprocessing, feature selection, train test split, model training, and evaluation. The flow of the proposed model is shown in Fig. 1.

The first step is dataset preprocessing, which involves loading the dataset and handling missing values. Then, the numerical columns in the dataset are normalized or standardized to bring all the features to the same scale. The categorical features are then one-hot encoded into multiple binary columns. And, the boolean column is transformed using Binarizer which converts boolean values into binary numbers. Feature selection is the second step, where relevant features are selected from the dataset. This phase is crucial since it aids in simplifying the model and guards against overfitting.

The third phase involves dividing the data into training and testing sets in a 3:1 ratio. The training data are used to train the model, while the test data are used to evaluate it. The SVM classifier is set up and trained using training data, and measures like accuracy, precision, recall, and F1 score are used to gauge the model's effectiveness. Also, a confusion matrix is displayed to show how the model is performing. An SVM classifier-based machine learning model is composed of these five phases.

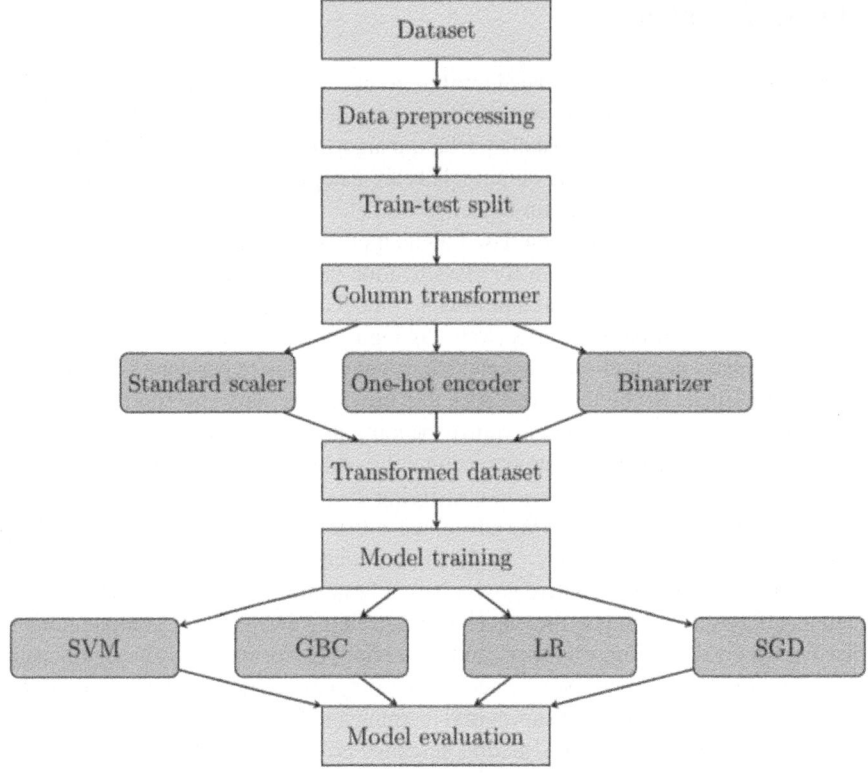

Fig. 1. Flow of the proposed model

4   Results and Discussion

4.1   Dataset Used

For training this model, this paper has used the dataset available from "Mental Health in Tech Survey" [12]. This dataset, which examines respondents' attitudes towards mental health and the prevalence of mental health illnesses in the IT workplace, is from a study conducted in 2014. This dataset contains data in 1259 rows and 22 columns that was obtained anonymously from a group of tech employees.

This dataset contains various information of interest of a person including their age, gender, whether they are self-employed, do they have family history of someone having high stress, have they ever had a treatment for their mental health issues, how often they are intervened by their professional work, whether their job is remote or not, do they work for a tech company, does their company provide additional benefits to them, do they have a wellness program for their employees, can they seek help from counsellors, are they offered enough paid leaves, are they physically/mentally healthy.

Figures 2 A and 2 B show the plot of the respondents' age and gender distribution, respectively. The range of the ages of respondents lie between 18 and 50 years. The gender ratio of the respondents is roughly 3.5:1 for males and females respectively.

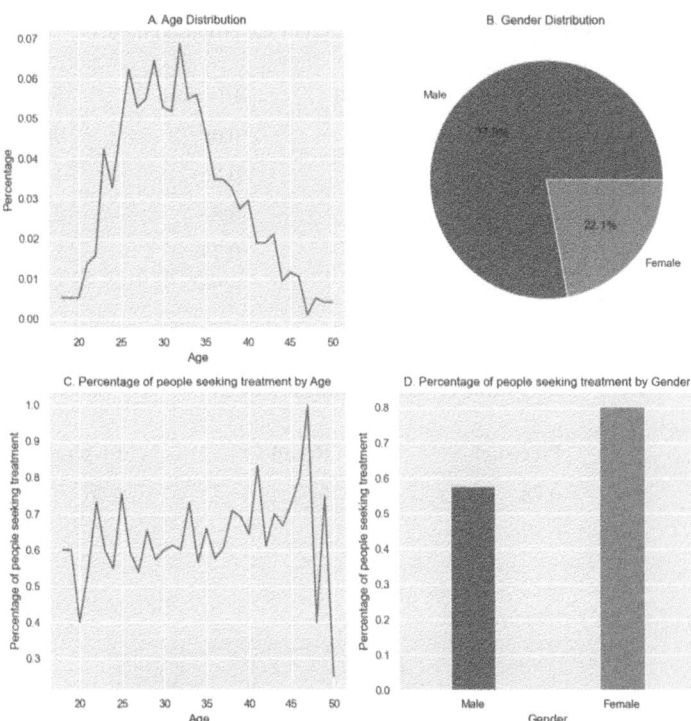

Fig. 2. Analysis of the age and gender columns in the dataset

Figures 2 C and 2 D compares the percentages of people seeking treatment by age and gender of the respondents, respectively. On an average, at least 60% of the people across all age groups seek treatment. Nearly 57% of the male respondents seek treatment, whereas, 80% of the female respondents seek treatment.

4.2 Performance Evaluation

This dataset after processing is then trained using an SVM model with a linear kernel, of which the results are presented in the Table 1 below. The output labels predicted from this binary classifier are False and True. False represents that the person doesn't require a mental health treatment, whereas, True indicates that they do need a treatment.

As represented in Table 2, the accuracy obtained from this model is 81%. The precision with which the model predicts the negative label is 54%, while for the positive label, it is 91%. The overall precision of the model is 78%. The recall score obtained is at 97%. The F1 score for prediction of the negative and positive labels are 68% and 86% respectively. The overall F1-score of the model is 86%.

The confusion matrix in the Fig. 3 above shows that the performance for true labels is optimum (97%). However, there is a high percentage of false positives (46%), though there is room for improvement by obtaining and training on a larger dataset with reduced biases.

Table 1. Performance evaluation of the proposed model

	Precision	Recall	F1-score	Support
False	0.91	0.54	0.68	90
True	0.78	0.97	0.86	147
Accuracy			0.81	237
Macro avg	0.84	0.76	0.77	237
Weighted avg	0.83	0.81	0.79	237

Table 2. Key performance metrics

Accuracy	Precision	Recall	F1-score
0.81	0.78	0.97	0.86

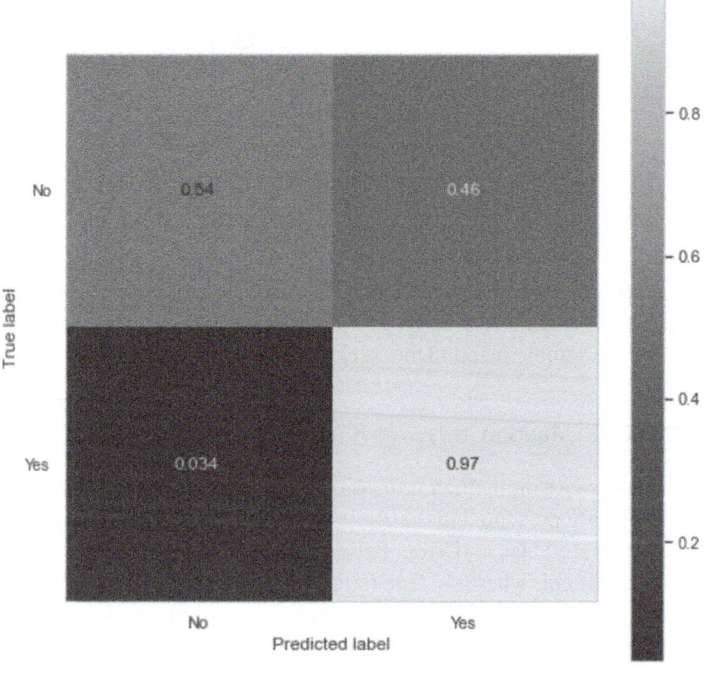

Fig. 3. Confusion matrix of the SVM model

4.3 Performance Comparison with Existing Models

Different existing classifier models, viz. Gradient Boosting Classifier (GBC), Logistic
Regression (LR), and Stochastic Gradient Descent (SGD) classifier has also been trained

on the processed dataset in order to determine their correctness. GBC is an ensemble of decision trees that sequentially trains weak models on residuals, known for its high accuracy and robustness to noisy data [13]. LR is a statistical method for binary classification that models the probability of an instance belonging to a particular class using a logistic function, extendable to multi-class classification using techniques like one-vs-rest or softmax [14]. SGD classifier is a linear classifier that optimizes the parameters of the model using stochastic gradient descent algorithm on a random subset of training data, known for its efficiency and ability to handle large datasets [15]. The normalized and processed independent and dependent datasets were then incorporated into the algorithm model.

The findings listed in the Table 3 as well as in the Fig. 4 suggest that the LR algorithm obtains the optimum performance level with an 80% accuracy value other than SVC. The other models' accuracy is also found to be in the range of 70% and 80%. As previously mentioned above, there are two possible outcomes from the model - True or False, where True represents that the person should get a professional counselling and False represents otherwise.

Table 3. Performance metrics obtained from different models

Model	Accuracy	Precision	Recall	F1-score
SVM	0.81	0.78	0.97	0.86
GBC	0.79	0.80	0.88	0.84
LR	0.80	0.79	0.93	0.85
SGD	0.66	0.88	0.53	0.66

The data visualization process also reveals that mental health illnesses are frequently present in, or at least have been present in, IT employees. Male respondents who were in the majority provided the proportion of data and the outcomes. Data on female respondents can be collected for future research, and the findings can be compared. The analysis's findings will likely be utilized as a guide for future study, as well as a representation of the respondents' average mental health, particularly that of IT professionals. The model can also be developed as a support tool for diagnosing mental health issues.

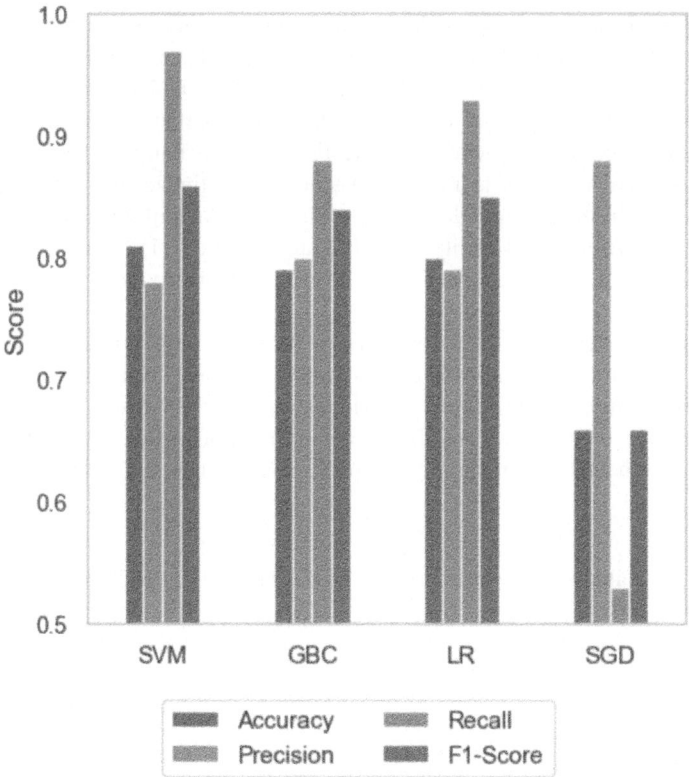

Fig. 4. Comparison of key performance metrics of different models

5   Conclusion

In this paper, a mental health classifier model has been proposed that allows patients to directly contact a specialist, even anonymously, and utilize a self-assessment tool to evaluate and monitor their mental health. The model is intended to increase the quality of mental health issue diagnosis and treatment, increase access to mental health care, and reduce the stigma attached to asking for help. The dataset used in the proposed model has been conducted on tech workers only. An analysis of the data suggests that there is high gender bias, non-uniform age distribution, and invalid values in the user responses. This led to sub-optimal predictions. A more comprehensive and larger dataset with greater diversity is needed to be able to make more accurate predictions. The proposed model has the potential to revolutionize the way mental health care is provided and accessed by patients. The model will improve access to mental health services, reduce the stigma attached to asking for help, and enhance the precision of mental health issue diagnosis and treatment. The model could also be integrated with existing healthcare systems and provide more efficient and effective mental health management.

References

1. Feng, X., Hu, M., Guo, W.: Application of artificial intelligence in mental health and mental illnesses. pp. 506–511 (12 2022)
2. Tan, C., Koo, A.C., Rahmat, H., Siew, W., Cheang, A., Amir Sharji, E.: A quantitative study exploring the acceptance of the ehealth model for mental wellness among digital workers. F1000Research **11**, 111 (01 2022)
3. Hennemann, S., Witthoft, M., Bethge, M., Spanier, K., Beutel, M.E., Zwerenz, R.: Acceptance and barriers to access of occupational e-mental health: cross-sectional findings from a health-risk population of employees. Int. Arch. Occup. Environ. Health **91**(3), 305–316 (2018)
4. Vaishnavi, K., Kamath, U.N., Rao, B.A., Reddy, N.V.S.: Predicting mental health illness using machine learning algorithms. J. Phys.: Conf. Series **2161**(1), 012021 (jan 2022)
5. Srividya, M., Mohanavalli, S., Bhalaji, N.: Behavioral modeling for mental health using machine learning algorithms. J. Med. Syst. **42**, 1–12 (2018)
6. Moulya, S., Pragathi, T.R.: Mental health assist and diagnosis conversational interface using logistic regression model for emotion and sentiment analysis. J. Phys.: Conf. Series 2161(1), 012039 (jan 2022)
7. Iyer, S., Shetty, D., Badgujar, P., Nambiar, A., Jewani, K.: A proposal for virtual mental health assistant. In: 2021 7th International Conference on Advanced Computing and Communication Systems (ICACCS) **1**, 953–957 (2021)
8. Winata, G.I., Lovenia, H., Ishii, E., Siddique, F.B., Yang, Y., Fung, P.: Nora: The well-being coach. ArXiv abs/2106.00410 (2021)
9. Podrazhansky, A., Zhang, H., Han, M., He, S.: A chatbot-based mobile application to predict and early-prevent human mental illness. In: Proceedings of the 2020 ACM Southeast Conference (2020)
10. Cortes, C., Vapnik, V.: A training algorithm for optimal margin classifiers. Mach. Learn. **20**(3), 273–297 (1995)
11. Pedregosa, F., et al.: Scikit-learn: machine learning in Python. J. Mach. Learn. Res. **12**, 2825–2830 (2011)
12. Mental health in tech survey (Nov 2016), https://www.kaggle.com/datasets/ osmi/mental-health-in-tech-survey
13. Friedman, J.H.: Greedy function approximation: a gradient boosting machine. Ann. Stat. **29**(5), 1189–1232 (2001)
14. Hosmer, D., Lemeshow, S., Sturdivant, R.: Applied Logistic Regression. Wiley Series in Probability and Statistics, Wiley (2013). https://books.google.co.in/ books?id=64JYAwAAQBAJ
15. Bottou, L.: Large-scale machine learning with stochastic gradient descent. In: Lechevallier, Y., Saporta, G. (eds.) Proceedings of COMPSTAT'2010. pp. 177–186. Physica-Verlag HD, Heidelberg (2010)

Enhanced Cloud Virtualization Based on Offloading Dynamic Switching Scheduling Algorithm Using Optimal Time Desk Task Allocation Model

S. Rajkumar[✉], M. B. Rahuraman, M. Rahul David, V. Sai Senthil Ganapathy, and S. N. Sanjay

Sona College of Technology, Salem, India
{rajkumar.s,rahuraman.19it,rahuldavid.19it,
saisenthilganapathy.19it,sanjay.19it}@sonatech.ac.in

Abstract. Cloud computing provides decentralized process to provide various services, having lot of task running depending of service request the server reach more energy consumption, time to complete task, cause burden. So many virtualized scheduling system task takes virtual allocation to complete the task. But the increasing time leads time consumption to complete the workload in the server. To resolve this problem, we propose an Offloading Dynamic Switching Scheduling Algorithm (ODSSA) based on optimal time desk task allocation model to increasing the cloud computing process. Initially the task loading process timing consumption was estimated based on Expected time completion matrix model. Then the time slot is constructed using Min Max Queuing Theory (MMQT). Depends on ETCM and MMQT dynamic switching scheduling algorithm is applied to balance the scheduling task to complete the task using optimal time desk task allocation model. This balances virtual task allocation when the task requires the virtualization task is allocated or other wise to terminate the idle machine to reduce the time and energy consumption. The proposed system achieves high performance compared to the other system as well to allocate the task to reduce the time consumption.

Keywords: Cloud computing · Scheduling · Virtual Machine · Time Consumption · Task Allocation · Dynamic Priority

1 Introduction

Cloud computing is an innovative way of providing shared resources on the Internet. It provides a computer environment where various resources, infrastructure, development sites and software as a service are provided to customers virtually, with effective time-based fees [1]. Cloud important features are Low cost, scaling, reliability and application-based computing. Cloud is a parallel distribution system with many virtual machines that provide integrated dynamically linked resources according to customer service level agreements between consumers and service providers through the internet [2].

P. Dassan et al. (Eds.): ICICSCNT 2023, CCIS 1970, pp. 46–57, 2024.
https://doi.org/10.1007/978-3-031-75957-4_5

The prime purpose of cloud computing is to provide distributed access and storage across a wide range of computing power [4]. Therefore, these services must be scalable, reliable, and capable of operating with autonomous support. Cloud users can specify the amount they need using the Quality of service (QoS) parameters of service listed in the Service Level Agreement (SLA) set by the provider [5].

In this research, an offloading dynamic switching scheduling algorithm based on optimal time desk task allocation model to increasing the cloud computing process. The objective of this research is to develop an efficient task loading process timing consumption was estimated based on Expected time completion matrix model. Then the time slot is constructed using min max queuing theory. Depends on ETCM and MMQT dynamic switching scheduling algorithm is applied to balance the scheduling task to complete the task using optimal time desk task allocation model. This balances virtual task allocation when the task requires the virtualization task is allocated or other wise to terminate the idle machine to reduce the time and energy consumption. The proposed system achieves high performance compared to the other system as well to allocate the task to reduce the time consumption.

2 Related Work

In cloud Task scheduling is a challenge in a diverse data center in cloud computing environments. Cloud performance depends on the task scheduling strategy [6]. The task scheduling algorithm will schedule the task resources required for use on the cloud platform. Although many methods for task scheduling have been proposed, they only consider the very small scope of optimal scheduling [7].

Cloud workflows are difficult to plan due to the large volume of workflows, flexibility and random cloud resources. In addition, the time and cost of implementation in the cloud pricing model are two important scheduling issues [8]. Cloud workflow scheduling is a multi-purpose optimization problem that simultaneously improves both implementation time and implementation cost [9]. Ant Colony Optimization (ACO) scheduling technique that upgrades the limited set of public and private computer resources to suit the time and cost constraints of a hybrid cloud computing environment. MOSACO uses laborious and cost effective single-purpose upgrade strategies to reduce task completion times and costs [10]. The entropy optimization model improves user service quality and resource provider benefits.

Efficient virtual machine management is important for saving energy, maximizing profits, and preventing Service Level Agreement (SLA) violations. Describe how each scenario uses prediction algorithms to make the most efficient and cost effective VM deployments [11]. In addition, each type describes the predictors such as element management methods and evaluation parameters, simulation software, workload data, and the power to reveal details of the VM deployment methods.

The cloud computing model, clients are charged in view of the assets and the Quality of Service (QoS). The normal QoS-based multi-objective assignment booking issue of NP-culmination issue is because of the NP-fulfilment of the undertaking planning issue and the enormous review space made by the different issue occasions [12]. It can't successfully accomplish the around the world ideal arrangement with the many existing solutions.

The main problems of cloud computing is task scheduling. When considering QoS requirements, task scheduling and resource management in the cloud are often intractable optimization problems. Much work under task scheduling focuses only on deadline issues and cost optimization, eschewing the importance of availability, robustness, and reliability [13]. The main is to develop optimization algorithms to achieve efficient resource allocation and scheduling in cloud environments [14]. Scheduling the tasks of these sharable resources is an important aspect of cloud computing and an area that attracts many researchers [15]. Workflow scheduling in cloud architecture is even more important because it consists of related tasks and is considered NP-hard. Implemented various traditional met heuristic scheduling techniques and evaluated their performance based on two parameters, Flow time and Make span.

The task scheduling of cloud computing is a modern system technology and is an NP-hard optimization problem. In a cloud environment, it is more important to maintain load balancing between processing units. Cloud computing is a virtualization of building applications, hosting applications, sharing resources, and delivering services over networks [16]. Task scheduling is gaining attention by ensuring the quality of service to users, reducing costs, and optimizing the scheduling of incoming jobs in a cloud location. One of the major challenges of task scheduling issues in a cloud computing is maximum time consumption [17]. Previous Fuzzy Flexible Job Shop Scheduling Problem (FFJSP). Introducing pathway re-linking technology for simulating object exchange through black/white holes. A hybrid stage that joins inclusion based heuristics and way of relinking innovation is fused into the calculation to extend the search space [18]. Cloud computing offers a variety of services to meet the needs of our customers and works in virtualized environments. Cloud environments include many geographically dispersed data centres. The main challenge facing cloud environments is data centre energy consumption [19]. To assign a task to a virtual machine, need an appropriate scheduling mechanism.

Dynamic work flow algorithm (DWFA) limits asset reaction time and, generally speaking, work process undertakings. Intended to further develop load adjusting by modifying the algorithm to help load adjusting [20]. Diminish asset reaction time by copying administration hubs across various locales. A calculation's speedy change to the arrangement of decisions brings about quicker asset reaction times than different algorithms.

3 Proposed System

In this research, an offloading dynamic switching scheduling algorithm (ODSSA) based on optimal time desk task allocation model to increasing the cloud computing process. Initially the task loading process timing consumption was estimated based on Expected time completion matrix model. Then the time slot is constructed using min max queuing theory. The proposed algorithm contains an agent container that has a dynamic agent. The proposed approach generates the required number of agents at runtime and stores them for scheduling. The number of agents generated depends on the number of available computers. Let assume Identical Machine (Im) set of jobs Js which has to be schedule, the proposed algorithm create set of agents A with the size equal to the size of machines.

Depends on ETCM and MMQT dynamic switching scheduling algorithm is applied to balance the scheduling task to complete the task using optimal time desk task allocation model.

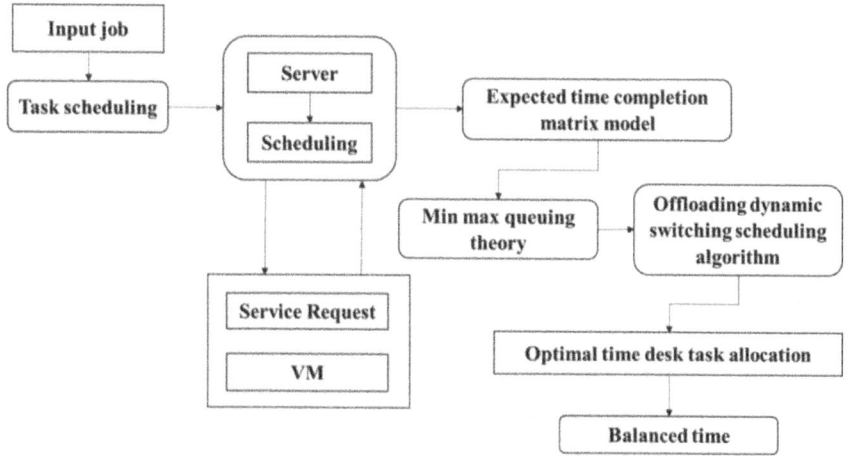

Fig. 1. Architecture diagram For Proposed System

This balances virtual task allocation when the task requires the virtualization task is allocated or other wise to terminate the idle machine to reduce the time and energy consumption. Figure 1 shows the Architecture diagram For Proposed System.The proposed system achieves high performance compared to the other system as well to allocate the task to reduce the time consumption. To gain computing power without deploying and maintaining their infrastructure. Because the cloud is built up as a large cloud, it consumes more energy. The displacement system in which multiple Virtual Machines (VMs) are housed in a small number of active physical machines is called VM Consolidation (VMC). Therefore, this technique is a practical way to balance CCS's power consumption with other QoS requirements.

3.1 Expected Time Completion Matrix Model

The proposed algorithm contains an task execution mean rate estimation that has a dynamic priority level. The proposed approach generates the required number of VM at runtime and stores them for scheduling. The number of agents generated depends on the number of available computers.

Algorithm steps

Begin

Step 1: Initialize the Agent Container A_c

Step 2: Read input job set J_s and Identical Machine (I_m)

Step 3: For each Identical Machine (I_m)

J_{si}= Randomly select job J_i from J_s

J_{si}=$\int(1n) R$ and J_s // n denotes number of jobs

Step 4: Generate Agent A_i

$A_i = J_{si}, M_i . A_i = \{J_{si}, M_i\}$

Stop

Let assume Identical Machine (I_m) set of jobs J_s which has to be schedule, the proposed algorithm create set of agents A with the size equal to the size of machines.

3.2 Min Max Queuing Theory

The proposed mathematical models for the problem described in, S 1, 2, 3 n single motion tasks and m 1, 2, m, scheduling suite. Provide a well-defined integer processing time Pi and outstanding date for each job. This allows all work to run at t0 time. There is no division of labor or union. Change m and n to match the above. The late Ti N of the work in the N table can be obtained as follows. Log in to S.T.S. from each label T, this method is based on the label interest or theme.

$$T_iN = Max\left[0 \; c_iN - d_i\right] \tag{1}$$

where Ci N is the working time. Let Ni be the sequence of tasks planned on the $j \in N$ machine.

The scheduling problem above is to find the assignment of all work to different machines and to plan all the work on all machines. At that point N = N1, N2, N2 .... N minimizes the total tardiness. Therefore, $P\|\sum T_i$ on any computer, even when the machine is idle. Therefore, the trial is for a valid schedule only. This is the equivalent of handling the own work. It starts at t = 0, which is assigned to the first place on each machine, and starts executing tasks as soon as the work in progress is completed. Mixed integer schemes $P\|\sum T_i$. The problem is to create mixed integer linear programming model. No machine codes are required for binary variables. Therefore, the proposed model is

$$\min Z = \sum_{i=0}^{S} T_i \tag{2}$$

Here, min Z is the work delay. H, h = 1... n,

$$\sum_{i=0}^{S} y_h \leq m \tag{3}$$

yh = binary variable, take the value 1 for a work, (h) 'is the main work of the (m) machine h = 1.... N, in the above model is always reduced slowly. This guarantees that the first jobs on the machine will be up to m jobs. Additional controls are required to ensure that each job is started on one machine or before other work.

3.3 Offloading Dynamic Switching Scheduling Algorithm

The actual processing time of the job is the linear increase function of the start time. During this operation, the machine performs only one maintenance function and the maintenance time is fixed. When the repair work is completed, the machine will return to its original state and resume working. The goal is to determine the optimal schedule that minimizes the sum of the maximum completion time (production time) and work completion time. To generate offloading level the task Ji is executed based on maximum priority. By assigning the devise factor D with N number of machines to create slot to schedule the task on each assigned sequence Si get visit the VM to execute the task without waiting time.

Procedure:

Input: Number of Task, ETMA

Output scheduling query to reduce time factor

Step1: Assign VM to task At j number

Step2: Estimate the priority task by using

$$\text{divisive issue D} = \int \frac{size(J)}{size(M)}$$

Step3: for each model mod

Arrange job set $J = \int_{i=1}^{size(J)} f(x)$

F(x) is the function of sorting order.

Split job set J according to D to produce Js $\int \frac{J}{D} =$ Split Machine set according

to D to produce Ms $= \int \frac{M}{D}$ For each division Di of D

For each machine Mi from Ms

Schedule the maximum priority in max window available Mibased on

Fitness selection task f(x). Mi $= \int f(x) \times js(i)$

Create priority sequence task on max support Si $= \Sigma si + Miji)$

End

Step4: stop

The assortment of jobs is founded on three factors: instruction, lowest priority and maximum importance. In each repetition, once a job finishes its allocated jobs, the similar is completed for the outstanding unvisited jobs. It permits active distribution and selection of VM machines. Lastly, get the set of purpose rows that perform the Desk table.

3.4 Optimal Time Desk Task Allocation

Time desk contain the virtualized execution task information completed status and time state. Based on the executed weightage the sequence of the scheduling principle is allotted to the completed slot weather the machine is idle to release the process of execution.

The time slot principle, make span process the allocation by desk information achieved by utilizing the sequence of information to give the priority execute the task.

Procedure:

Input: Time consumption T, service S, VM.

Output: Optimized VM service execution Task

Step1: create desk Information for sequence priority switching for each sequence Si

from SS Compute service make span Mk =

$$\int_{i=1}^{N} \sum Ti(j) \in Si$$

$$\int_{i=1}^{N} \frac{\sum Ti(j) \in Si}{Max(Ti(j) \in Si) \times N} \times 100$$

Queuing allotment retain units RU = St End

Step2: For all sequence Si→Sequence task St (Ex→Ts) Estimate the slot VM Desk information

Scheduled weight (sw) = maxi support MS→Ex(Ts)

Sw←Ts

End.

Step3: select the optimal desk VM to set task

Optimal solution Os = Max (Sw)×SS.

Step4: End task execution.

The Make span allocation give scheduled task of execution by depending the waiting task to switch the time based allocation in optimal virtualization state. This allocation reduce the VM utilization, time consumption, to improve the cloud processing.

4   Result and Discussion

The proposed approach implemented and verified in different jobs and machines and at different times at different jobs. Microsoft visual frame work is developed with cloud AWS server for VM migration, up to 1000 tasks with random queuing service request is applied to simulate the result tested with prosed system. The various testing parameters is applied to shows comparison with existing system.

The comparison results on Energy performance efficiency the proposed approach compared with different types of algorithm performance present in Table 1. The proposed technique has improves the energy performance than ACO, DWFA, and FFJSP approaches.

Figure 2 defines energy gain performance the proposed algorithm provide high gain compared to previous methods. The proposed algorithm Multi agent approach achieves 7.2 Joules for 50 jobs similarly the existing algorithm ACO, DWFA, and FFJSP approaches results are 4.2J, 5.2J, and 6.8J respectively.

Table 1. Simulation parameter tool

No. of. Service	ACO in J	DWFA in J	FFJSP in J	ODSSA in J
10	1.2	1.8	2.5	2.8
20	1.8	2.8	3.8	4.2
30	2.5	3.3	4.5	4.7
40	3.5	4.3	5.8	5.9
50	4.2	5.3	6.8	7.2

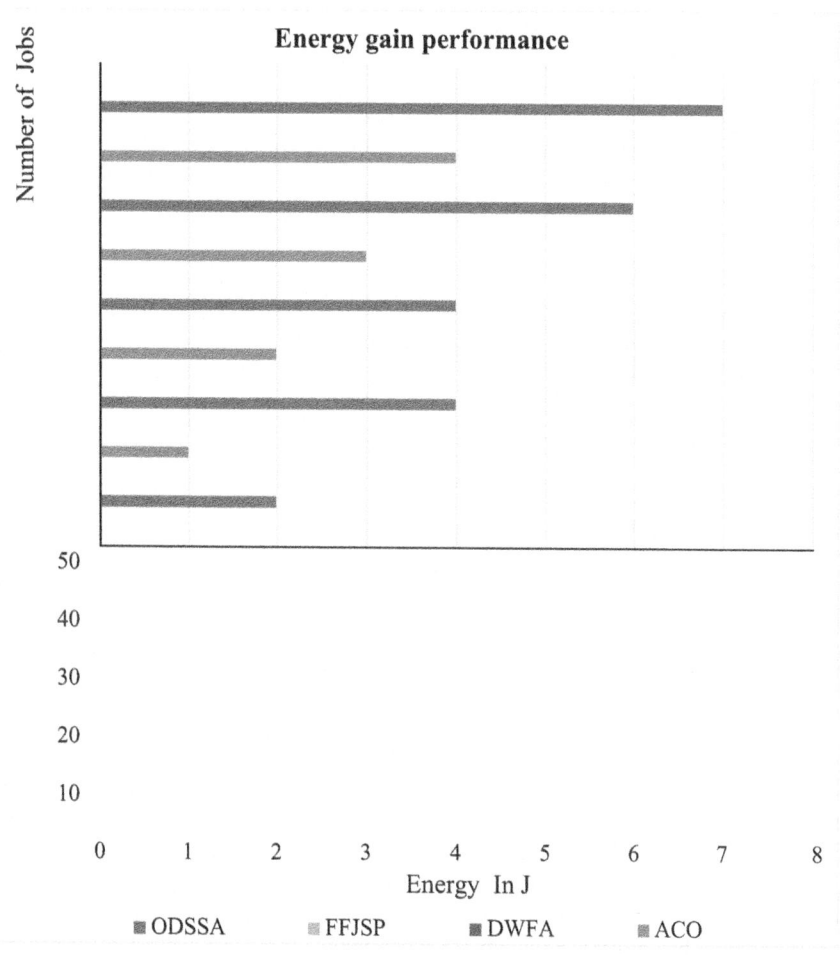

Fig. 2. Analysis the Energy gain performance

The comparison results on makespan time complexity performance the proposed approach compared with different types of algorithm performance present in Table 2.

Table 2. Exploration of makespan time complexity performance

No. of. Service	ACO in J	DWFA in J	FFJSP in J	ODSSA in J
10	45	39	35	32
20	55	48	39	36
30	59	52	42	39
40	62	55	50	45
50	68	60	58	50

The proposed technique has improves the energy performance than Heuristic, ACO, DWFA, and FFJSP approaches.

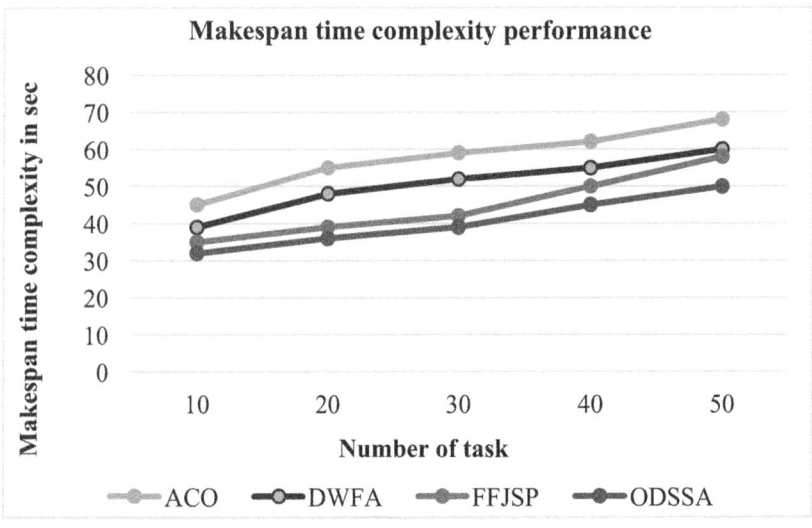

Fig. 3. Analysis of make-span time complexity performance

Figure 3 represents the analysis of make-span time complexity performance for Job scheduling comparing the previous and proposed methods. The proposed The proposed algorithm Multi agent approach achieves 50 s for 50 jobs similarly the existing algorithm ACO, DWFA, and FFJSP approaches results are 68 s, 60 s, and 58 s respectively.

The comparison results on end-to end delay performance the proposed approach compared with different types of algorithm performance present in Table 3. The proposed technique has improves the energy performance than ACO, DWFA, and FFJSP approaches.

Figure 4 described Delay performance exchanges generated in various ways show the final output response of the service in the proposed order and have a longer duration.

Table 3. Exploration of End to end delay performance

No. of. Service	ACO in J	DWFA in J	FFJSP in J	ODSSA in J
10	2.6	0.5	2.3	0.1
20	2.8	0.7	2.5	0.3
30	3.2	0.9	2.7	0.5
40	3.7	1.3	2.9	0.8
50	3.9	1.6	3.1	1

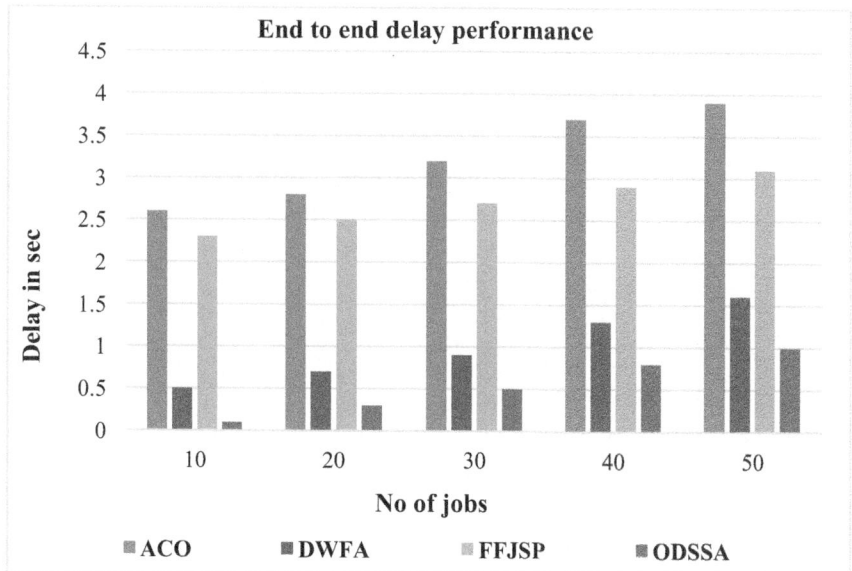

Fig. 4. Analysis the End to end delay

Table 4. Exploration of scheduling performance

No. of. Service	ACO in J	DWFA in J	FFJSP in J	ODSSA in J
10	30	35	40	48
20	36	42	49	55
30	45	52	62	69
40	54	61	70	75
50	60	68	74	80

The comparison results on scheduling performance efficiency in a different types of algorithm performance present in Table 4. The proposed technique has improves the scheduling performance than ACO, DWFA, and FFJSP approaches.

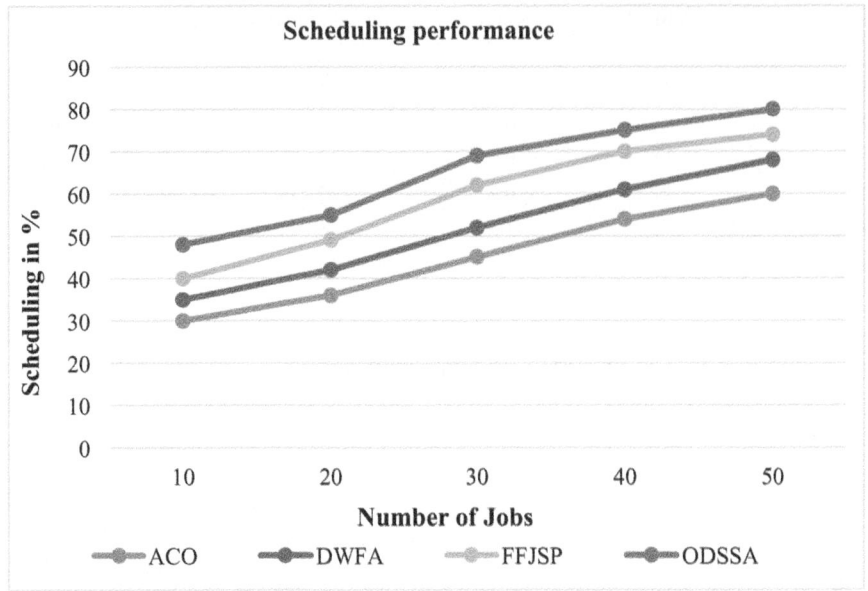

Fig. 5. Analysis the Scheduling Performance

Figure 5 describes the scheduling performance based on the number of jobs, and scheduled the tasks comparing the existing and proposed approaches, in the proposed method Multi Agent algorithm shows the better scheduling performance.

5 Conclusion

To conclude that Offloading dynamic switching scheduling algorithm based on optimal time desk task allocation model to increasing the cloud computing process. Initially the task loading process timing consumption was estimated based on Expected time completion matrix model. Then the time slot is constructed using min max queuing theory. Depends on ETCM and MMQT dynamic switching scheduling algorithm is applied to balance the scheduling task to complete the task using optimal time desk task allocation model. The proposed system achieves high performance compared to other system a well in scheduling state to optimize the task balancer.

References

Moreno-Vozmediano, R., Montero, R.S., Huedo, E., Llorente, I.M.: Efficient resource provisioning for elastic cloud services based on machine learning techniques. J. Cloud Comput. 8(1), 5 (2019)

Tesfatsion, S.K., Klein, C., Tordsson, J.: Virtualization techniques compared: performance, resource, and power usage overheads in clouds. In: Proceedings of ACM/SPEC International Conference on Performance Engineering, pp. 145–156. Association for Computing Machinery, Berlin, Germany (2018)

Zhang, F., Liu, G., Fu, X., Yahyapour, R.: A survey on virtual machine migration: challenges, techniques, and open issues. IEEE Commun. Surveys Tuts. **20**(2), 1206–1243 (2018)

Marimuthu, M., Akilandeswari, J., Chellaiah, P.R.: Identification of trustworthy cloud services: solution approaches and research directions to build an automated cloud broker. Computing **104**, 43 (2021)

Karthick, K., et al.: Iterative dichotomiser posteriori method-based service attack detection in cloud computing. Comput. Syst. Sci. Eng. **44**(2), 1099–1107 (2023)

Gamsız, M., Özer, A.H.: An auction based mathematical model for energyaware virtual machine allocation in clouds. In: Proceedings 4th International Conference Computer Science Engeneering (UBMK), pp. 1–6, (2010)

Tan, X., Leon-Garcia, A., Wu, Y., Tsang, D.H.K.: Online combinatorial auctions for resource allocation with supply costs and capacity limits. IEEE J. Sel. Areas Commun. **38**(4), 655–668 (2020)

Sakthivel, S.: UBP-trust: user behavioral pattern based secure trust model for mitigating denial of service attacks in software as a service (SaaS) cloud environment. J. Comput. Theor. Nanosci. **13**(10), 7649 (2016)

Liu, X., Cheng, B., Wang, S.: Availability-aware and energy-efficient virtual cluster allocation based on multi-objective optimization in cloud datacenters. IEEE Trans. Netw. Service Manag. **17**(2), 972–985 (2021)

Karthikeyan, D., Mohan Raj, V., Senthil Kumar, J., Suresh, Y.: Intrusion detection using ensemble wrapper filter based feature selection with stacking model. Intell. Autom. Soft Comput. **35**(1), 646–659 (2022)

Filho, M.C.S., Monteiro, C.C., Inácio, P.R.M., Freire, M.M.: Approaches for optimizing virtual machine placement and migration in cloud environments: a survey. J. Parallel Distrib. Comput. **111**, 222–250 (2018)

Kasthuripriya, S., et al.: LFTSM-local flow trust-based service monitoring approach for preventing the packet during data transfer in cloud. Asian J. Inf. Technol. **15**(20), 3927 (2016)

Liu, X., Cheng, B., Wang, S.: Availability-aware and energy-efficient virtual cluster allocation based on multi-objective optimization in cloud datacenters. IEEE Trans. Netw. Serv. Manag. **17**(2), 972–985 (2020)

Sha, J., Ebadi, A.G., Mavaluru, D., Alshehri, M., Rajabion, L.: A method for virtual machine migration in cloud computing using a collective behavior-based metaheuristics algorithm. Concurr. Comput. Pract. Exp. **32**, e5441 (2019)

Dhiyanesh, B., Sakthivel, S.: Secure data storage auditing service using third party auditor in cloud computing. Int. J. Appl. Eng. Res. **10**, 37 (2015)

Abdullahi, M., Ngadi, M.A., Dishing, S.I., Abdulhamid, Si.M., Ahmad, BIe.: An efficient symbiotic organisms search algorithm with chaotic optimization strategy for multi-objective task scheduling problems in cloud computing environment. J. Netw. Comput. Appl. **133**, 60 (2019)

Marahatta, A., Pirbhulal, S., Zhang, F., Parizi, R.M., Choo, K.-K.R., Liu, Z.: Classification-based and energy-efficient dynamic task scheduling scheme for virtualized cloud data center. IEEE Trans. Cloud Comput. **9**(4), 1376–1390 (2021)

Sakthivel, S.: F2C: an novel distributed denial of service attack mitigation model for Saas cloud environment. Asian J. Res. Soc. Sci. Humanit. **6**(6), 192–203 (2016)

Liu, Y., Zhao, Y., Dong, J., Li, L., Wang, C., Zuo, D.: I-Neat: an intelligent framework for adaptive virtual machine consolidation. Tsinghua Sci. Technol. **27**(1), 13–26 (2022)

Lo, H.-Y., Liao, W.: CALM: survivable virtual data center allocation in cloud networks. IEEE Trans. Serv. Comput. **14**(1), 47–57 (2021)

A Quick and Effective Loan Eligibility Analysis Model Using LightGBM

C. Gowdham[2], R. Ashwin[1(✉)], and T. S. Senthil Raj[1(✉)]

[1] Department of Computer Science, Hindustan Institute of Technology and Science, Chennai, Tamil Nadu, India
ashwinrk2001@gmail.com, senthilsdd@gmail.com
[2] School Of Computing & Information Technology, REVA University, Bangalore 560064, India
gowdhamchinnaraju@gmail.com

Abstract. In recent times, it is extremely common for the people to make use of bank credits to satisfy their day to day needs. This is increasing rapidly. There are lot of schemes are provided by the banks. Lending money is one of the most crucial bank programmes. Banks generally provide loans to customers based on their requirements. But, due to some situation, some clients cannot repay their debts on time or they will make it delay because of lack of finances. There are so many people who take advantage of this and misuse the facilities that are provided by the bank. To overcome this issue, Banks should utilize certain methods to help in predicting the status of loan repayment. The financial system needs an appropriate modelling system to handle a variety of problems. For any bank, one of the most challenging tasks is the task of anticipating loan late payers. However, the banks will undoubtedly be able to minimise their non-profit assets in order to decrease their loss as a result of the expected bankruptcies. As a result, it enables the repayment of approved debts to proceed with no losses and contributes to the credit report. This demonstrates the significance of researching this technique to foresee lending decisions. The proposed method utilizes machine learning methods for predicting default on loans. This is because they provide greater accuracy on prediction challenges.

Keywords: Machine learning · Deep learning · Predictive analysis · Data analytics · Neural networks

1 Introduction

The system of loans controlled by lenders is one of the key elements that significantly influence both the financial and economic well-being of our nation. Banks all over the globe have performed the evaluation of danger to banks. Multiple methods are used to calculate severity levels because it is very challenging to determine creditworthiness. Financial management of risks is also regarded as one of the crucial tasks performed by the field of banking. The information of previous consumers from different banks, whose advances on a number of limits have been verified, is used in the suggested method. Since we are aware that the main operation of all banks is the movement of funds, an artificial intelligence algorithm is created on the information in order to produce the intended outcomes. Forecasting the success of financing is the primary objective of this examination.

Rewards gained from advancements that the bank communicates are the primary component which makes a contribution to its helpful assets. Both the candidate and banker staff will find the technique of financing approval estimation to be much simpler. This method's primary goal is to offer a quick, easy, and straightforward way to select the qualified candidates.

The bank should be responsible to make sure their assets are perfect. The execution of this system helps to find and assure that applicant is safe or not for the loan. This is performed with the help of this automatic loan approval prediction system. The bank may profit from a number of things, including giving individuals a deadline to determine whether or not their borrowing request will be approved. This prediction system is so productive because it gives lenders the freedom to focus more on the bank's significant resources rather than on the weak candidates. This approach will cut down on the amount of time it takes for the potential borrower to complete his or her credit registration. The solution associated with a specific Lending identification is forwarded to various banking divisions. Due to this, the necessary action on application can be taken by them. Hence, it is benefit for all others department to perform other formalities.

When completing a digital framework, financial businesses and banking institutions must streamline the lending verification process (continuously) according to information supplied by clients. It covers a variety of details, such as gender, relationship status, studying, count of family members, earnings, credit value, and financial standing. The problem is created to bring together the consumer groups which are eligible over a given amount of credit in order to electronically store this link, and it enables those to precisely focus on these customers. In such scenarios, a fractional informational collection is provided. One of the most common real-life complexities that is faced by all companies in their lending operations is the Approval of Loan. The automation of the financing clearance processes may lower human workload, which in turn accelerates up customer service. The improvement in consumer fulfillment and the cost reductions from businesses are crucial factors to take into account. However, the bank can only gain from it if it has a reliable method in place for determining what customer loans to approve and which to deny. This is crucial for reducing the potential danger.

The method has some serious flaws, like how it assigns various quantities to every component. But in everyday circumstances, financing approval may occasionally be dependent solely on a single, significant factor, which is impossible for this system to handle.

- The proposed system architecture is implemented based on comparative analysis of various algorithms applicable for loan eligibility analysis using machine learning.
- The suggested system takes into account Light gradient boost model, the naive Bayesian approach, a random forests classification algorithm, and Logistic regression approach.
- A dynamic web application that collects customer records dynamically.
- Training of the system considers existing loan eligibility data gather from bank.

- Further testing with each machine learning algorithms, the proposed system identifies the best performing system to provide the eligibility result and criteria.

The remainder of the essay is structured as an extensive literature review in Part II. In Part III, the choice of system resources and issue determinations are covered. In Part IV, the structure of the system design and specific system design stages are covered. The paper's conclusion discusses potential future development.

2 Background Study

M. Alaradi et al. 2020 This paper generates a high performance predictive method for loan approval prediction with the help of decision trees. Several studies using different tree-methods have been conducted. The decision tree, which is very simple and easy to comprehend, is at the bottom of the list, and random forests are at the top. With regard to simpler decision trees, the efficacy of this research is inadequate. Due to the complicated and strongly associated feature set, many of the important factors affecting lending decisions were not taken into account. So, it ended up with an inappropriately simplistic tree. The boosting method, however, produced outstanding efficiency, importance, as well as understanding and as evidenced by the significance of the chart's precision ratings of [98.75%], [100%], [92.85%] for minority class forecasting precision and [97.0%] for categorization performance on the experimental sample. Therefore, a boosting-based decision-tree forecasting framework was suggested to arrive at judgments regarding the suitability of those seeking loans based on their individual characteristics.

M.S. Park et al. 2021 the number of data increases rapidly, and hence it is not possible to make assumptions by the existing economic analysis method which is not satisfied and the complexity arises to create a new analysis model. Hence for the bankruptcy prediction, the demand for utilizing the machine learning techniques is increased because of its high performance. Standard tree-based theories, that possess the ability to assess the characteristic significance of each model on their own, are where the method is first applied. This demonstrated that the attribute significance assessed by LIME might serve as a suitable indication of the significance of the features assessed by model based on trees. The projected liquidation chance from the model is also used to study the reliability of feature significance. This suggests that it may be feasible to handle loan qualification criteria fairly by using measurements of important characteristics as a foundation.

Ugochukwu.E 2022 In all industries, involving housing, privacy, computer science, and the financial services industry, machine learning methods are regarded as the modernizing power. One of the most challenging processes in the financial sector is granting loans. The dataset gets processed and assessed with the Python language tools on Kaggle's Jupyter Notebook cloud-based platform. Our study's findings demonstrated that the Random forest method provides exceptionally fast precision. It has the greatest grade of 95.55%, while the score for logistic regression is the least at 80%. In terms of precision-recall and reliability, the efficacy of the suggested Method is significantly better than that of the other two of the three loan forecasting approaches noted in the research.

Miraz Al Mamun 2022 Machine Learning (ML) algorithms are helpful for separating the patterns from a common loan-approved dataset and predicting the relevant loan applicants. It makes use of the previous data from the customers to make the study, including their age, income type, loan annuity, last credit bureau report, Type of organization they work for, and length of employment. In order to identify the most deserving features, that is, the element which has the most impact on the prediction output, ML methods such as Random Forest, XGBoost, Adaboost, Lightgbm, Decision tree, and K-Nearest Neighbor were employed. The above algorithms are compared and evaluated against one another with the help of standard metrics. Among these, the highest accuracy of 92% is obtained by the Logistic Regression.

Murthy et al. 2020. The profit that is gained from loans is the main source of income of bank assets. The main aim of banks is investing their assets in safe customers. In recent days, the banks will only sanction loan after so many process of checking and evaluation. However, it is not possible to find whether the chosen customer is safe or not. Hence, there is a necessity to use different methods in banking sector for the correct selection of customer who pays loan on time. This method utilizes random forest algorithm for the categorization of data. Random forests algorithm develops a method from trained dataset and this method is used on test data and the required result is obtained.

3 System Design

In this modern world, getting loans from banks is very common. The main business of the bank is lending money. The main benefit that the bank can get from loan is the interest on it. On the other hand, it is capable of being given to only a handful of individuals due to the banks' inadequate balances. Choosing who can receive the loan and who constitutes the best option for the bank is part of the usual procedure. Loan eligibility prediction is considered as a categorization issue. It includes the prediction of whether a loan would be approved or not. In such complex scenarios, the discrete values should be predicted based on a given set of independent variables. The profit or a loss of the bank depends mainly on loans. In other words, it relies on whether the borrowers are making their loan payments or are in violation. The bank may be able to reduce its bad debts after bankruptcy proceedings are anticipated. This increases the significance of this research. Every feature associated with the approval process can be ranked by the Credit Estimation Systems, and identical characteristics may be handled according to their dimensions on fresh test results. In most cases, the potential borrower is given a timeframe by which to learn whether or not he or she's credit will be confirmed.

The candidate's information is carefully verified by bank staff before the loan is approved for the appropriate candidate. Examining each applicant's specifics in depth takes a lot of time. A bank's credit risk is estimated via a machine learning neural network algorithm. The Feed-forward Back Propagation Neural Network anticipates loan rejection. The technique by which two or more classifiers merge to create a combined model for more accurate forecast. Additionally, the random forest method is used. It employ both the bagging and boosting approaches. Classifiers have the duty to improve the efficiency of the information and offer greater effectiveness. In addition to multi-class grouping, this article presents various grouping techniques for binary categorization. The fresh approach of grouping is called COB, and it very successfully performs segmentation. With noise and classificational outlier information, it is, however, additionally weakened. The group-based method improves the learning data collection results, it is eventually determined.

- Inspecting the details of all applicants is a time consuming one.
- The possibility of human error may occur when inspecting all details manually
- There occur some chances of assigning loan to ineligible applicant.
- The detailed methodology and steps involved are explained over here in Section IV.

4   Methodology

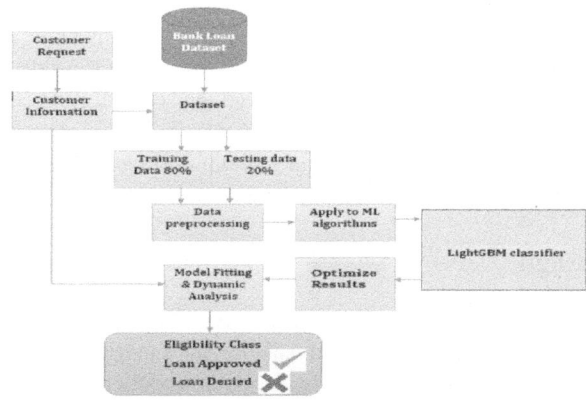

Fig. 1. System architecture of Proposed Loan eligibility analysis model

Figure 1 depicts the suggested creditworthiness analytical model's system structure. According to the information given by the prospective borrower, the banking management must streamline the loan approval process (in real-time). While filling out an application form, information including Debt Amount, Sexuality, Relationship Position, Revenue, Financial History, Learning, The total number of family members, as well as some other specifics are required. They developed an approach to help recognize the different candidate kinds who are eligible for different loan amounts and then target them in particular in accordance to simplify things better. This is regarded as an issue of category because all that must be grouped before deciding whether the loan's current status is positive or negative. A loan application's probability of being approved or denied can be determined by the algorithm much more quickly.

To achieve the desired outcome, the suggested method employs an entirely distinct algorithm. Python is used because it has all of the required tools and libraries, making it one of the most commonly used and well-liked programming languages in AI and ML.

4.1   Platform Configurations

Numerous tools are included in it, including matplotlib for data display and experimental investigation of data and pandas for input screening. Furthermore, it makes use of sklearn, a component of Scikit-learn that contains numerous grouping, the theory of regression, and categorization techniques that are frequently employed in machine learning as well as AI. The specified collection of data containing all of the details about our clients is then used by the framework for implementing the above method. The methods are applied sequentially in an orderly manner. Following that, the data is analyzed, categorizes, and passed into an algorithm for training it. The accuracy percentage is given shortly after every algorithm has been calculated.

4.2 Model Analysis

To produce an accurate outcome, the predictive system undergoes training via a variety of algorithms. Utilizing a 70% training sample, 15% assessment set, and 15% assurance set, the XGB Classifier Algorithm, LGBM Classifier Algorithm, Decision Tree Algorithm, and Random Forest Classifier Algorithm are used. It is found that the XGB Classifier Algorithm and LGBM Classifier provide good precision. Following the evaluation procedure that it learns from the data used for training sets, the algorithm makes a prediction regarding whether the present applicant is likely to be approved for a loan. Hence, the result shows that, if the detection of the capable borrower is good, then it more beneficial to the organization.

- The loan approval time they spend will be as brief as possible.
- Human mistake won't occur because the complete process will be computerized.
- For those who meet the criteria, funding will be authorised right away.

4.3 Data Collection

Both the training set and the testing set are supplied with the information that was amassed for forecasting debt failure customers. The training set and testing group are typically separated using 80:20 ratios. The decision tree-based data model that was created will be used to the training set, and based on how well the test takers performed; test set estimates is carried out. Loan-id, Gender, Dependents, Education, Self-Employed, Applicant Income, Coapplicant Income, LoanAmount, Loan_Amount _term, Credit_history, etc. represent a few of the characteristics in the dataset.

4.4 Data Pre-process

The collected data will sometimes have missing values which may cause inconsistency. So, for obtaining best results, the data need to be pre-processed. As a result, the system's performance is increased. The factors should be converted, and the anomalies should be removed. To resolve these issues, the chart function is utilized.

4.5 Building a Model

Following pre-processing, the data collection is thoroughly evaluated in order to evaluate it and gain an acute awareness of its characteristics. A exploratory approach to data analysis is used to create controlled and unsupervised learning models after it is finished. Before the modelling process begins, numerous theories are first generated by studying the information at hand. EDA is used to validate and assess the theories that have been put forth. The related Univariate Evaluation Methods are frequently used to carry out the Qualitative Information Evaluation. The Bivariate Study is carried out in order to determine the relationship between each factor in the data measure and the objective component. It discusses the findings of every category in the unstructured data inventory.

4.6 Model Training

The algorithm has now been trained using the training information set, and it can forecast results using the test information. Two paths, such as train and declarations, make up the trained data collection. This training portion trains the algorithm, which aids in making predictions for the testimony portion. In this manner, the forecast is assessed because it offers the best explanation for the testimony portion (which is missing from the test sample).

4.7 Light GBM Method

It is claimed that LightGBM is a gradient enhancement architecture that relies on decision trees in order to boost the accuracy of models while limiting memory consumption. It uses two significant methods, exclusive feature binding (EFB) and gradient-based single-side averaging. In all GBDT (Gradient Boosting Decision Tree) systems, the histogram-based method is the most common technique utilized, and this fulfills its requirements. The properties of the LightGBM Method are made by the two GOSS and EFB approaches that are described below.

They collaborate to improve the device's capabilities and set it apart from other GBDT models. Gradient-based LightGBM One Side Sampling Method:

In the computation of knowledge benefit the numerous data points play a variety of responsibilities. The benefit of data will be greater for the cases with greater gradients (i.e., under-trained examples). For preserving the precision of data gain calculation, GOSS uses examples that have substantial gradients (e.g., bigger than a pre-set cut off and within the higher percentiles) and only drops cases with little gradients at chance. When the value of data benefit has a broad spectrum, this eventually yields gain prediction that becomes more precise than evenly sampling at random with an identical goal testing pace.

LightGBM Pseudocode

1. Initialize Database, Load dataset opted for analysis
2. Split dataset into training datra and validation sets
3. Define LightGBM parameters to make analysis
4. Split the dataset and train the model uisng LightGBM regressor to evaluate prediction
5. Formulate accurate predictions on validation set 15%, Training set 70%, Testing set 15%
6. Once the model Fits, evaluate performance of model using metric scores

Step 1: Load the dataset
input = load_dataset*('data.csv')
Step 2: Split dataset into training and validation sets
train_set_, val_set_ = split_dataset(input)
Step 3: Call LightGBM parameters
params = {
    '.objective.': '.binary'.,
    '.boosting_type.': '.gbdt.',
    '.metric.': 'auc'.,
    '.num_leaves.': 31,
    '.learning_rate.': 0.05,
    '.feature_fraction.': 0.9,
    '.bagging_fraction.': 0.8,
    '.bagging_freq'.: 5,
    '.verbose.': -1
}
Step 4: Deploy the model and train LightGBM algorithm
model = lgb.train(Init_params, train_set., num_boost_round.=100,
valid_sets.=[val_set], early_stopping_rounds.=10)
Step 5: Analyze the data and make prediction
prediction_result= model.predict_(val_set.)
Step 6: Final_metrics_score = Classification_performance(val_set, predictions)

4.8 Predicting the Outcome

The first step is importing all the necessary python modules and then imports the database for both TESTING and TRAINING. When it's finished, make sure no NULLVALUES are displayed. Fill the database with the appropriate coding if NULLVALUES departs. The exploratory data analysis procedure is completed for all ATTRIBUTES in the table, and all plots are then mapped with the MATPLOTLIB tool. After that, construct the LIGHT GBM Model for the coding. Finally, we the proper output is predicted by using the predicted model.

5 Results and Discussion

Figure 2 Shows the system interactive front-end system collects required information from the consumer.

Figure 3 Shows the system front-end interactive web page of proposed model with instructions given in the screen. Based on the input arrived from the consumer details, the notification message on loan eligibility is shown.

Figure 4 Shows the system loan rejection status collected from the backend analysis of various eligibility criteria.

Figure 5 Shows the system home loan rejection notification window after the critical analysis done by the proposed lightGBM technique.

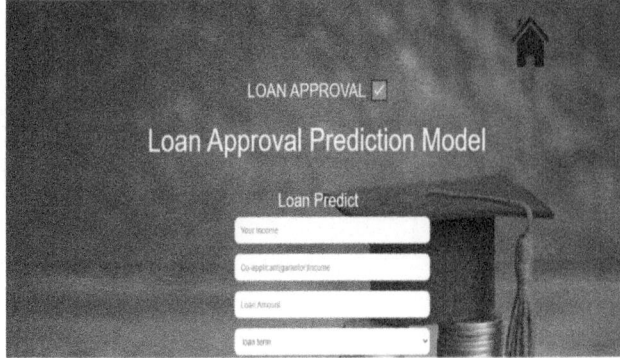

Fig. 2. Interactive front-end

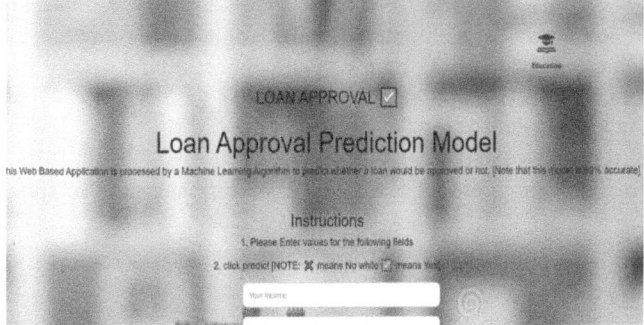

Fig. 3. Notification window

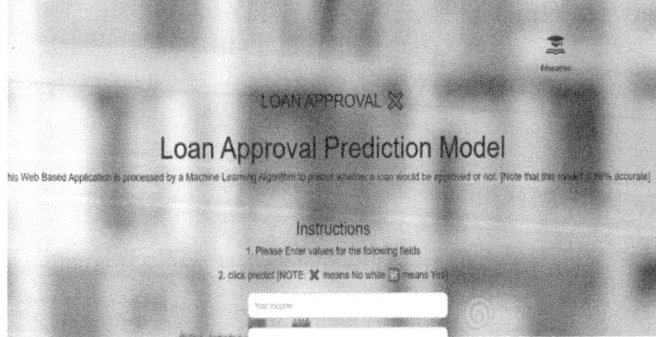

Fig. 4. Education Loan rejection notification

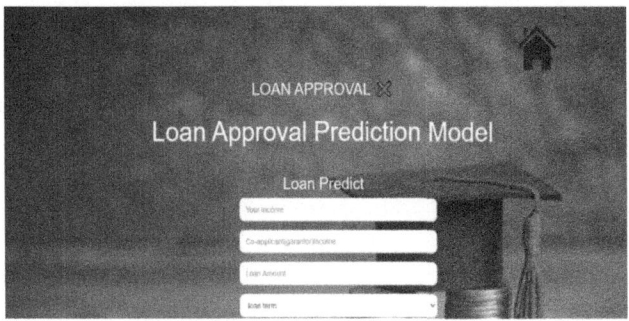

Fig. 5. Home Loan rejection notification

6 Conclusion

One of a bank's most challenging tasks is forecasting who will default on credit. However, the anticipation of loan defaulters will undoubtedly assist banks in minimizing their non-profit assets to reduce their loss. As a result, it contributes to the bank statement and enables the restoration of sanctioned loans to continue without incurring losses. The significance of investigating this loan approval prediction method is made clear by this. The suggested method uses deep learning methods to forecast failures of loans. This is because they solve prediction challenges with greater precision. Comparing the various existing approaches, LightGBM model achieved better accuracy of 97% comparing existing state-of-art approaches.

References

1. Zie̦ba, M., Tomczak, S.K., Tomczak, J.M.: Ensemble boosted trees with synthetic features generation in application to bankruptcy prediction. Expert Syst. Appl. **58**, 93–101 (2020)
2. Barboza, F., Kimura, H., Altman, E.: Machine learning models and bankruptcy prediction. Expert Syst. Appl. **83**, 405–417 (2020)
3. Chen, T., Guestrin, C.: XGBoost: a scalable tree boosting system. In: Proc. 22nd ACM SIGKDD Int. Conf. Knowl. Discovery Data Mining, pp. 785–794 (2020). https://doi.org/10.1145/2939672.2939785
4. Ke, G., et al.: LightGBM: a highly efficient gradient boosting decision tree. In: Proc. Adv. Neural Inf. Process. Syst., pp. 3146–3154 (2019)
5. Ancona, M., Ceolini, E., Öztireli, C., Gross, M.: A unified view of gradient-based attribution methods for deep neural networks. In: Proc. NIPS Workshop Interpreting (NIPS). Zürich, Switzerland: ETH Zürich, 2020. [Online]. Available: https://scholar.google.co.kr/scholar?start=10&hl=en&as_sdt=0,5&cluster=7129422820232184089
6. Gilpin, L.H., Bau, D., Yuan, B.Z., Bajwa, A., Specter, M., Kagal, L.: 'Explaining explanations: an overview of interpretability of machine learning. In: Proc. IEEE 5th Int. Conf. Data Sci. Adv. Analytics (DSAA), Oct. 2020, pp. 80–89
7. Carvalho, D.V., Pereira, E.M., Cardoso, J.S.: 'Machine learning interpretability: a survey on methods and metrics. Electronics **8**(8), 832 (2019)
8. Ribeiro, M.T., Singh, S., Guestrin, C.: Why should I trust you?': explaining the predictions of any classifier. In: Proc. 22nd ACM SIGKDD Int. Conf. Knowl. Discovery Data Mining, pp. 1135–1144 (2020)

9. Kim, M., Kim, J., Park, K.: 'A study on financial characteristic of delisting companies by Kosdaq. Rev. Accounting Policy Stud. **16**, 125–142 (2019)
10. Bae, Y.I., Song, S.H., Hong, S.K., Yu, S.Y.: 'The comparative analysis of financial factors that influence on corporate's survival and bankruptcy: before and after foreign exchange crisis in Korea. IE Interfaces **21**(4), 385–393 (2019)

Lifting Wavelet Transform Based Data Steganography for Health Care and Military Applications

Bini M. Issac[1,2]([⊠]) [iD] and S. N. Kumar[2,3] [iD]

[1] Department of CSE, Amal Jyothi College of Engineering, Kanjirappally, India
binimissac@amaljyothi.ac.in
[2] APJ Abdul Kalam Technological University, Thiruvananthapuram, Kerala, India
[3] Department of EEE, Amal Jyothi College of Engineering, Kanjirappally, India

Abstract. Steganography is the practice in which a message or information is hidden within another seemingly innocuous message or file, without anyone else being aware of its existence. Its importance lies in its ability to provide secure and covert communication between parties without raising suspicion from third parties who may be monitoring or intercepting the communication. Steganography has been used throughout history to transmit confidential information, such as military secrets or espionage, without being detected. In the modern era of digital technology, steganography is more important as the use of electronic communication has become widespread particularly in the healthcare sector for teleradiology applications. To securely transmit sensitive information like confidential patient data for teleradiology applications, we can apply steganography together with cryptography. For sensitive data transmission in military applications also, we can employ this method. In this work, AES encryption algorithm is applied to the confidential patient medical data and then it is embedded it into a cover image using Lifting wavelet transform and transferred through the cloud network. At the receiver side, the decrypted confidential data is extracted and we have compared the method using different wavelet transforms and the results obtained shows that haar and bs3 transforms have better PSNR and MSE values compared to other transforms.

Keywords: Steganography · AES encryption · Lifting wavelet · teleradiology

1 Introduction

Steganography provides a layer of security for sensitive information by hiding messages in seemingly innocuous files such as images, videos, or audio files [1] that cannot be achieved through encryption alone. Even if an attacker intercepts the communication and breaks the encryption, they may not be aware that hidden information exists, making it more difficult for them to decipher the message. Steganography is also used for digital watermarking, copyright protection, and tamper detection, among other purposes [17]. Overall, the importance of steganography lies in its ability to provide secure and covert communication, which is essential in many fields, including military, law enforcement, business and healthcare applications.

P. Dassan et al. (Eds.): ICICSCNT 2023, CCIS 1970, pp. 69–81, 2024.
https://doi.org/10.1007/978-3-031-75957-4_7

Steganographic techniques are divided into two groups: frequency-domain techniques and spatial domain techniques. The cover image itself contains the hidden messages when using spatial domain approaches. The benefits of spatial domain approaches include straightforward implementation, high payload, easy control, and stereo image quality. Frequency domain techniques are another common method of data concealment. In frequency domain techniques, before embedding the hidden message, the image is transformed into frequency domain coefficients. The most popular are Fast Fourier Transforms (FFT), Discrete Cosine Transform (DCT), and Discrete Wavelet Transforms (DWT) [18].

Cryptography is a technique utilized to safeguard confidential information. The method involves encrypting a secret message using a specific encryption key. Even if a third party is aware of the existence of a secret message, they cannot comprehend or interpret it unless they possess the decryption key. The primary benefits of cryptography are to ensure the confidentiality and security of the message content. However, this technique does not conceal the message itself, which is considered suspicious. Additionally, the privacy of data transmission is entirely dependent on the secrecy of the encryption key.

The paper organization is as follows. Section 2 discusses some of the related works. Section 3 discusses about the proposed approach using AES algorithm and Lifting wavelet transform. Section 4 focusses on the performance metrics considered for the evaluation and in Sect. 5, we conclude our work.

2 Related Work

This section presents an overview of the related studies of steganographic methods.

In [2] Kumar et al. proposed a method combining the AES encryption and LSB cryptography for providing double layer of security to the data transmitted and they evaluated the PSNR and MSE values. Bansal et al. [3] proposed a method combining cryptography and steganography for hiding secret information. They used Elliptic Curve Cryptography (ECC) for encoding the secret information and Least Significant Bit Inversion method is used for embedding the encrypted information to a cover image. Khari et al. [4] suggested a way to enhance the security of data in the Internet of Things (IoT) field by combining cryptographic and steganographic algorithms. They introduced an Elliptic Galois cryptography protocol that encrypts secret data obtained from medical sources. Additionally, they utilized a Matrix XOR encoding steganography method to conceal the data that is encrypted within a simple image. To improve the process, they also used an optimization algorithm known as Adaptive Firefly for optimizing the selection of cover images.

Maksoud et al. [5] introduced a sound encryption system based on a fractional-order chaotic system. Saleh et al. [6] proposed a merged technique for data security using steganography and cryptography using the AES algorithm for encrypting the messages and then for hiding the encrypted message using PVD_MPK and MALDIP-MPK algorithms. Jero et al. [7] proposed a technique for associating patient information with biomedical signals. They employed the Curvelet transform to conceal patient details within the ECG signal. The Curvelet transform can divide the ECG signal into various sub-bands of frequency, and a quantization method is utilized to embed the confidential data of patient into coefficients that have values near zero in the high-frequency sub-band.

Wahab et al. [8] proposed a data compression algorithm that utilized the RSA (Rivest–Shamir–Adleman) algorithm to improve security when employing lossy and lossless compacting steganography techniques. This approach can effectively reduce the size of transmitted data, facilitating faster transmission over slow internet connections and for conserving storage space on different medias. Huffman coding is used for compressing the text initially, while the cover image is compressed via DWT, which incorporates lossy compression to reduce its dimensions and LSB technique is used for embedding the encrypted data. Wazirali et al. [9], proposed a method incorporating huge amount of data into an image without sacrificing the image quality. The authors proposed a spatial steganography method that employs genetic algorithms. By utilizing Least Significant Bits (LSB) matching between the carrier and the steganographic image, they were able to increase the embedding capacity and decrease distortion. Seyyedi et al. [10] suggested a method that involves partitioning a cover image into 8 × 8 blocks and applying an integer wavelet transform through a lifting scheme to every block. To ensure high security and authentication, a symmetric RC4 encryption method is used to encrypt secret messages. The authors then performed Tree Scan Order in the frequency domain to identify suitable locations to embed the confidential messages. As a result of this process, the secret messages are embedded in the cover image with minimal degradation of its quality.

Shahadi et al. [11] introduced an audio steganography approach which is lossless based on Integer-to-Integer Lifting Wavelet Transform (Int2Int LWT) and LSB substitution method. And for increasing the security level, they proposed an encryption with adaptive key. By utilizing the Int2Int LWT-based speech steganography algorithm, complete recovery of the embedded secret messages is possible on the receiver side. Wavelet-based Secure Steganography with Scrambled Payload was proposed by Reddy et al. (WSSSP) [12]. By dividing the XD band into an upper and lower band for embedding the payload, they transformed the cover image using the Daubechies Lifting Wavelet Transform (LWT). The payload was split into four equal blocks, and the Haar LWT was used to create wavelet transform bands on the alternate blocks (F1 and F2). In the spatial realm, the remaining blocks remained. The stego-objects were mixed together by the authors using Decision Factor-Based Manipulation (DFBM). Lastly, a stego-object was subjected to the Daubechies Inverse LWT2 to produce a stego-image in the spatial domain.

Shirafkan et al. [13] introduced a method for steganography based on lattice vector quantization and DWT which can provide hiding and security to the data and using error correction coding Reed-Solomon encoding. The possibility of successfully mining concealed data was significantly decreased by this method, which employed the 3-level lifting wavelet transform, blocking, and embedding in specific coefficients utilizing the lattice vector quantization. A steganographic technique based on wavelet and modulus function was suggested by Lai et al. [14]. An image is first divided into blocks of a specific dimension, and then each block is separated out into a single-level wavelet. Then, using a greater number of wavelet coefficients, the capacity of the concealed secret data is determined. Finally, steganography based on modulus function embeds hidden information.

Luen Lai et al. [15] proposed an adaptive data hiding technique in their study, which operates in the frequency domain. The original image is divided into sub-blocks measuring 8x8 pixels, and each block undergoes Haar Discrete Wavelet Transform (HDWT) to produce LL1, HL1, LH1, and HH1 bands. As the human eye is not so sensitive to the edge region, large amount of data can be embedded when the LL1 band is complex. For analyzing the complexity of the LH1, HL1, and HH1 bands, a data hiding capacity function was utilized. If these sub-bands are found to be complex, the LL1 band undergoes further decomposition, and additional data bits being embedded in the resulting LH2, HL2, and HH2 bands. Hasoon et al. [17] presented a technique for concealing speech within audio in their study. The characteristics of the secret speech are extracted using Linear Predictive Coding (LPC), and these parameters are then embedded into the audio signal. By applying DWT to audio frames, the signal is split into high and low frequencies, and the embedding parameters are inserted into the high-frequency band. The resulting stego-audio is imperceptible from the original audio and the proposed method enables hiding speech and audio of equal duration.

3 Proposed Approach

This paper proposes a data steganography model comprising of AES (Advanced Encryption Standard) is for encrypting the text data and then lifting wavelet transform for embedding the encrypted text data within a cover image, at the sender side. At the receiver side, the reverse process is applied to extract the encrypted text data and then decryption is performed. Figure 1 shows the steps of encryption and embedding of confidential data within a cover image at the sender side and Fig. 2 shows the steps of extraction of the encrypted data and decryption of the confidential data.

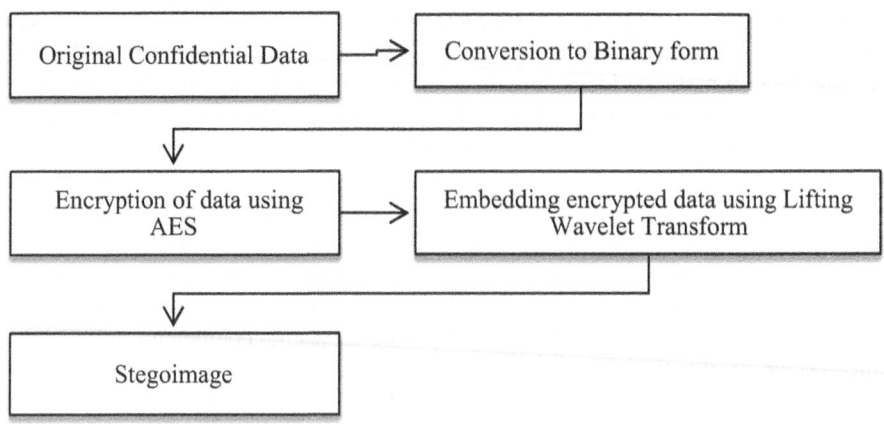

Fig. 1. Encryption and embedding at Sender side

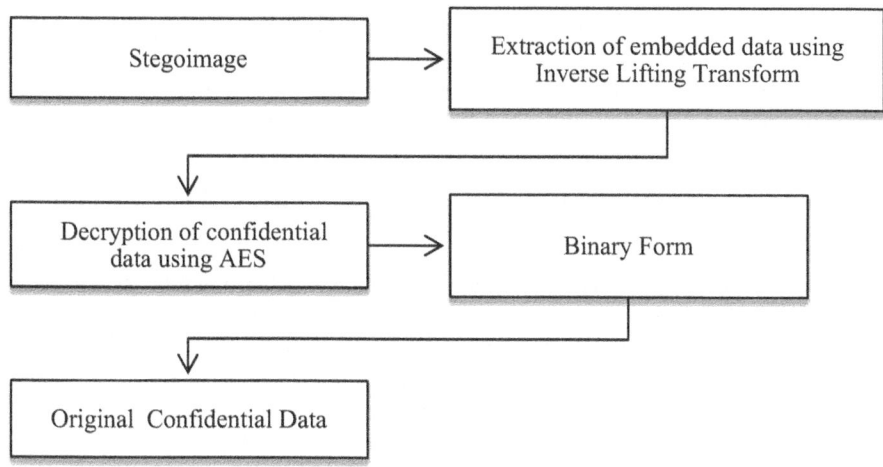

Fig. 2. Extraction and decryption at Receiver Side

3.1 Advanced Encryption Standard (AES) for Confidential Data Encryption and Decryption

AES is a widely used symmetric-key encryption algorithm for encrypting data [19]. In our proposed method, for the encryption of confidential data, AES encryption is used. This algorithm uses a block cipher, which means that it encrypts data in fixed-size blocks, typically 128 bits. It operates on plaintext data using a secret key that is shared between the sender and the recipient. The algorithm has several rounds of encryption, with each round using a different subkey that is derived from the main key.

The various step in the AES algorithm is as follows:

1. Key expansion: The secret key is expanded into a set of round keys, which is used for each round of encryption.
2. Initial round: The plaintext block is combined with the first-round key using a bitwise XOR operation.
3. Main rounds: The plaintext block is processed through a series of transformation rounds, each consisting of four steps:

   - SubBytes: Each byte of the block is replaced with a corresponding byte from a fixed substitution table (called the S-box).
   - ShiftRows: The rows of the block are shifted by a certain number of bytes.
   - MixColumns: Each column of the block is multiplied with a fixed matrix to create a new column.
   - AddRoundKey: The block is combined with the round key for the current round using a bitwise XOR operation.

4. Final round: The final transformation round is similar to the main rounds, but without the MixColumns step.
5. The resulting encrypted block is the ciphertext.

To decrypt the ciphertext, the same process is applied in reverse, using the same secret key but with the round keys applied in reverse order. Overall, AES is a widely used and well-regarded encryption algorithm that provides strong security for sensitive data.

3.2 Lifting Wavelet Transform for Embedding Encrypted Text into the Cover Image

The lifting wavelet transform [20] is a type of DWT that uses a lifting scheme to decompose a signal into its wavelet coefficients. The lifting scheme is a method of constructing a wavelet transform using only lifting steps, which are simple linear operations on the signal values. The lifting wavelet transform consists of two types of steps: prediction and update. The prediction step estimates the values of the high-frequency coefficients based on the values of the low-frequency coefficients, while the update step updates the low-frequency coefficients based on the estimated high-frequency coefficients. For reconstructing the original image from the wavelet coefficients, the lifting wavelet transform is applied in reverse, using the inverse prediction and update steps. The basic steps in lifting wavelet transform are:

1. Start with the input signal.
2. Apply the prediction step to estimate the high-frequency coefficients.
3. Apply the update step to update the low-frequency coefficients.
4. Repeat steps 2–3 for the desired number of levels of decomposition.
5. The resulting wavelet coefficients are the output of the transform.

Compared to the traditional DWT, which uses a pair of high-pass filters and low pass filters to decompose a signal into its wavelet coefficients, the lifting wavelet transform has several advantages, including reduced computational complexity, reduced memory requirements and increased flexibility.

The following are the various stages in the manipulation of images using lifting wavelet transform:

1. Choose a lifting scheme: There are many different lifting schemes that can be used for image processing, each of which has its own strengths and weaknesses.
2. Decomposition of the image: Once we chose a lifting scheme, we can use it to decompose the image into several components. For example, if we are using the Haar wavelet transform, we apply a series of high-pass and low-pass filters to the image to extract different levels of detail.
3. Apply lifting steps: After decomposition, the image is processed in a sequence of lifting steps. Each lifting step involves applying a linear transformation to the image that modifies its coefficients in a specific way. The specific lifting steps depend on the chosen lifting scheme.
4. Reconstruction of the image: After processing the image with the lifting steps, we can reconstruct the image from its transformed coefficients. The image that is reconstructed will not be the same as original image, but it will capture useful information or features that may be used for further analysis or processing.

3.2.1 Embedding of Encrypted Text into a Cover Image

The detailed procedure for applying the lifting transformation to the cover image for the embedding of encrypted text is as follows. First the input image and the encrypted test data is read and loaded into MATLAB software. The text data is represented as an array of 8-bit characters and converted to binary representation. Then the input image is converted from the RGB color space to the YCbCr color space. Only the Cb channel is extracted for further processing. Then the Haar wavelet is constructed using the liftwave function. This wavelet is used for the DWT. After that DWT is applied to the Cb channel, producing four sets of coefficients: LL, HL, LH, and HH. Now, the binary text data is hidden in the HH and HL coefficient sets. If the text data is too large to be embedded in the available coefficient space, an error message is displayed. Otherwise, the number of characters in the text data is encoded as the first size_length bytes in the HH coefficient set. The remaining binary text data is then encoded in the HH and HL coefficient sets, with HH given priority. If the text data cannot fit entirely in the HH coefficient set, the remainder is encoded in the HL coefficient set. The modified HH and HL coefficient sets are reshaped into flat vectors using the reshape function. If the text data fits within the available coefficient space, the flat vectors are updated with the binary text data. Each byte of the text data is represented as eight bits in the binary array. After that the flat vectors are restored to their original matrix dimensions. The modified coefficient sets are used to reconstruct the steganographic image using the inverse DWT, ilwt2. Now, the steganographic image is converted back to the RGB color space using the ycbcr2rgb function and saved to an output file. Now the final output image will look similar to the original input image but with the hidden text data embedded in it.

The liftwave function in MATLAB is used to generate lifting wavelet filters with various properties. If we give, 'haar' it specifies that we want a lifting wavelet filter for the Haar wavelet, which is a simple and widely used wavelet. The second argument, 'Int2Int', specifies the type of lifting scheme to be used. 'Int2Int' stands for integer-to-integer lifting, which means that the lifting steps only involve integer arithmetic, making the transform more efficient. The resulting haar_wavelet object is a MATLAB data structure that contains the lifting steps for the Haar wavelet transform, to perform the wavelet decomposition and reconstruction of images and signals. For example, the following code creates a lifting transform corresponding to the haar wavelet transform.

$$\text{haar\_wavelet} = \text{liftwave} \left(\text{`haar'}, \text{`Int2Int'} \right); \qquad (1)$$

Similarly, by varying the wavelet transforms, we have tested the proposed approach with the following wavelet transforms- bior3.5, bs3, coif3, db2, db4, db6, haar.

3.2.2 Extraction of Encrypted Text from the Cover Image

Following steps are applied for the extraction of the encrypted text from the cover image. Firstly, the input image is read and converted to the YCbCr color space. Then the blue-difference chroma (Cb) channel is extracted from the YCbCr image. After that, a Haar wavelet is constructed using the 'liftwave' function. The Haar wavelet is then applied to the Cb channel of the image using the 'lwt2' function, which performs a two-dimensional lifting wavelet transform. This produces four sets of coefficients: approximation (LL), horizontal detail (HL), vertical detail (LH), and diagonal detail (HH) and the HH and HL coefficients are flattened into 1D arrays. The number of characters embedded in the image is also read from the first size_length bytes of the HH coefficients. Now the characters are converted to its binary representation and the binary text is read from the HH and HL regions. After that, the bits from the HH coefficients are extracted. If the binary text couldn't be fully extracted from the HH coefficients, the remaining bits are read from the HL coefficients. The binary values are then converted to an ASCII string. Finally, the extracted text is the encrypted confidential message which can then be given to AES algorithm for decryption.

4   Results and Discussions

The proposed method is implemented in MATLAB 2020a. First, the data to be embedded for example, in this case the confidential patient information is first converted into binary, then it is encrypted using AES encryption algorithm. Then the resulting encrypted text is embedded into the cover image and the cover image is transmitted. So, this provides an extra layer of security to the data transmitted which is useful in the healthcare sector for tele radiology applications. Below given is the confidential data about a patient that has to be sent by hiding it in a cover image.

Confidential data sent: The patient is suspected to have abnormality in the white matter and grey matter distribution.

The confidential data is first converted into binary form and then AES encryption algorithm is implemented and the resulting ciphertext is then embedded to a cover image and then transmitted. At the receiver side, the ciphertext is extracted from the stego-image and then decryption algorithm is implemented to retrieve the confidential data back. Figure 3 shows the binary data corresponding to the confidential patient data after applying lifting transform function to haar wavelet transform and Fig. 4 shows the encrypted form of the confidential data after AES encryption for the lifting haar wavelet transform.

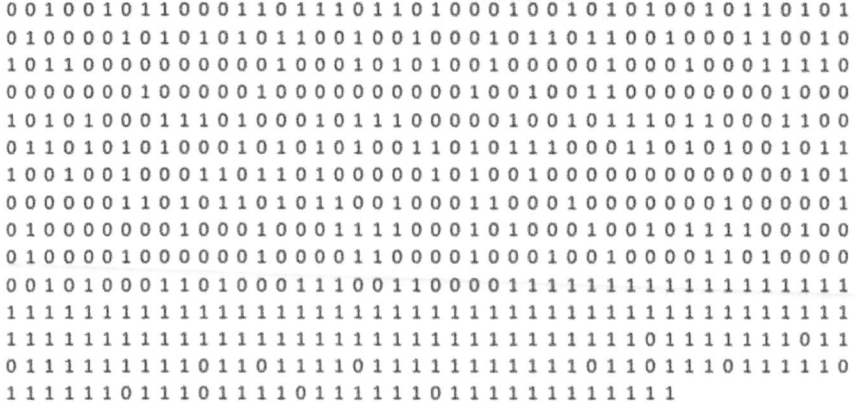

Fig. 3. Conversion of input data into binary form

accihbidjixxxxxedefjchxxxxxxxxfcgddxxacddfabcghxxxxxedcifedxxxxxxxxfchgbxxacddfabcghxxxxxedcifedxxx
xxxxfchgbxxacddfabcghxxxxxedefjchxxxxxxxxfcgddxxacddfabcghxxxxxedefjchxxxxxxxxfchgbxxaccihbidjixxxx
xedefjchxxxxxxxxfcgddxxacddfabcghxxxxxedefjchxxxxxxxxfcgddxxacddfabcghxxxxxedefjchxxxxxxxxfchgbxxac
cihbidjixxxxxedefjchxxacddfabcghxxxxxedcifedxxxxxxxxfchgbxxaicbiagefexxxxxedefjchxxxxxxxxfcgddxxacdd
fabcghxxxxxedcifedxxxxxxxxfcgddxxacddfabcghxxxxxedefjchxxxxxxxxfcgddxxacddfabcghxxxxxedefjchxxxxxx
xfchgbxxacddfabcghxxxxxgdgabgjxxxxxxxxfcgddxxaccihbidjixxxxxedefjchxxxxxxxxfcgddxxaccihbidjixxxxxede
fjchxxxxxxxxfcgddxxaccihbidjixxxxxedefjchxxacddfabcghxxxxxedcifedxxxxxxxxfchgbxxaccihbidjixxxxxedefjch
xxxxxxxxfcgddxxacddfabcghxxxxxedefjchxxxxxxxxfchgbxxacddfabcghxxxxxedcifedxxxxxxxxfcgddxxacddfabcg
hxxxxxedefjchxxxxxxxxfchgbxxacddfabcghxxxxxedefjchxxxxxxxxfcgddxxaccihbidjixxxxxedefjchxxxxxxxxfchgb
xxacddfabcghxxxxxedefjchxxxxxxxxfchgbxxaccihbidjixxxxxedefjchxxacddfabcghxxxxxedcifedxxxxxxxxfchgbx
xacddfabcghxxxxxedefjchxxxxxxxxfcgddxxaicbiagefexxxxxedcifedxxxxxxxxfcgddxxacddfabcghxxxxxedcifedx
xxxxxxxfcgddxxacddfabcghxxxxxedefjchxxxxxxxxfcgddxxacddfabcghxxxxxedcifedxxxxxxxxfcgddxxacddfabcgh
xxxxxgdgabgjxxxxxxxxfchgbxxaccihbidjixxxxxedefjchxxxxxxxxfcgddxxaccihbidjixxxxxedefjchxxaccihbidjixxxxx
edcifedxxxxxxxxfchgbxxaccihbidjixxxxxedefjchxxxxxxxxfchgbxxaccihbidjixxxxxedcifedxxxxxxxxfchgbxxacddfa
bcghxxxxxedcifedxxxxxxxxfcgddxxacddfabcghxxxxxedcifedxxxxxxxxfcgddxxacddfabcghxxxxxedcifedxxxxxxxf
chgbxxacddfabcghxxxxxedefjchxxxxxxxxfchgbxxaccihbidjixxxxxedefjchxxxxxxxxfcgddxxacddfabcghxxxxxede
fjchxxacddfabcghxxxxxedcifedxxxxxxxxfchgbxxaccihbidjixxxxxedcifedxxxxxxxxfchgbxxaccihbidjixxxxxedefjch

Fig. 4. AES encryption of output data

The histogram is a graphical representation showing the frequency of occurrence of gray levels of a digital image used in digital image processing. The number of pixels for each tonal value is plotted on a graph in a histogram. For a good steganographic method, there should be little to no histogram change between the sample cover image and the stego-image. Figure 5 shows the stego-image and its histogram after applying the haar lifting wavelet transform and Fig. 6 shows the stego-image and its histogram after applying the bs3 lifting wavelet transform.

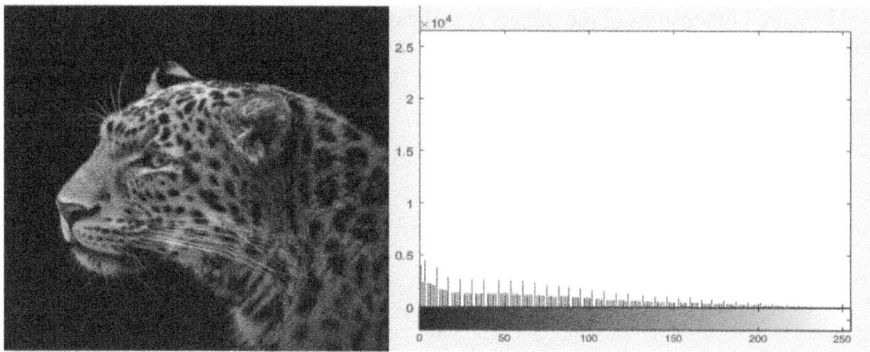

Fig. 5. Stego-image and corresponding histogram after application of lifting haar wavelet transform to encrypted data

In steganographic approaches, it is important to ensure that Peak Signal to Noise Ratio (PSNR) and the Mean Square Error (MSE) values are as high and low respectively as possible, which indicates that the encrypted stego-image is of high quality and has minimal distortion or noise. These metrics are used for evaluating the performance of different encryption techniques and to compare the quality of the decrypted data

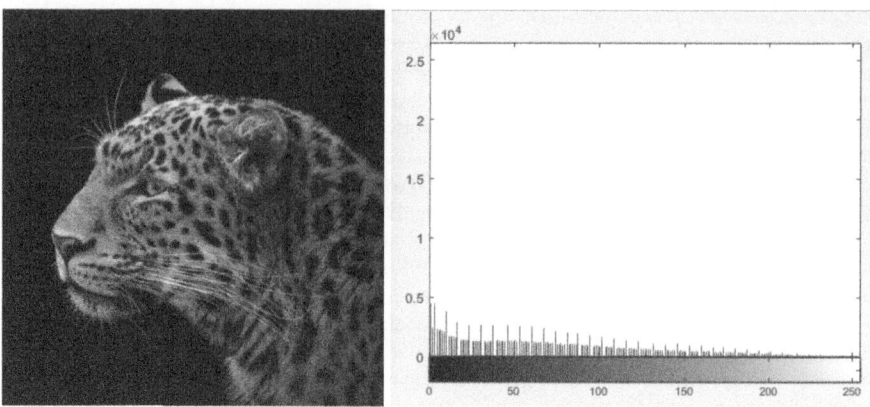

Fig. 6. Stego-image and corresponding histogram after application of lifting bs3 wavelet transform to encrypted data

obtained from different encryption methods. We have analyzed the MSE, PSNR and hiding capacity of the proposed approach for the bior3.5, bs3, coif3, db2, db4, db6 and haar lifting wavelet transforms and the results are discussed below.

The average squared difference between the original and encrypted images is what MSE calculates. It provides us with a gauge of the inaccuracy the data embedding method caused in the cover image. A lower MSE number signifies that the encrypted image is more similar to the original image, which improves the quality of the image.

$$MSE = \frac{\sum_{U,V} [I_1(u, v) - I_2(u, v)]^2}{U * V} \qquad (2)$$

Here U,V are Dimensions of the image, I_1 is the original image and I_2 is the stego-image (Fig. 7).

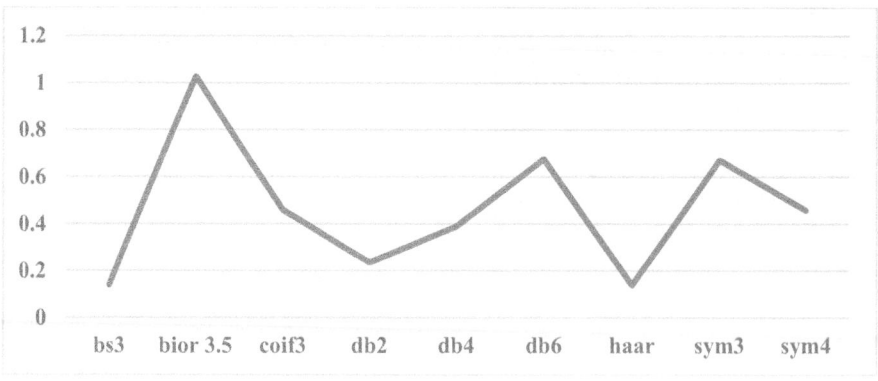

Fig. 7. Mean Squared Error comparison of different lifting wavelet transforms

After applying the AES encryption and lifting transform to the nine wavelet transforms, it is found that haar wavelet lifting transform and bs3 have better values compared to other lifting transforms and the value obtained is 0.13887 where the bior3.5 have MSE equal to 1 which means that the error rate is high with bior3.5. It means that that the stego-image quality is better when using the haar and bs3 wavelet lifting transform with AES encryption for transmission of confidential patient data. This method may be also employed in military application also for securing sensitive data.

PSNR measures the degree of distortion that happened in the cover image due to embedding. It is defined as the ratio between the maximum possible value of a signal and the power of distortion noise (MSE). It is measured in dBs. Obtaining a higher value of PSNR indicates that a better-quality embedding is achieved.

$$PSNR = 10log_{10}\left(\frac{R^2}{MSE}\right) \tag{3}$$

where R = 255 for an 8-bit grayscale image.

From the results obtained, it is found that the haar transform and bs3 have maximum values for PSNR and the value obtained is 55.29 which indicates that they performed well on the proposed method whereas the other transforms achieved less values (Fig. 8).

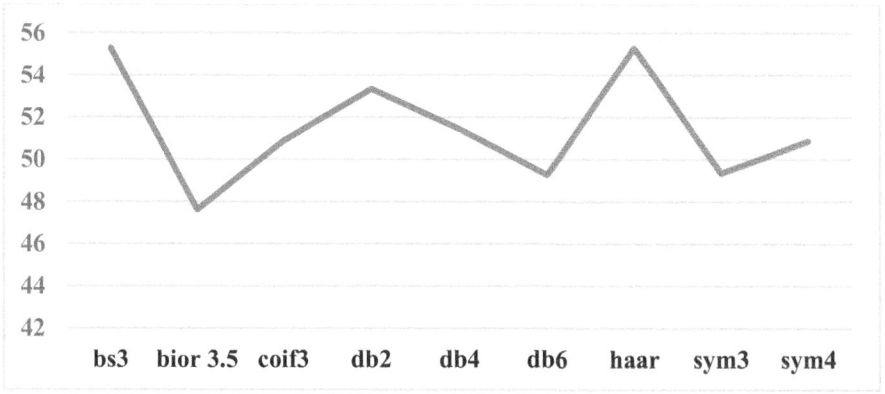

Fig. 8. Peak Signal to Noise Ratio comparison of different lifting wavelet transforms

Hiding capacity is another critical factor in encryption techniques that are used to hide or protect confidential data. Hiding capacity refers to the amount of data which can be hidden or embedded within a cover object, such as an image or audio file, without significantly affecting the cover objects quality or raising any suspicion. In the context of encryption, hiding capacity is essential because it determines the amount of confidential data which is hidden within a cover object. The larger the hiding capacity, the more data that is hidden within the cover object. This is useful in scenarios where a significant amount of data needs to be encrypted and protected, such as in military or intelligence operations. From our results, we observed that the hiding capacity is independent of the type of the wavelet transform used and the value obtained is 1798 bytes in this particular scenario.

The proposed work employs steganography as an additional security measure for the data, in addition to cryptographic techniques. Steganography conceals encrypted messages in a manner that makes it difficult to detect their presence. Modern digital steganography uses conventional cryptographic methods for data encryption. The data is then inserted into redundant data within an image, using a specific algorithm. In the proposed work, the data is encrypted and embedded within a cover image block and then uses lifting wavelet transform to enhance security.

In the medical field, steganography is used to securely store and transfer medical data such as patient records, images from medical scans, and other sensitive information. By embedding this data within other files, such as images, the data may be hidden from prying eyes, making it more difficult for unauthorized individuals to access it.

In military applications, steganography is used to securely transmit sensitive information, such as military orders or intelligence reports, without the risk of interception by enemy forces. By embedding this information within seemingly innocuous files, such as images or videos, the information may be hidden from view and transmitted securely.

5 Conclusion

By combining steganography with cryptography, we can encrypt the message to ensure that only the intended recipient can decrypt it, while also hiding the fact that the message even exists by embedding it within a seemingly innocent file. This makes it more difficult for someone to intercept or detect the message, as they would need to both crack the encryption and find the hidden message. Overall, the combination of steganography with cryptography can help to provide greater security and privacy for sensitive information, making it more difficult for unauthorized individuals to access or intercept the message. In our work, we combined cryptography with steganography for providing an additional layer of security to the confidential medical data that is transferred through the network. Our results show that out of the different lifting wavelet transforms used, haar and bs3 lifting wavelet transforms have better results compared to the others.

References

1. Kaur, S., Singh, S., Kaur, M., et al.: A systematic review of computational image steganography approaches. Arch Computat. Methods Eng **29**, 4775–4797 (2022). https://doi.org/10.1007/s11831-022-09749-0
2. Kumar, M., Soni, A., Shekhawat, A.R.S., Rawat, A.: Enhanced digital image and text data security using hybrid model of LSB steganography and AES cryptography technique. In: 2022 Second International Conference on Artificial Intelligence and Smart Energy (ICAIS), Coimbatore, India, pp. 1453–1457 (2022). https://doi.org/10.1109/ICAIS53314.2022.9742942
3. Bansal, R., Badal, N.: A novel approach for dual layer security of message using Steganography and Cryptography. Multimed. Tools Appl. **81**, 20669–20684 (2022). https://doi.org/10.1007/s11042-022-12084-y
4. Khari, M., Garg, A.K., Gandomi, A.H., Gupta, R., Patan, R., Balusamy, B.: Securing data in Internet of Things (IoT) using cryptography and steganography techniques. IEEE Trans. Syst. Man Cybern. Syst. **50**(1), 73–80 (2019)
5. Abd El-Maksoud, A.J., et al.: FPGA implementation of sound encryption system based on fractional-order chaotic systems. Microelectron. J. **90**, 323–335 (2019)

6. Saleh, M.E., Aly, A.A., Omara, F.A.: Data security using cryptography and steganography techniques. Int. J. Adv. Comput. Sci. Appl. **7**(6) (2016)
7. Jero, S.E., Ramu, P.: Curvelets-based ECG steganography for data security. Electron. Lett. **52**(4), 283–285 (2016)
8. Wahab, O.F., Khalaf, A.A., Hussein, A.I., Hamed, H.F.: Hiding data using efficient combination of RSA cryptography, and compression steganography techniques. IEEE access. **9**, 31805–31815 (2021)
9. Wazirali, R., Alasmary, W., Mahmoud, M.M., Alhindi, A.: An optimized steganography hiding capacity and imperceptibly using genetic algorithms. IEEE Access **7**, 133496–133508 (2019)
10. Seyyedi, S.A., Sadau, V., Ivanov, N.: A secure steganography method based on integer lifting wavelet transform. Int. J. Netw. Secur. **18**(1), 124–132 (2016)
11. Shahadi, H.I., Jidin, R., Way, W.H.: Lossless audio steganography based on lifting wavelet transform and dynamic stego key. Indian J. Sci. Technol. **7**(3), 323 (2014)
12. Reddy, H.M., Raja, K.B.: Wavelet based secure steganography with scrambled payload. Int. J. Innov. Technol. Explor. Eng. **2**, 121–129 (2012)
13. Shirafkan, M.H., Akhtarkavan, E., Vahidi, J.: A image steganography scheme based on discrete wavelet transform using lattice vector quantization and reed-solomon encoding. In: 2015 2nd International Conference on Knowledge-Based Engineering and Innovation (KBEI), 5 November 2015, pp. 177–182. IEEE (2015)
14. Zhiwei, K., Jing, L., Yigang, H.: Steganography based on wavelet transform and modulus function. J. Syst. Eng. Electron. **18**(3), 628–632 (2007)
15. Lai, B.L., Chang, L.W.: Adaptive data hiding for images based on harr discrete wavelet transform. In: Chang, L.W., Lie, W.N. (eds.) Advances in Image and Video Technology, pp. 1085–1093. Springer, Heidelberg (2006). https://doi.org/10.1007/11949534_109
16. Hasoon, J., Al-Saad, S.: Audio hiding based on wavelet transform and linear predictive coding. Iraqi J. Comput. Inf. **42**(1), 30–37 (2016)
17. Sajedi, H., Yaghobi, S.R.: Information hiding methods for E-Healthcare. Smart Health. **1**(15), 100104 (2020)
18. Kour, J., Verma, D.: Steganography techniques–a review paper. Int. J. Emerg. Res. Manag. Technol. (2014). ISSN 2278–9359
19. Abdullah, A.M.: Advanced encryption standard (AES) algorithm to encrypt and decrypt data. Cryptogr. Netw. Secur. 16, 11 (2017)
20. Acharya, T., Chakrabarti, C.: A survey on lifting-based discrete wavelet transform architectures. J. VLSI Signal Process. Syst. Signal Image Video Technol. **42**, 321–339 (2006)

To Enhance the Efficiency and Features of Text to Image Generation Using Neural Network Models

S. Muruganandam, V. Jeevana[✉], S. Harish Kumar, and G. V. Harish

Department of Computer Science and Engineering, Rjalakshmi Engineering College, Thandalam, Chennai, India

{muruganandam.s,jeevana.v.2019.cse,harishkumar.s.2019.cse, harish.gv.2019.cse}@rajalakshmi.edu.in

Abstract. Generation of photorealistic images have multiple utilization in the field of photo editing, fashion, product, game designing, painting and so on. Individuals involved in these fields are in need of visualizing their own ideas. There isn't any specific solution that serves them. Existing works result in limited features and low-quality images with less accuracy. It's important for developers to visualize their ideas before moving into the development phase. The proposed system is designed in such a manner to mitigate all these issues. It generates image with the help of a pre trained model. The accuracy is improved by ranking, denoising and upscaling the generated 2D image. By combining different models, a new algorithm is proposed for generating the high-resolution 2D image. To improve the features of the text to image generation, character 3D modelling and video generation are included. The 3D modelling feature displays 3D model by taking a 2D character or human image as input. The Video feature generates the video by taking list of text prompts as input. Best models are chosen from the survey to implement video generation and 3D model construction. The proposed system will stand as a unique solution for users to envision their thoughts and ideas. This aids several developers and designers, by simplifying work and effectively utilizing time.

Keywords: Decoder · denoising · encoder · pre-processing · upscaling

1 Introduction

Visualizing one's idea always require artistic skill or an artist. Animators, story writers, game designers are exhausted by using separate software for specific set of tasks. They often accompanied by an artist, to visualize their creative ideas in the form of an art, before moving for implementation. Recreating ideas in the form of art by the artist is a trial-and-error process. Here, the artist and the designer or developer needs to work together until they both get satisfied. This process consumes lot of effort and time. There are few issues like developer might not capable of explaining the ideas they visualize, artist might fail in recreating the ideas, some artist might find difficulties in connecting

P. Dassan et al. (Eds.): ICICSCNT 2023, CCIS 1970, pp. 82–94, 2024.
https://doi.org/10.1007/978-3-031-75957-4_8

with the developer and developers might not satisfied with the work done by the artist. Hence proposed system is developed by considering these issues. It can be used by anyone to visualize their ideas.

The main goal of it is to develop an application to aid all kinds of designers and developers. It generates the corresponding 2D image with the given text prompt as input. With the help of several models and techniques, high-resolution 2D image is produced. Further for the animators, 3D character modelling of a particular character or human image and video generation is developed with an efficient model from the survey. Video generation is developed by passing list of prompts as input and serves as the best solution the creators like story writers and cartoon developers. These features help in eliminating the need of artist, accompanying animators, story writers and designers. The proposed solution is developed as an application to make ease of the work of animators, creators and designers. It stands as a unique and cutting-edge solution for visualizing and developing their ideas.

2   Related Work

The majority of the models in the current systems correspond to the generation of the images using deep learning methods like Convolutional Neural Networks (CNN). These models, however, can only produce photos with low resolution. A few studies discuss 3D picture generating methods that employ MaskRCNN, convolutions, image encoders, and decoders.

Transformers were used to model text and image as a single piece of data [1]. The CLIP (Contrastive Language–Image Pre-training) model made it possible to apply zero shot transfer using NLP (Natural Language Processing) to numerous datasets and is performed very well in large scale [2]. Pointwise technique performed better than any previous work because it learned the ranking function [6]. The computation in data relevance picture ranking worked well for huge images and transfer learning was used to create a deep rating system for photos [8]. Based on the visual similarity of the image with the provided input text, similarity scores were computed [25].

An image denoising approach based on CNN in which the image quality was assessed using a quality analysis method [20]. Deep CNN performed the mapping between the low- and high-resolution pictures [7]. To efficiently extract the huge feature set, numerous blocks with multiple layers were linked [9]. Direct mapping was performed using a range of resolutions because only one image was used and the outcome was not accurate [18]. Intrusion detection technique with most efficient intelligent algorithm [13] suggested the image resolution was transformed into a low or high representation by the upscaling factor and using a sparse approach, individual images of remarkable resolution were created [19]. A discrete and temporal cellular neural network is employed for image upscaling [21].

Gradient descent was used to optimise the initialised 3D model in order to lessen the loss in 2D rendering for 3D creation [4]. Both the image generator model and the shape reconstruction networks are trained on the unpaired image and shape dataset for the same item to improve the 3d image reconstruction [5]. Colour theories were used to gather the data for 3D creation [10]. Convolutions were performed using the Pytorch3D library, Mask RCNN, graph convolutions, and image encoders and decoders to develop an automated approach of creating 3D models [11]. Cluster header algorithm was used to classify nodes [12]. CNN was employed to predict how a 2D representation of an object would look in 3D [14]. To create a Colour Filtered Aperture, an objective lens with red, green, and blue colour filters was employed using disparity settings and depth alignment for 3d creation [15]. Image-based 3D human shape prediction was made possible by deep neural networks' substantial improvement in representational power [17]. To build an ill-defined framework for the 3D images without any inconsistency, a theory of warm and cold colours was applied to provide depth data [22]. With the use of decoders and two integrated encoders, fine-grained output was produced in the form of a 3D model [24].

3   Proposed System Architecture

The proposed architecture deals with the input-output flow as shown in Fig. 1. Firstly, the text prompt is taken from the user and the image is generated using DALL E algorithm. DALL E performs the pre-processing of the given text. With the help of pre-processed text tokens, 2D images is generated. The number of 2D images to be generated is passed as an argument to the model. All the generated images are then passed to the CLIP model for ranking. It selects the best image based on the similarity score of the image with respect to the text prompt. The selected image is then passed through a non-local means denoising function. It helps in removing the noise from the selected image. The selected denoised image is then passed into the Super Resolution LapSRN (Laplacian Pyramid Super-Resolution Network) model. This helps in generating the final 8× resolution 2D image.

Secondly, the human image or 2D images of characters for designing the game or story is supplied into the PIFuHd model for generating 3D model.

Lastly, a list of prompts is provided as input to the Stable Diffusion video generator model. It processes each prompt and combines the generated image in the form of a video.

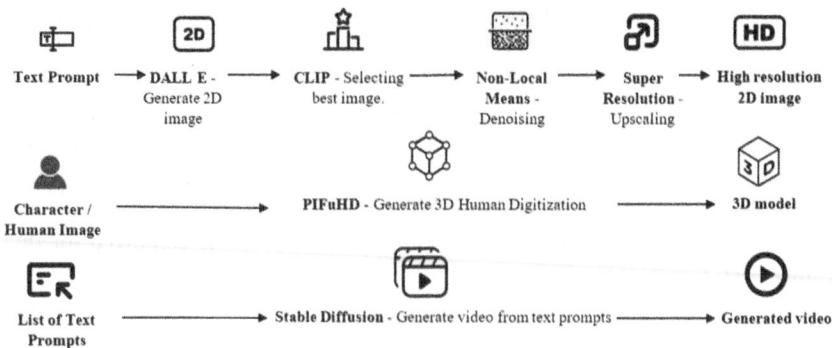

Fig. 1. Architecture of the Proposed System

3.1 Generation of 2D Image

DALL E is the well-known pre-trained model for construction of image from the text. Comparing with other existing methods, this produces the results efficiently in terms of time. The model is trained with several input-text pairs, and it includes two modules. The first one was about learning the vocabulary. Discrete Variational Auto-Encoder is used to compress the given training images into smaller tokens of 32 × 32 grid. These 1024 image token (integers) are used for reconstruction of an image. The second module performs the learning of the prior distribution. It concatenates the 1024 tokens together with the 256 tokens. These tokens are used to train the transformers autoregressively. The two-step process involved in this is, 1024 image tokens is generated by the transformer from the 256 text encoded tokens. Then full stream of tokens is taken by the transformer to map from the embedding space to the image space.

The whole process works as shown in the Fig. 2. When the input text is provided as input, the tokenization step is being carried out. The input text is divided into tokens. These tokens then would be passed into several stages and at last contains only the required tokens in the form of list. DalleBart processor helps in performing such a pre-processing step of the text prompt. These text tokens help in generating the images accordingly. DallBart processor consist of DalleBart encoder and DalleBart decoder. VQGAN (Vector Quantized Generative Adversarial Network) encoder and decoder helps in tokenizing and constructing the images.

Sometimes, the image generated by the model would not accurately matches with the given text prompt. To solve this issue an additional ranking algorithm is introduced as the next stage. This basically ranks the images with the help of similarity metric. This further improves the accuracy and results of DALL-E.

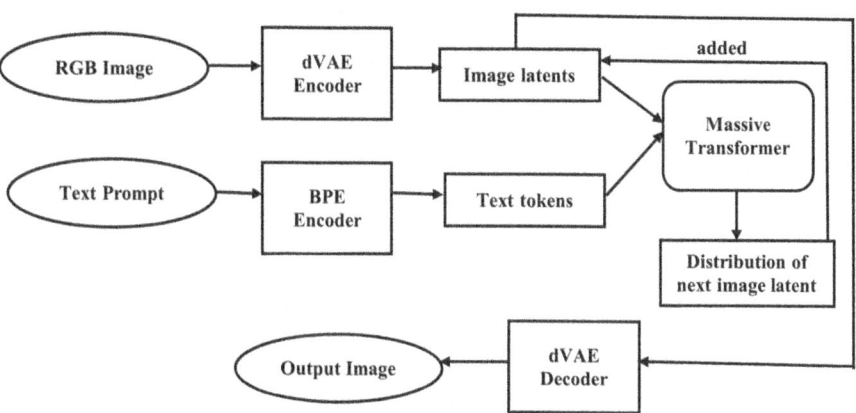

Fig. 2. Image Generation using DALL E

3.2 Ranking of Generated 2D Image

CLIP is a pre-trained model, which is also used for image generation from the text. The CLIP and DALL E differs by the methodology used to generate images. CLIP uses similarity scores for generating the images. Such technique is used here as a ranking algorithm for getting the most accurate image from DALL E.

It mainly focuses on to jointly pre-train language model and image model. Image encoder helps in encoding an image into small grid like tokens. Text encoder helps in generating tokens for the text. A matrix is constructed to find similarity score between caption and image pair. With these values, the model is trained to maximize the values at the correct pair position, especially along diagonal. Once the pre-training process is done, representation of an input image is acquired from the visual model and it is applied to several prompts that are acquired from language model. The text representation with the maximum similarity value will be chosen as the best prompt for the image being used as input.

The same model is planned to apply in a slightly different manner as shown in Fig. 3. Here the input text and some randomly generated list of images would be given as input to the model. The model calculates the similarity score of each text to image pair. Then the list of score is sorted in descending order to get the maximum similarity score at the index 0. Then the corresponding image would be used for further process. Through this methodology, the accurate image can be generated with high similarity score from DALL E. The process doesn't stop with this. Next few techniques will be applied to the selected image for increasing the quality of the image.

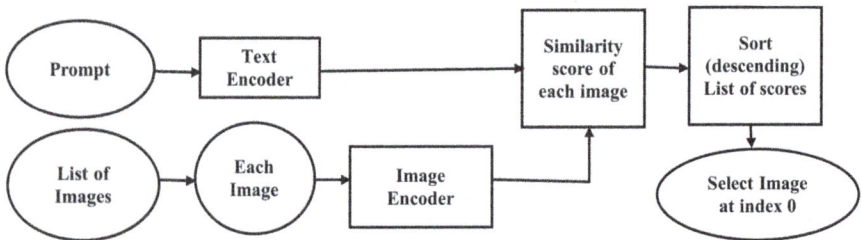

Fig. 3. Ranking Process using CLIP

3.3 Denoising of the Selected Image

Non-Local Means Denoising [3] helps in removing the noise in the image generated by the DALL E. A very simple model, where the average colour of the similar pixel is taken to replace the particular colour of the pixel in the image. It is based on the probabilistic theory that the standard deviation of a noise can be divided by three, if the similar nine pixels are averaged. Finding the similar pixels for the pixel to be denoised is the complicated process. This is done by just navigating the window across all the pixels of the image. Such filter is termed as non-local means.

Though several pre trained model is analysed and surveyed, they are not efficient in terms of time. Many diffusion models take much time in generating a denoised image. Hence this OpenCV approach for denoising the image is efficient in terms of time and quality, for generating the denoised image.

3.4 Upscaling of the Denoised Image

Upscaling is performed by a pre-trained model named super resolution. This used four deep learning algorithms namely EDSR, ESPCN, FSRCNN and LapSRN [23]. The upscale ratio of the first three algorithms is about 2, 3 and 4 times. The last one upscales about 2, 4 and 8 times of an original image. LapSRN is chosen for upscaling an image as it produces high resolution image. It stands as the solution for contrasting the strategies of an image at the start and the end. It is named as Laplacian pyramids. The upscaling is done in the shape of pyramid till the end of the image. Feature extraction and the image reconstruction are considered as the two basic path of an image. Depending on the number of times the image being resolution, the number of pyramids is utilized. $2\times$ uses one pyramid, $4\times$ used two pyramids and $8\times$ used three pyramids. Feature extraction is comprised of many convolutional layers and only one transposed convolutional layer for $2\times$ scaling. Then the output from this is linked to two different layers. For reconstructing convolutional layer is used and for extracting the features another convolutional layer is used. Image reconstruction works by combing upscaled image and the predicted residual image. Predicted residual image is taken from the feature extraction branch. This produces high-resolution image.

3.5 Algorithm for Generation of High-Resolution 2D Image

Input: Text Prompt
Output: High-Resolution 2D Image
Decide on the number of images to be generated
Create an empty list named resultant_images
For i in number of images
3. Generate image using DALL E with prompt and random int as seed value
4. Append the generated image to the resultant_images list
End for
5. Create an empty list named scores
For image in resultant_images
6. Calculate the similarity score between prompt and image by passing prompt and image as input to CLIP model
7. Append the similarity score value to the scores list
End For
8. Sort the scores list in descending order
9. Select the image at index 0
10. Pass the selected image to Non-Local Means Denoising function
11. Pass the denoised image to Super resolution (LapSRN) model
12. Display and save the upscaled image

3.6 Generation of Character Modelling (3D Model)

By analysing various methodologies and pre trained models for generating 3D model as per the Table 1, PIFuHD stands as the perfect solution in terms of both time and quality. PIFu takes an input and as a result it generates feature embedding of the input image.

More PIFu modules are stacked one above the another to generate a high-resolution image. Feature of the high-resolution image as input is extracted as the initial step. Then these features as well as the 3D embeddings are combined to calculate the occupancy of probability field. Maps of front and back sides in image space are fed as an input to improve the quality of the model being constructed.

Table 1. Comparison of various Methodology for 3D generation

Methodology	Advantage	Disadvantage
CNN	The 3D model can be predicted well from a 2D photograph	It was only used to foretell how a 2D image of an object will seen in 3D
PIFuHD	Reconstruction was done effectively	It's suitable only for human 3D Model generation
ShapeNet	It showed good results on complex and pure background images	Requires high computational power
DreamFusion	It doesn't require any dataset for 3D generation	It requires much time to generate the output
Real time hardware	It doesn't require large memory	Hardware requirement

The PIFu is a function that commonly predicts the binary occupancy value in the continues camera space. It uses the probability formula that is, if it is inside the mesh surface then it assigns value to be 1 and 0 otherwise. Occupancy of the 3D point is calculated by just first finding image feature embedding and then orthogonal projection. Convolutional Neural Network and Multilayer perceptron are used for the functions. It basically contains two modules, and they are coarse level and fine level. The coarse level integrates the global geometric information by down sampling the image into 128 × 128 resolution. Fine level adds more subtle information by producing the feature of 512 × 512 image. Fine level takes the 3D embeddings of the input image to process. The 3D embeddings are the front and back faces of the image. These image feature outputs are used for reconstruction of the image as shown in Fig. 4.

Dream fusion and other pytorch available libraries doesn't produce much realistic 3D models. This PIFuHD generates more realistic 3D model for the human images as input.

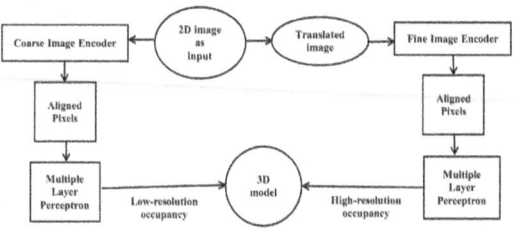

Fig. 4. 3D model generation using PIFuHD

It works well for human digitization compared to others. The 3D human models can be directly used by animators and game developers for developing their own creative ideas. This also models the particular posture of the human as image into the more realistic one. This stands as the solution for recreating a 3D model for a particular character without wasting much time in 3D tools.

3.7 Generation of Video

Stable diffusion [16] is a text to image generation model in which the generating process is broken into many steps as shown in Fig. 5. Each step involves generating an image from the given input text. The process involves a model called diffusion, where the generated image is blended in each step gradually. This process mainly helps to reduce the noises that will rise during the process of image generation. The diffusion process applies noise to the generated image at each step to make the image stabilized. This process helps the model to prevent the generated image from converging and help to generate large number of images from the given input text.

In addition, this model normalizes the generated image and matches with the given input text. This helps to maintain the image quality and relevant to the given input text. This model also ensures the quality of the generated image using feedback system to evaluate and to make the changes to the model. Finally, this model ensures and help us to make the generation process normalized and produces high image quality that exactly matches with the user's input text.

After generating a number of images from the given input texts with the help of this model. The video is generated by running the generated images in one-by-one order.

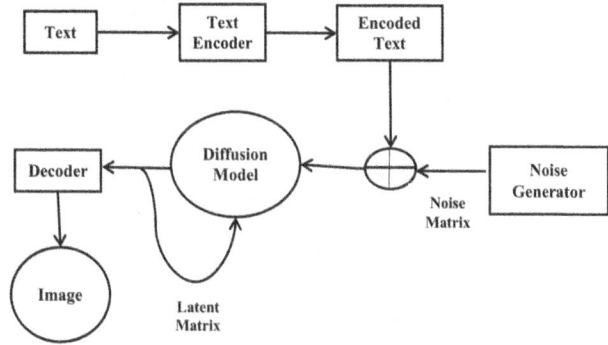

Fig. 5. Image generation using Stable diffusion for Single Prompt

4 Proposed Algorithm

Text to Image Generation with additional features
Input: Select 2D or 3D character modelling or video
Output: output from the corresponding model

If input is 2D:
Pass prompt to 2D image generator
Display and save the generated 2D image
Else if input is 3d character modelling:
Pass 2D image to 3D model generator
Display and download the generated 3D object
Else:
Pass the list of prompts to video generator
Display and download the generated video
End if

5   Results and Discussions

The proposed system is implemented and the results are discussed below.

5.1   Comparing the Outputs of DALL E and Proposed Algorithm for 2D Image Generation

Table 2 shows the output from DALL E and proposed model for the provided text prompts. By comparing the output of DALL E and proposed methodology from the table, it is proved that the proposed model works well in terms of resolution and accuracy of the image for the provided input prompts.

Table. 2. Comparison of Generated 2D image with DALL E outputs

TEXT PROMPT	OUTPUT (DALL E)	OUTPUT (PROPOSED METHODOLOGY)
A writer writing story in the forest covered with snow		
Tortoise flying near the moon		
Aliens having food at the restaurant		

5.2 Accuracy of DALL E and Proposed Algorithm

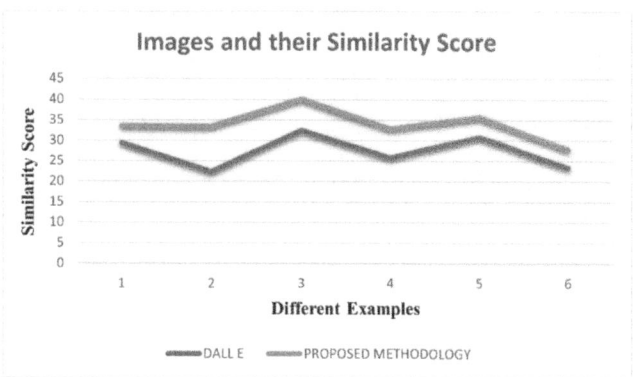

Fig. 6. Similarity Score of DALL E and Proposed Methodology

The above Fig. 6 depicts that for different samples of text as input, the similarity score of the existing model (DALL E) with respect to the generated image is lesser than the proposed model. Hence the accuracy of the image is improved by the proposed algorithm.

5.3 Resolution of Existing and Proposed Algorithm

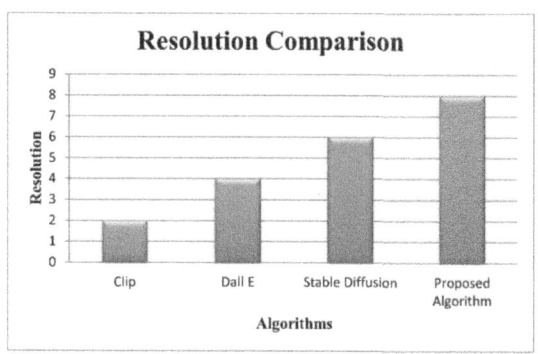

Fig. 7. Resolution of Existing and Proposed Algorithms

The chart in Fig. 7 represents the resolution of the final image generated by existing and proposed algorithms. Proposed algorithm generates a 2D image of the highest resolution. Performance enhancement is also implemented efficiently in the proposed algorithm.

5.4 Output of 3D Character Modelling

The Fig. 8 shows the output of the character modelling with 2D image of character as input. Image at the right side of the figure shows 3D model, while at the left side shows 2D character image.

Fig. 8. Output of Character 3D Modelling

5.5 Outputs of Video Generator

The Fig. 9 shows the snapshots of the video generator module. Here, the two prompts namely "A lion sitting in a snow near a boy" and "boy riding lion in snow" are given as input. These snapshots run as combined in the form of video.

Fig. 9. Snapshots of Video Generator

6 Conclusion and Future Enhancements

Visualizing one's idea plays an important role in several fields. This project aims to make easier the work of animators, designers and developers. This innovative approach is put forth by integrating various models to improve the accuracy, efficiency and features of the text to image generation. The image is generated using Dall E. The proposed system improves the accuracy of the generated image by ranking, denoising and upscaling it via CLIP, non-Local means denoising and Super resolution methodologies. This system also includes features like generation of video from the text prompts. It further includes 3D character modelling generator which takes a 2D image as an input. These features are implemented with the help of Stable diffusion and PIfuHD models respectively. The developed solution includes features and improves efficiency of text to image for users to visualise their ideas.

Additional visualizing and editing features like colour editing, texture creation, picture smoothing, story generation, deep motioning will be added as a part to the proposed system in the future. 3D modelling feature will be implemented to all kind of images as it is now only limited to humans or characters.

References

1. Ramesh, A., et al.: Zero-shot text-to-image generation (2021). arXiv:2102.12092
2. Radford, A., et al.: Learning transferable visual models from natural language supervision (2021). arXiv:2103.00020
3. Buades, A., Coll, B., Morel, J.M.: Non local means denoising. IPOL J. (2011)
4. Poole, B., Jain, A., Barron, J.T., Mildenhall, B.: DreamFusion: text-to-3D using 2D diffusion (2022). arXiv:2209.14988
5. Kaya, B., Timofte, R.: Self-Supervised 2D image to 3D shape translation with disentangled representations. IEEE (2020)
6. Prakash, C., Sarkar, A.: Ranking with deep neural networks. IEEE (2018)
7. Dong, C., Loy, C.C., He, K., Tang, X.: Image Super Resolution using Deep Convolutional Networks (2014). arXiv:1501.00092
8. EI-Batrawy, H., Atwan, A., Soliman, H., Elmogy, M.: Image ranking relevancy based on semantic web using deep learning technique. IEEE (2020)
9. Liang, J., et al.: SwinIR: Image Restoration Using Swin Transformer. Computer Vision Lab, ETH Zurich, Switzerland KU Leuven, Belgium. (2021). arXiv:2108.10257
10. Yamada, K., Suzuki, Y.: Real-time 2D-to-3D conversion at full HD 1080P resolution. IEEE (2009)
11. Mehta, M., Kothawade, S., Kudale, S., Dole, S.: Automated 2D image to 3D model construction: a survey. Int. Res. J. Eng. Technol. **7**, 1452–1457 (2020)
12. Muruganandam, S., Renjit, J.A.: A node quality based cluster header selection algorithm for improving security in MANET. In: Emerging Research in Computing, Information, Communication and Applications. Springer, Heidleberg (2021). https://doi.org/10.1007/978-981-16-1338-8_11
13. Muruganandam, S., Srinivasan, N., Sivaprakasam, A.: An intelligent method for intrusion detection and prevention in mobile adhoc networks. Int. J. Intell. Syst. Appl. Eng. **10**(3), 154–160 (2022)
14. Stigeborn, P.: Generating 3D objects using neural networks. DiVA portal (2018)
15. Deshpande, R.R., Renu Madhavi, C., Bhatt, M.R.: 3D image generation from single image using color filtered aperture and 2.1D sketch-a computational 3D imaging system and qualitative analysis. IEEE Access **9**, 93580–93592 (2021). https://doi.org/10.1109/ACCESS.2021.3089938
16. Rombach, R., Blattmann, A., Lorenz, D., Esser, P., Ommer, B.: High-Resolution Image Synthesis with Latent Diffusion Models (2021). arXiv:2112.10752
17. Saito, S., Simon, T., Saragih, J., Joo, H.: PIFuHD: multi-level pixel-aligned implicit function for high-resolution 3D human digitization (2020). arXiv:2004.00452v1
18. Schulter, S., Leistner, C., Bischof, H.: Fast and Accurate image upscaling with super resolution forests. IEEE (2015)
19. Sharmila, T., Leo, M.: Image upscaling based convolutional neural network for better reconstruction quality. IEEE (2016)
20. Ghose, S., Singh, N., Singh, P.: Image denoising using deep learning: convolutional neural network. IEEE (2020)
21. Otake, T., Konishi, T., Aomori, H.: Image resolution upscaling via two layered discrete – time cellular neural network. IEEE (2006)

22. Yamada, K., Suehiro, K., Nakamura, H.: Pseudo 3D image generation with simple depth models. IEEE (2005)
23. Lai, W.S., Huang, J.B., Ahuja, N., Yang, M.H.: Fast and accurate image super-resolution with deep laplacian pyramid networks (2018). arXiv:1710.01992v3
24. Zhang, Y., Liu, Z., Liu, T., Peng, B., Li, X.: RealPoint3D: an efficient generation network for 3D object reconstruction from a single image. IEEE Access **7**, 57539–57549 (2019). https://doi.org/10.1109/ACCESS.2019.2914150
25. Zhu, S., Luo, Q., Hu, J.: Relevance-quality ranking based image retrieval. IEEE (2013)

Diabetic Retinopathy Image Recognition Using AI Technique in Rasberry PI

P. Latha[1](✉), V. Sumitra[1], K. Yogeshwari[2], P. Shyamala[2], S. Swathi[2], and T. Arulmurugan[2]

[1] R.M.K. Engineering College, Tiruvallur, India
pla.ece@rmkec.ac.in

[2] Department of Electronics and Communication Engineering, R.M.K. Engineering College, Tiruvallur, India

Abstract. Regular eye exams are especially important because diabetes can interfere with eye health. Damaged or abnormal new blood vessels can cause vision loss due to high blood sugar, which may damage the eye. Various neural and layered architectures have been applied using CNN algorithms, trained on publicly available image datasets of diabetic retinopathy. Neural networks have been observed to capture lesion color and texture specific to each disease at the time of diagnosis, similar to human decision-making. Experimenting with different features of diabetic retinopathy as inputs to a convolutional neural network is recommended.

Keywords: Diabetic Retinopathy · CNN · Tensorflow

1 Introduction

The primary cause of blindness today is diabetic retinopathy (DR), which has emerged as a significant health issue for a large number of diabetic patients [1–3]. Ophthalmologists use detailed fundus images to diagnose DR lesions [2, 3]. However, there hasn't been an effective cure for this condition so far. Early detection and intervention to prevent disease progression are the best-known treatments.

2 Existing System

Segmenting diabetic retinopathy (DR) lesions is an ideal approach for aiding ophthalmologists in diagnosis [4]. While countless investigations have been conducted in this field, most past attempts emphasized network designs instead of the clinical connection of lesions [4, 5]. We found that the characteristic DR lesions of certain arteries were occluded, reflecting an interrelated pattern by pre-examining the pathogenic origin of DR lesions [4–7]. Based on these observations, we suggest a relation transformer block (RTB). Our network's outstanding experimental findings may be credited to GTB and RTB, which study intra-class relationships.

P. Dassan et al. (Eds.): ICICSCNT 2023, CCIS 1970, pp. 95–110, 2024.
https://doi.org/10.1007/978-3-031-75957-4_9

3 Objectives

Diabetic retinopathy is a major factor in our healthcare ecosystem [6]. Many patients are proactively managing their condition around the globe. Diabetic retinopathy is difficult to diagnose. As a consequence, this project aims to easily identify the type of diabetic retinopathy through deep learning (DL) [7].

4 Proposed Design

Importance of AI and DL
     The construction explains why it is decisive in this application.
Introduction to Domain (AI)
AI is broadly utilized in different industries, including healthcare, finance, transportation, and entertainment, to automate routine tasks, enhance decision-making, and generate novel services and products. Artificial Intelligence is used in many different industries, such as speech recognition, computer vision, robotics, and natural language processing. Though AI has the potential to importantly progress our lives in numerous ways, it also increases important ethical and societal complications, such as the possibility of misuse, bias and discrimination, loss of jobs, and privacy and security apprehensions. To maximize advantages while minimizing hazards, AI must be developed and applied ethically and transparently.
Introduction to Domain (DL)
By replicating the composition and operations of the human brain, deep learning purposes to progress artificial intelligence that can perform sophisticated tasks like natural language processing, speech and image recognition, and decision-making [9]. Artificial neurons are usually layered in these models, with each layer processing and enhancing the incoming data to produce progressively abstract demonstrations. To rise accuracy over time, these models employ a training method known as backpropagation. The weights of the associates among neurons are modified by backpropagation in accordance with errors in the model's predictions. Deep learning has accomplished important breakthroughs in numerous fields, such as computer vision, natural language processing, and robotics. For instance, it has allowed the improvement of self-driving cars, precise medical image analysis, and speech recognition systems that can interpret human speech with great precision [10].

5 Methodology

Data collection
This dataset includes roughly 1,000 training and 200 test image recordings of optical coherence tomography (OCT) characteristics, which are then separated into two classes:

- No Diabetic Retinopathy
- Severe Diabetic Retinopathy

Overview

One has preprocessed the dataset. The test image undergoes comparable processing as well [10–12]. A dataset consisting of approximately two classification images of diabetic retinopathy is created, and each image may be used as a software test image [11]. The program can classify the diabetic retinopathy classifier image in the dataset once the model has been trained efficiently [12]. The trained model is used to evaluate the test image and forecast the DR following successful training and preprocessing.

CNN Model steps:

Conv2d:

Deep learning operations of the Conv2d type are regularly used by convolutional neural networks (CNNs). Typically, image data is used for this two-dimensional convolutional activity. Applying filters to each point in the input image results in small matrices, usually 3×3 or 5×5. Features like edges, textures, and shapes that were in the innovative image are highlighted in the feature map that is created. Backpropagation is used to modify the filter weights during the training phase in order to decrease the error among the expected and actual outputs [14–16]. This allows the network to pick up on characteristics and patterns in the input image that are pertinent to the problematic it is attempting to solve. Conv2d is a key building block for numerous deep learning applications. CNNs can accomplish high fidelity in a variation of computer vision tasks and extract progressively complex features from images by layering multiple Conv2d layers with dissimilar filters, adding activation functions, and pooling layers in among.

MaxPooling2D layer

Image processing activities use a MaxPooling2D layer. MaxPooling2D works by choosing the main value inside each of the rectangular windows that propagate over the input feature map. These windows are usually 2×2 or 3×3. This method is efficient in cutting the spatial dimensions in half while retaining the most important components of the feature map. By reducing the convolutional neural network's sensitivity to small variations in the input image, including shifts and rotations, MaxPooling2D increases the network's robustness. It also helps avoid overfitting since the number of parameters in the network is reduced and the network is prevented from memorizing the training data.

Flatten layer

In the deep learning models applied for computer vision tasks such as image classification, there is a layer known as "flatten layer". This layer creates a 1D feature vector from the output of a MaxPooling2D or Conv2D layer. This is done by keeping the stack size constant while reshaping the output tensor of the preceding layer to one dimension. For example, the Flatten layer changes the shape tensor which is the output of a Conv2D layer to another shape tensor. This flattened demonstration can be passed to dense layers or other layer types that expect a 1D feature vector. A flatten layer is a basic and effective method to change the output of a CNN into a format that can be used by other types of layers in the network and is commonly used in many deep learning structures.

Dense layer

During training, backpropagation modifies the dense layer's weights to reduction the difference between the expected and actual outputs. To add nonlinearities and permit the network to recognize more intricate patterns, an activation function is also applied to the dense layer's output. A neural network's last layer for classification tasks is often a dense layer. The output is then passed through a softmax activation function to provide probability distributions over classes. It can also be utilized as an intermediate layer in neural networks to learn different demonstrations of input data and perform feature extraction. Dense layers are powerful tools for learning problematic relationships between input and output data. Overfitting can occur when the number of neurons in a layer exceeds the quantity of the training data. Thus, it is important to use training data to regulate the perfect amount of neurons for a dense layer.

Dropout layer

In neural networks, overfitting is avoided with the help of a dropout layer. This technique provides a quick and simple solution to improve generalization and lessen neuronal codependency in networks. A fraction of neurons is randomly chosen by the dropout layer during training, and their output is set to zero depending on a predefined probability (usually between 0.1 and 0.5). This procedure runs independently for each input example, and a dropout mask (a set of neurons set to zero) is randomly created for each iteration. During inference, the dropout layer is bypassed, and all neurons are used.

Image Data Generator

While training the deep learning model, the DataGenerator utility class from the Keras deep learning package preprocesses image input and creates batches of images in real-time [13]. It offers many image enhancement techniques such as rescaling, rotating, zooming, panning, mirroring, and shearing, which help increase the variety of training data and improve model robustness. The ImageDataGenerator takes input data and applies the specified augmentation techniques to generate batches of data. These batches can be directly fed into Keras models for training, allowing you to create more robust and diverse training datasets.

Training Process

Training deep learning models can be difficult and time-consuming, requiring careful consideration of data pretreatment methods, hyperparameter adjustment, and optimization algorithms [14]. Nonetheless, these models can perform very well on a variety of tasks when used in conjunction with a thoughtful strategy.

Epochs

The term "epoch" describes a neural network's complete run through the training dataset. The weights and biases of the neural network are modified following each batch of samples during training. The network considers an epoch to be finished when it has processed all training data and modified the weights and biases appropriately [15, 16]. Prior to training starting, the number of epochs must be set as a hyperparameter. Computational resources, dataset size, and model complexity are some of the variables that affect the optimal number of epochs. To find the best time to terminate training, model performance is usually tracked using a different validation dataset. Training typically ceases when performance on the validation dataset no longer improves or begins to decline.

System Architecture

The architecture in Fig. 1 represents how the obtained data will be examined. The processed and compared data is subsequently used for the final deployment. The Django framework was used for the final deployment.

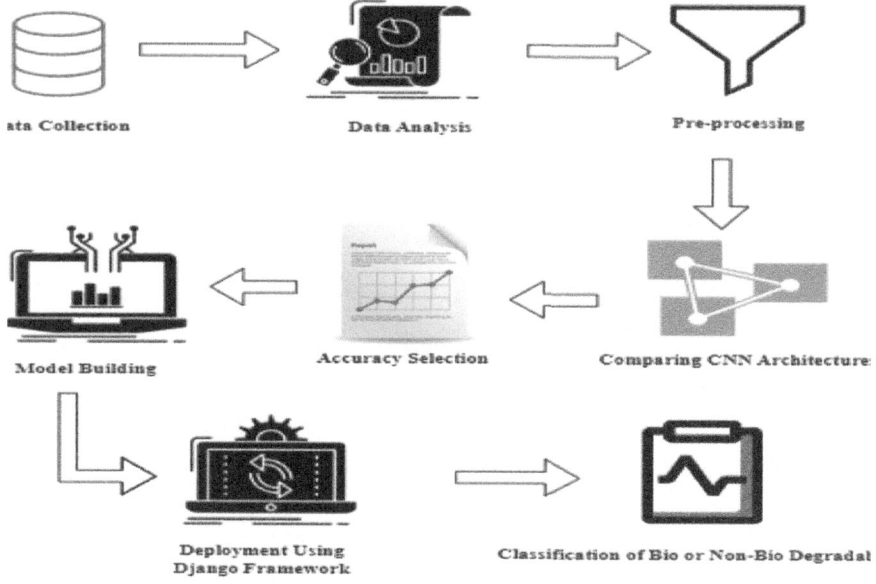

Fig. 1. Architecture of the System

6   Experiments and Results

Types of CNN:

- AlexNET
- LeNET

AlexNET:
AlexNet has 8 levels. The first two layers are basic: an 11 × 11 kernel size and 96 filters in the Conv2D layer, and a stride of 2 in the MaxPooling layer with a pool size of 3 × 3. [17]. Subsequently, the network uses distinct filter counts and sizes for its Conv2D and MaxPooling layers. The last three layers are a compression of information. Except for the final layer, which utilizes a softmax activation function for classification, the other layers use ReLU activation functions. In the fully connected layers, dropout regularization is used to prevent overfitting. AlexNet, an early deep learning model, was the first to use convolutional neural networks for large-scale image identification tasks [16–19].

Architecture of AlexNet:
(See Fig. 2).

LeNET
LeNet is made up of seven layers: two convolutional layers, two pooling layers, and three densely connected layers. The network begins with a MaxPooling layer that has a 2×2 pool size and a stride of two, and a Conv2D layer that has six filters and a 5× 5 kernel. The filter widths and number of filters utilized in the subsequent Conv2D and MaxPooling layers vary. For a class 10 handwritten digit identification test, the last three

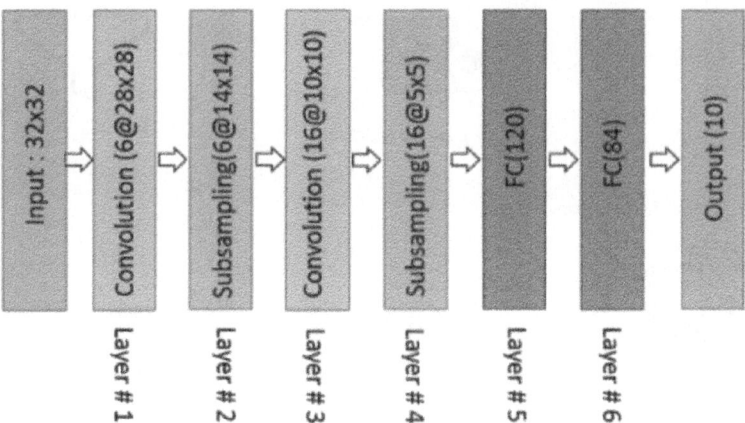

Fig. 2. Architecture of AlexNet

layers are dense layers; the first two levels contain 120 neurons, and the last layer has 10 neurons [18]. All layers of LeNet use the Tanh activation function, while the last layer uses the Softmax activation function for classification. There is no regularization through dropouts in the original LeNet architecture.

The architecture of LeNet:

(See Fig. 3).

Module Description

   Import The Given Image From The Dataset:

   In deep learning, images are typically imported from a dataset using the data generators provided by deep learning frameworks such as Keras or PyTorch. We first import the ImageDataGenerator class from Keras [19]. The training and validation sets are then given different data generators using the flow_from_directory technique. Using the designated augmentation techniques (in this case, only rescaling), this procedure reads the images from the supplied directory and preprocesses them. Finally, class_mode specifies the type of labels for the images (in this case, binary classification). Once the data generators are generated, they can be used to feed the images into a Keras model for training or assessment [20].

   To Train The Module By Given Image Dataset:

Model: "model"

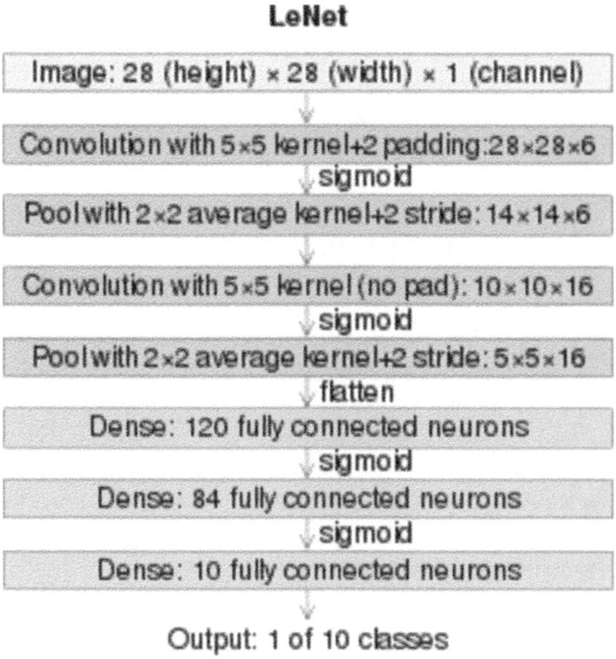

LeNet

Image: 28 (height) × 28 (width) × 1 (channel)

Convolution with 5×5 kernel+2 padding:28×28×6
sigmoid

Pool with 2×2 average kernel+2 stride: 14×14×6

Convolution with 5×5 kernel (no pad): 10×10×16
sigmoid

Pool with 2×2 average kernel+2 stride: 5×5×16
flatten

Dense: 120 fully connected neurons
sigmoid

Dense: 84 fully connected neurons
sigmoid

Dense: 10 fully connected neurons

Output: 1 of 10 classes

Fig. 3. Architecture of LeNet

batch_normalization_3 (Batch	(None, 49, 49, 256)	1024
activation_3 (Activation)	(None, 49, 49, 256)	0
conv2d_4 (Conv2D)	(None, 49, 49, 64)	16448
batch_normalization_4 (Batch	(None, 49, 49, 64)	256
activation_4 (Activation)	(None, 49, 49, 64)	0
conv2d_5 (Conv2D)	(None, 49, 49, 64)	36928
batch_normalization_5 (Batch	(None, 49, 49, 64)	256
activation_5 (Activation)	(None, 49, 49, 64)	0
conv2d_6 (Conv2D)	(None, 49, 49, 256)	16640
batch_normalization_6 (Batch	(None, 49, 49, 256)	1024
activation_6 (Activation)	(None, 49, 49, 256)	0
conv2d_7 (Conv2D)	(None, 49, 49, 64)	16448
batch_normalization_7 (Batch	(None, 49, 49, 64)	256
activation_7 (Activation)	(None, 49, 49, 64)	0
conv2d_8 (Conv2D)	(None, 49, 49, 64)	36928
batch_normalization_8 (Batch	(None, 49, 49, 64)	256
activation_8 (Activation)	(None, 49, 49, 64)	0
conv2d_9 (Conv2D)	(None, 49, 49, 256)	16640
batch_normalization_9 (Batch	(None, 49, 49, 256)	1024
activation_9 (Activation)	(None, 49, 49, 256)	0

CNN Model Summary details

Working Process of Layers in CNN Model:

The layers of a convolutional neural network (CNN) collaborate to extract and modify the features of the input images. During training, the network uses backpropagation and stochastic gradient descent to determine the optimal values for the weights and biases in each layer (or a variant thereof). By fixing these values, the network can learn and manipulate the features in the input images to make correct predictions on new unseen data. The input layer of the CNN stores image data. A 3D matrix assigns this image data, which needs to be reshaped to be a single column.

Diabetic retinopathy

(See Fig. 4).

Trainned data for Severe:

```
====== Images in:  dataset/Train/Severe
images_count:   100
min_width:      1024
max_width:      1024
min_height:     679
max_height:     768
```

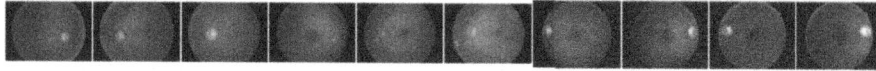

Fig. 4. Trained data for severe DR

Normal

(See Fig. 5).

Trainned data for No_DR:

```
====== Images in:  dataset/Train/No_DR
images_count:   100
min_width:      1024
max_width:      1024
min_height:     679
max_height:     768
```

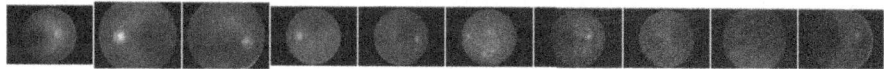

Fig. 5. Trained data for No DR

Convo Layer:

The convolutional layer is an important component of a CNN, in which many filters are used to extract various features from an input image, including edges, corners, and textures. During training, small matrices known as filters, kernels or weights are obtained. The multiple convolutional layers with many filters can help a CNN learn progressively more complex characteristics of an input image. Such features can then be used for tasks such as classification, object detection, and segmentation (Fig. 6).

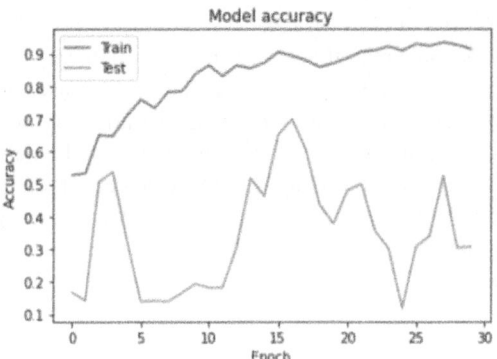

Fig. 6. CNN model trained dataset accuracy

Softmax/Logistic Layer:
The softmax function is one of the most common activation functions used in the output layer of the neural networks used in classification problems. It performs especially well when, in addition to classifying each input into one of a number of possible kinds, a procedure known as multi-class classification, is required. The raw output of the network is passed through the softmax function to obtain a probability distribution over all the classes and is used to calculate the performance metrics such as accuracy and cross-entropy loss. The softmax layer creates the assumption that the classes are mutually exclusive, which means that each input can only be a member of one class. A distinct output layer, such as a sigmoid, is used for tasks where multiple labels might apply to a single input (such as object detection or semantic segmentation) (Fig. 7).

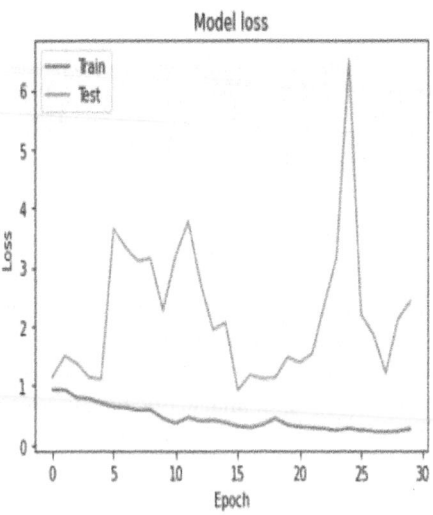

Fig. 7. CNN model trained dataset loss values

This module converts a trained deep learning model into a hierarchical data format (.h5 file). This file is used by the Django framework to provide a better user interface and, finally, on the Raspberry Pi to predict the output. Comparisons of other methods.

Manual Architecture Output

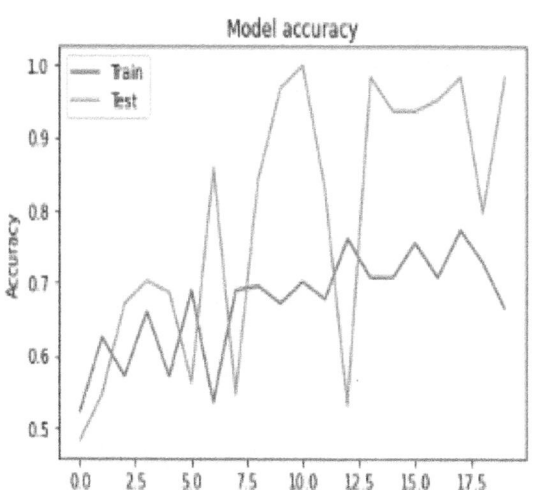

Fig. 8. Accuracy Graph of Manual Architecture

Alex Net Output

Lenet Output

The method used here is LeNet. From Figures 8, 9, 10, 11, 12, and 13 above, the LeNet model is more accurate than the manual and AlexNet models. LeNet's loss rate is much lower than that of the Manual Architecture and AlexNet.

No Diabetic:

(See Fig. 14).

Severe Diabetic:

(See Fig. 15).

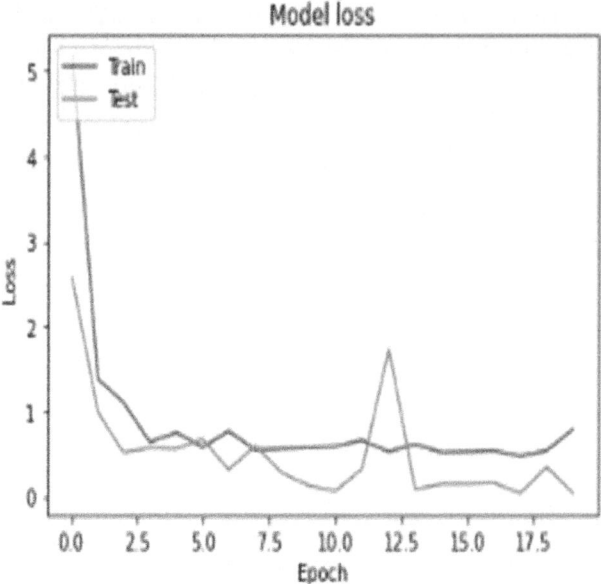

Fig. 9. Loss Graph of Manual Architecture

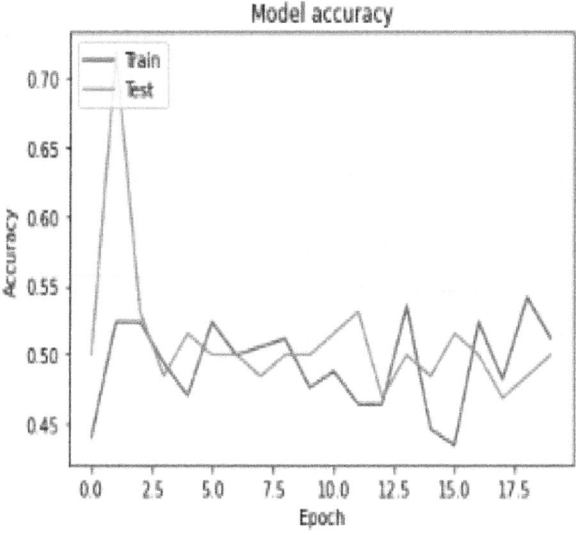

Fig. 10. Accuracy Graph of Alex's net

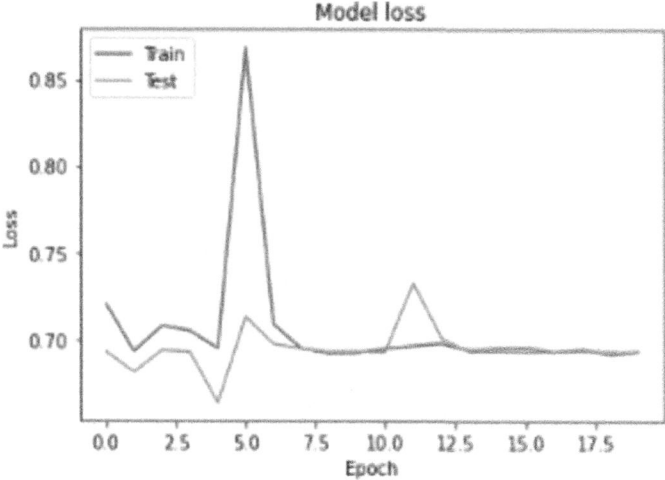

Fig. 11. Loss Graph of Alex's net

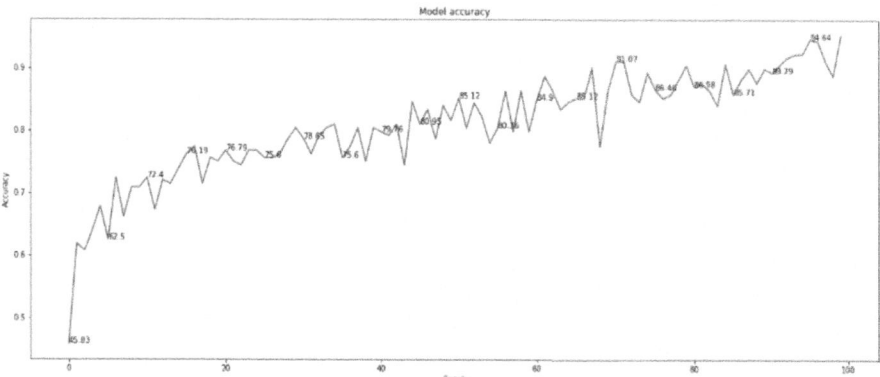

Fig. 12. Accuracy Graph of Lenet

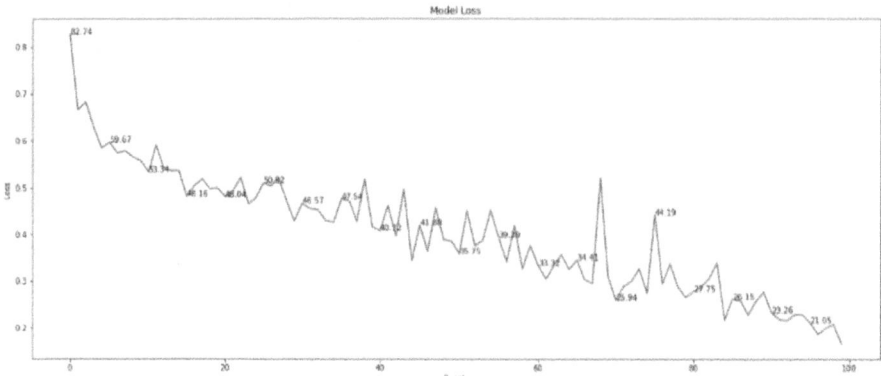

Fig. 13. Loss Graph of Lenet

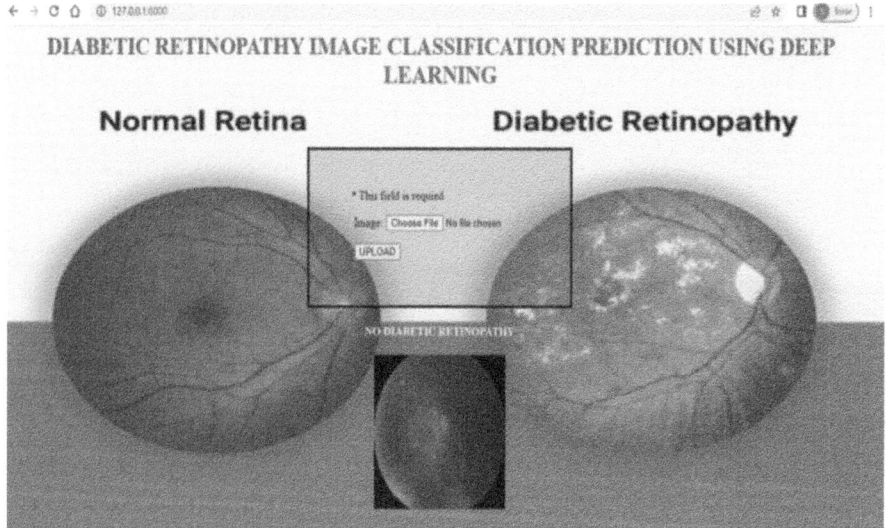

Fig. 14. Output as NO DIABETIC

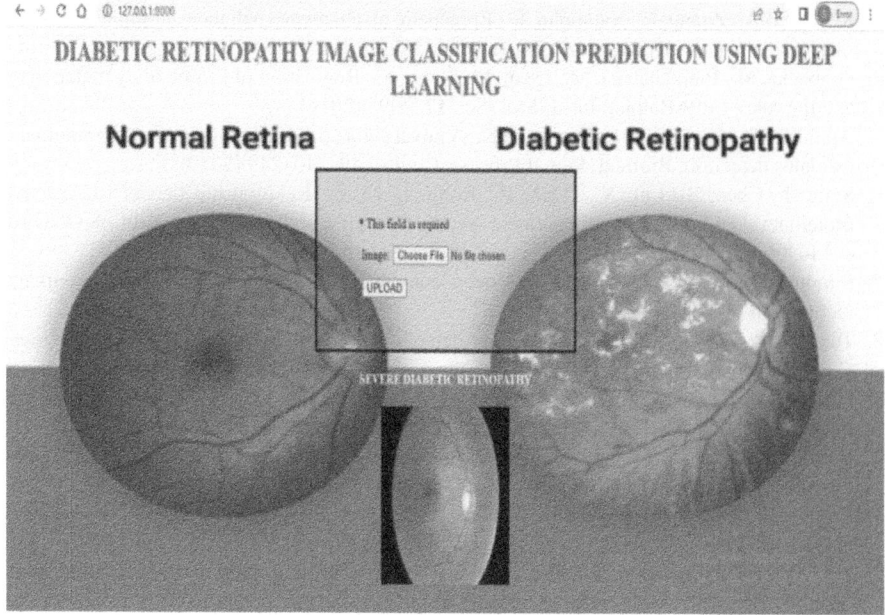

Fig. 15. Output of Severe Diabetic

7 Conclusion and Discussion

It focused on how CNN models can be used to predict disease patterns in diabetic retinopathy using a given dataset (trained dataset) and images from previous datasets. This yields some of the following insights for predicting diabetic retinopathy. One important advantage of the CNN classification system is its capacity for automatic image classification. Eye illnesses are typically irreversible because they are detected too late by patients.

8 Future Scope

1. Any cloud-based system can use this architecture.
2. Real-time data collection is possible by connecting to an embedded system.

References

1. Leasher, J.L., et al.: Global estimates on the number of people blind or visually impaired by diabetic retinopathy: a meta-analysis from 1990 to 2010. Diab. Care **39**(9), 1643–1649 (2016)
2. Cheung, G.C.M., Ting, D.S.W., Wong, T.Y.: Diabetic retinopathy: global prevalence, major risk factors, screening practices, and public health challenges: a review. Clin. Exp. Ophthalmol. **44**(4), 260–277 (2016)

3. Reddy, M.A., Zhang, E., Natarajan, R.: Epigenetic mechanisms in diabetic complications and metabolic memory. Diabetologia **58**(3), 443–455 (2015)
4. Grabacka, M., Pierzchalska, M., Dean, M., Reiss, K.: Regulation of ketone body metabolism and the role of PPARalpha. Int. J. Mol. Sci. **17**, 2093 (2016)
5. Aliahmad, B., Khojasteh, P., Kumar, D.K.: A novel color space of fundus images for automatic exudates detection. Biomed. Signal Process. Control **49**, 240–249 (2019)
6. Song, J., Chen, S., Liu, X., Duan, H., Kong, J., Li, Z.: Relationship between C-reactive protein level and diabetic retinopathy: a systematic review and meta-analysis. PLoS ONE **10**, e0144406 (2015)
7. Simonyan, K., Zisserman, A.: Very deep convolutional networks for large-scale image recognition (2014). arXiv:1409.1556
8. Tu, Z., Xie, S.: Holistically-nested edge detection. In: Proceedings of IEEE International Conference on Computer Vision (ICCV), pp. 1395–1403 (2015)
9. Feng, Y., Mo, J., Zhang, L.: Exudate-based diabetic macular edema recognition in retinal images using cascaded deep residual networks. Neurocomputing **290**, 161–171 (2018)
10. Behl, T., Kotwani, A.: Omega-3 fatty acids in the prevention of diabetic retinopathy. J. Pharm. Pharmacol. **69**, 946–954 (2017). https://doi.org/10.1111/jphp.12744
11. Erginay, A., Jeulin, C., Klein, J.C., Massin, P., Ordonez, R., Walter, T.: Automatic detection of microaneurysms in color fundus images. Med. Image Anal. **11**(6), 555–566 (2007)
12. Seki, M., et al.: Involvement of a brain-derived neurotrophic factor in early retinal neuropathy of streptozotocin-induced diabetes in rats: therapeutic potential of the brain-derived neurotrophic factor for dopaminergic amacrine cells. Diabetes **53**, 2412–2419 (2004)
13. Girshick, R., Gupta, A., He, K., Wang, X.: Non-local neural networks. In: Proceedings of IEEE/CVF Conference on Computer Vision Pattern Recognition, pp. 7794–7803 (2018)
14. Wong, T.Y., et al.: Guidelines on diabetic eye care: the international council of ophthalmology recommendations for screening, follow-up, referral, and treatment based on resource settings. Ophthalmology **125**(10), 1608–1622 (2018)
15. Zhou, Y., et al.: Collaborative learning of semi-supervised segmentation and classification for medical images. In: Proceedings of IEEE/CVF Conference on Computer Vision and Pattern Recognition (CVPR), pp. 2079–2088 (2019)
16. Chen, X., Shi, F., Wu, B., Zhu, S., Zhu, W.: Automatic detection of microaneurysms in retinal fundus images. Comput. Med. Imag. Graph **55**, 106–112 (2017)
17. Clearside Biomedical I. Suprachoroidal injection of CLS-TA in subjects with macular edema associated with non-infectious uveitis (PEACHTREE). ClinicalTrialsgov. Bethesda (MD): National Library of Medicine (US), NCT02595398 (2015)
18. Guo, S., Kang, H., Li, N., Li, T., Wang, K., Zhang, Y.: L-SEG: an end-to-end unified framework for multi-lesion segmentation of fundus images. Neurocomputing **349**, 52–63 (2019)
19. Kontturi, L.S., Collin, E.C., Murtomaki, L., Pandit, A.S., Yliperttula, M., Urtti, A.: Encapsulated cells for long-term secretion of soluble VEGF receptor 1: material optimization and simulation of ocular drug response. Eur. J. Pharm. Biopharm. **95**, 387–397 (2015). https://doi.org/10.1016/j.ejpb.2014.10.005
20. Jazani, S., Nazar, M., Tavakoli, M.: Automated detection of microaneurysms in color fundus images using deep learning with different preprocessing approaches. In: Proceedings of SPIE Medical Imaging Information and Healthcare Research and Application, vol. 11318, Art. no. 113180E (2020)

Melanoma Skin Cancer Classification Using Transfer Learning

V. Subapriya(✉), B. Bala, and S. Aswin

Sathyabama Institute of Science and Technology, Chennai, India
subapriya.cse@sathyabama.ac.in

Abstract. Dermatologic issues may have psychological consequences that have a significant impact on patients' lives. Skin conditions are more than just an eyesore; they can cause psychological issues including anxiety, sadness, and others that have an impact on patients' life similar to that of arthritis or other incapacitating conditions. If skin conditions are not treated promptly, they can progress to more serious conditions and possibly be deadly. The need for early illness detection is therefore extremely important. There is a huge demand to create a system that can predict whether a person would get diseased or not, alternately, this system can prove to be of major help in hospitals and clinics.

Keywords: Dermatologic · Psychological · anxiety · eyesore · arthritis

1 Introduction

Skin disease is defined as the disease which affects the human skin. The symptoms are the rashes in the skin and change in texture of the skin. Skin diseases are of 2 types. One is temporary and other is permanent. The temporary ones usually can be cured in 2–3 months of time if proper medication is taken and the permanent ones are the one which take a longer period to time to get cured. Some of the skin disease are painful and other are painless. Mostly skin disease occurs through improper hygienic practices and some skin diseases are genetic based ones. However, in recent days there are more deaths due to skin diseases. The main reason behind these deaths are carelessness, many people aren't aware of the harmful effects of the skin diseases thus leading to more deaths.

2 Literature Review

Mobile based Skin disease detection application was developed since in the remote areas, there will be few dermatologists. Image processing is the technology that has been used and it has been found that this system achieves 80% efficiency, and the average time taken for testing and training is 0.78 s and 2.06 s respectively [1].

Human skin diseases are various types and some are fatal and some are minor ones. In this paper [2], classification of human skin diseases is studied and they are categorized under 5 types. The first one is Malignant melanoma and this is a dangerous skin disease

P. Dassan et al. (Eds.): ICICSCNT 2023, CCIS 1970, pp. 111–118, 2024.
https://doi.org/10.1007/978-3-031-75957-4_10

and this disease mostly occurs in white countries, the second one is the Genetic Diseases in which the skin diseases will be transmitted genetically, the third one is Leprosy and this disease is transmitted through open skin. This disease mostly occurs in undeveloped and rural areas. The fourth one is viral infectious diseases, which can be easily identified by clinical test and the last one is the most common diseases which is acne and Psoriasis.

In [3], Image processing is used for the implementation of the skin disease detection and breaking image processing, 3 methods were used, the first one is the histogram processing in which the histograms for variety of skin diseases is produces using MATLAB and after this step, Color Segmentation is done by segmenting objects in an RGB image and finally Texture analysis is used and it is a simple tool to segment the image into distinct characteristics.

CNN and traditional way of skin disease prediction systems are analyzed and compared. From the analysis is found that CNN requires larger dataset while the traditional method doesn't require large dataset and out of other CNN Architecture, CNN With Convolutional: 16 filters and 7*7, pooling produced accuracy of 98.32% sensitivity of 981.5%, specificity of 98.41% [4].

SVM based skin disease prediction system is implemented [5] and after the analysis is has been found that SVM is better than Neural networks. The reason why SVM is considered over neural networks is Also, SVMs are less prone to over fitting and SVM need very less memory for the storage of the predictive model. The achieved accuracy of the model is about 90%.

ANN is used for the implementation of the system. The reason being ANN is used is ANN can have the capability to learn the patterns, hence it can learn the symptoms of the particular diseases as they form a pattern [6].

In paper [7] various image processing techniques are analyzed and implemented to find whether the given input image has skin disease or not. The four image segmentation techniques that are implemented here are Edge detection, Morphology based image segmentation, Adaptive thresholding and Edge detection.

In this article, a skin disease prediction system is constructed using Convolutional Neural Networks (CNN) [8].The dataset has over 1200 images and this dataset is divided into 10 classes. The images are preprocessed and resized in order to fit into our model and these images will be used for validation purposes and also will be used for training purposes. The implemented system can detect up to 8 facial skin diseases. The accuracy achieved in this system is around 88%.

The preprocessed photos are used in paper [9] to extract the important skin characteristics. Color, diameter, various surface, and sub-surface structures are among the important skin characteristics. After extracting the key skin features, the skin changes are monitored over a function of time. The changes are then compared with the threshold value, and if the changes are higher than there is a high possibility of the skin disease to occur and if the change is less than the threshold, the skin condition is normal.

3   System Analysis

Tool: Anaconda software.
   Packages used:

- Keras
- Numpy
- Tensorflow

   Deployment: Flask.

3.1   Anaconda Software

Anaconda is software used to execute machine learning programs, data science programs, computing and analysis programs. The term conda means package and environment management packages and it has more than 150 packages. We are using anaconda for implementing python programs.

Flask
Flask is one of the popular Python framework and it is a web framework. Flask is mainly used for building web applications that uses python as an backend language. Using flask, any simple applications can be scaled and complex applications can be made.

4   System Design

Technology Used

Frontend – HTML, CSS
Backend – Python

HTML
HTML is mainly used to structure the web page. In this project HTML is used to design the input image upload web page.

CSS
CSS is mainly used for styling the web page CSS gives the web page a better look with the designs and styles.

Python
Machine learning is employed in this project, and Python is the preferred language for machine learning applications. Python is used for a variety of purposes. Python's simplicity and ease of use make it possible for others to understand it and, as a result, make it simple to create machine learning models. The second major factor is that Python is a perfect choice since it has many helpful libraries that make it possible for developers to perform complicated tasks with ease. Because machine learning mostly focuses on probability, optimisation, and statistics.

5 Modular Design

The general structure of the module is defined by the modular structure. The degree to which a system's components may be disassembled and reassembled is a common definition of the general idea of modularity.

The following modules make up the proposed system:

1) GUI & Data Preprocessing Module
2) CNN Implementation
3) Prediction

5.1 GUI and Preprocessing Module

This is the interface via which the user or patient communicates with the programme. This user interface (UI) was created in a straightforward manner to make it simple for people to engage with. These input parameters in this module are properly trained, and a model is produced. To achieve great accuracy, these input parameters need to be well educated. The training data's characteristics and the calibre of the labelled training data will impact how precisely the machine learns to recognise the result. If there are any missing values in our dataset, it might be quite problematic for our machine learning model. Since the dataset contains missing values, it is necessary to handle them. As a result, a data preprocessing module is needed to carry out tasks like cleaning the data and making it suitable for a machine learning model, which helps to increase the accuracy and efficiency of a machine learning model. We require clean, properly formatted data in order to successfully anticipate the case using machine learning. Our dataset is similarly split into a training set and a test set throughout the data processing stage. This is one of the key data pretreatment processes since by following it, we may improve the effectiveness of our machine learning model. Four phases make up data preparation.

1) Data cleaning is the initial phase, during which duplicate, improperly formatted, or damaged data is corrected or eliminated.
2) Data integration is the next phase, which unifies data from several sources into a single perspective.
3) The data are encoded, scaled, and sorted as necessary in the third phase, which is known as data reduction.
4) Data transformation, which involves transforming the data into the necessary format, is the last phase.

5.2 CNN Implementation

The first phase of CNN is the convolutional layer, which consists mostly of flattened convolutions and 1 * 1 convolutions. There are really 7 different types of convolutional layers. The next function is the Rectified Linear Activation Function, or ReLU. Using x as the input, it is a max function $(x, 0)$. ReLU's primary function is to compute the matrix after convolution, keeping all other values constant and setting any negative values in the matrix to zero. The size of the feature maps are decreased by pooling layers. As a result, it lessens the quantity of computation done in the network and the number of parameters to learn. The last several levels are full connected layers, which combine the

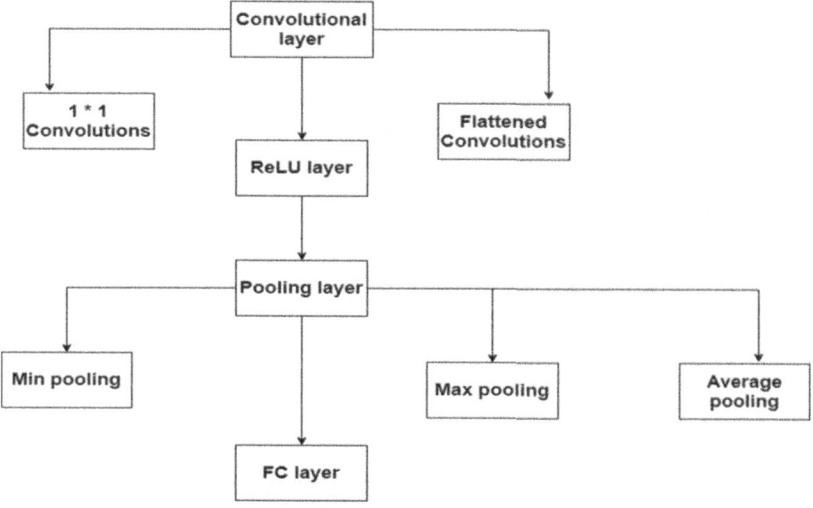

Fig. 1. Steps in CNN

data collected by earlier layers to create the final output, and the last layer is the FC (Fully Connected) layer (Fig. 1 and Table 1).

Table 1. Types of Convolutional layer

S.No	Types
1	Simple Convolution
2	1x1 Convolutions
3	Flattened Convolutions
4	Spatial and Cross-Channel convolutions
5	Depthwise Separable Convolutions
6	Grouped Convolutions
7	Shuffled Grouped Convolutions

The 2D convolutional layer is employed in this system, and then max pooling is applied after that. The advantage of max pooling over other strategies is that it requires fewer parameters to learn, which lowers the computational cost.

5.3 Prediction

This is the last module where the user-provided input picture is forecasted using the learned model. The word "malignant" will be displayed if the input image contains skin disease, indicating that immediate action must be taken. The word "benign" will be displayed if the input image does not contain skin disease and if the skin is normal, indicating that there is no need for concern and that the skin appears to be in perfect condition.

6 Coding

In this comparison of the two algorithms, CNN and SVM, it has been discovered that CNN provides more accuracy than SVM (Fig. 2).

```
MODEL_PATH = 'mdl_wts_xc.hdf5'

# Load your trained model
model = load_model(MODEL_PATH)
print('Model loaded. Check http://127.0.0.1:5000/')


def model_predict(img_path, model):
    img = image.load_img(img_path, target_size=(150, 150))

    # Preprocessing the image
    x = image.img_to_array(img)
    x = np.true_divide(x, 255)
    x = np.expand_dims(x, axis=0)


    preds = model.predict(x)
    return preds
```

Fig. 2. Comparison with the existing system

```
def upload():
    if request.method == 'POST':
        # Get the file from post request
        f = request.files['file']

        # Save the file to ./uploads
        basepath = os.path.dirname(__file__)
        file_path = os.path.join(
            basepath, 'uploads', secure_filename(f.filename))
        f.save(file_path)

        # Make prediction
        preds = model_predict(file_path, model)

        # Process your result for human
        res = preds[0]
        print(res)

        if res > .5:
            result = "Benign"
        else:
            result = "Maligant"
```

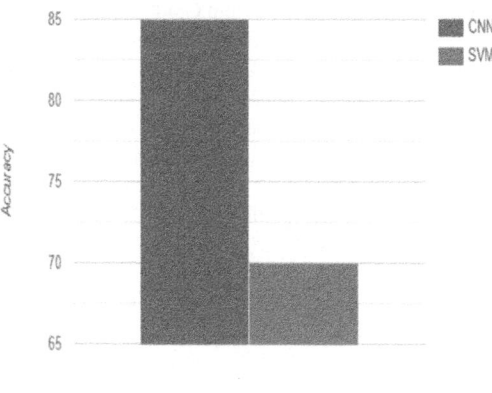

Fig. 2. (*continued*)

7 Conclusion

As a result, a web application is created to forecast skin diseases, and by employing CNN, 85% accuracy is attained. Because it automatically learns distinguishing characteristics for each class, CNN is able to carry out this project. It is also thought to be computationally effective. Clinicians and doctors will benefit much from this method.

Future Enhancement
The accuracy of the system may be increased by examining and studying more algorithms, and this web application can be set up on a cloud platform so that it is available to everyone.

References

1. Ajith, A., Goel, V., Vazirani, P., Roj, M.M.: Digital dermatology. In: International Conference on Intelligent Computing and Control Systems (2017)
2. Manoorkar, P.B., Kamat, D.K., Patil, P.M.: Analysis and classification of human skin diseases. In: International Conference on Automatic Control and Dynamic Optimization Techniques (ICACDOT) (2016)
3. Sahir, M., Das, M.M., Kib, M.A.: Feature based skin disease estimation using image processing for teledermatology. In: International Conference on Computer, Communication, Chemical, Material and Electronic Engineering (2018)
4. Goswami, T., Dabhi, V.K.: Skin disease classification from image - a survey. In: International Conference on Advanced Computing & Communication Systems (ICACCS) (2020)
5. Kumar, N.V., Kuma, P.V.: Classification of skin diseases using Image processing and SVM. In: International Conference on Vision Towards Emerging Trends in Communication and Networking (ViTECoN) (2020)
6. Islam, M.N., Gallardo-Alvarado, J., Abu, M., Salman, N.A.: Skin disease recognition using texture analysis. In: IEEE 8th Control and System Graduate Research Colloquium (2017)
7. Roy, K., Chaudhuri, S.S., Ghosh, S.: Skin disease detection based on different segmentation techniques. In: International Conference on Opto-Electronics and Applied Optics (Optronix) (2019)
8. Bakhshi, S.: Deep convolutional neural network for face skin diseases identification. In: 2019 Fifth International Conference on Advances in Biomedical Engineering (ICABME) (2019)
9. Dhinagar, N.J.: Early diagnosis and predictive monitoring of skin diseases. In: IEEE Healthcare Innovation Point-Of-Care Technologies Conference (HI-POCT) (2016)

Epidemic Outbreak Prediction Using SIR Model

Venkata Lohith Kumar Patibandla[1]([✉]), Shalini Patibandla[2], and M. Saravanan[1]

[1] Sathyabama Institute of Science and Technology, Chennai, Tamil Nadu, India
lohithkumarp2512@gmail.com, saravanan.cse@sathyabama.ac.in
[2] Sri Chandrasekarendra Saraswathi Viswa Mahavidyalaya, Chennai, Tamil Nadu, India
shalinichowdary145@gmail.com

Abstract. The pace at which viruses are being transferred by individuals is rapidly rising, which has resulted in the loss of human life. The majority of those who get this covid-19 virus are likely to have hereditary disorders. This study examined how long it will take a patient to recover from a virus. This will assess the length of time a patient will need to recover from a virus using Deep Learning techniques. The combination of DBSCAN clustering and the SIR model is utilized to estimate the time required for a patient to recover from a virus. The dataset is first subjected to analysis using the DBSCAN clustering algorithm, which groups the data based on age as the primary criterion. The resulting cluster output is then inputted into the SIR model to assess accuracy. However, the accuracy of the cluster output is not satisfactory when compared to previous results. Therefore, the cluster output will be rechecked using different clustering algorithms to obtain more reliable results. By utilizing the more precise results obtained from the cluster output as a parameter, and integrating natural death and death caused by the disease as additional parameters in the SIR model, the project can obtain the most accurate outcomes. These parameters will facilitate the creation of a dependable visual predictor. The SIR model outcomes are presented through a web interface, which enables users to input population size, infection rate, and a COVID-19 dataset. Utilizing a Flask application, the SIR model output is visualized on a webpage.

Keywords: COVID-19 · DBSCAN · SIR model · Flask application

1 Introduction

In the year 2019, the World Health Organization (WHO) announced that the widespread outbreak of coronavirus, commonly referred to as COVID-19, had reached pandemic proportions [1]. To stop the virus from spreading further, an international coordinated effort is required. An abnormally high percentage of the population is affected by the virus, which is defined as occurring over a large geographical area [2]. The 2009 H1N1 flu pandemic was the last pandemic to be officially confirmed worldwide.

In December 2019, an unknown pneumonia outbreak occurred in Wuhan, China. By the end of January 2020, it had infected 106 people in 19 other countries and 9,720 people in China, resulting in 213 deaths. Several research facilities identified the causative agent, COVID-19, a few days after the outbreak [3, 4]. The World Health Organization has

© The Author(s), under exclusive license to Springer Nature Switzerland AG 2024
P. Dassan et al. (Eds.): ICICSCNT 2023, CCIS 1970, pp. 119–130, 2024.
https://doi.org/10.1007/978-3-031-75957-4_11

designated COVID-19 as a severe and highly contagious respiratory illness caused by the SARS-CoV2 virus. The disease has caused significant global contamination since its emergence in 2019, with 79,890 cases and 3,354 fatalities in China as per the World Health Organization's daily report. This study will also develop software requirements specifications (SRS) [5].

Through the utilization of Deep Learning techniques, an analysis will be conducted on the duration of recovery from viral infections. Two distinct methods, namely DBSCAN and the SIR Model, have been employed to estimate the period of recovery for patients. Clustering, a fundamental technique in data analysis, plays a critical role in determining the data's structure. Essentially, clustering is the process of identifying subgroups in a dataset that exhibit high similarity, while the data in different groups have significant differences. This technique is classified as an unsupervised learning method, as it lacks a predetermined standard against which the accuracy of the grouping algorithm can be evaluated.

2 Literature Survey

Researchers and computer scientists have researched this problem statement in detail over the past few years to find a solution, and all their answers range from examining various cluster techniques to the analysis of epidemic outbreaks of different data collection.

The study by Marina Bagić Babac [6] suggested the model dynamical mathematical method is SIR, which provides a more accurate in predicting the covid-19 data set. The new virus was spreading more in Italy and the infection rate is increasing but entering the data (infected people) online is very less. The risk of the second wave is very less because of the first wave project. The result of this project is shown in the graph using the SIR model.

DBSCAN, or Density-Based Spatial Clustering of Applications with Noise, was the subject of a study conducted by Ashutosh S et al. [7] in April 2020. The researchers carefully identified the model's parameters and explored different approaches to enhance the accuracy of their findings. Their study distinguished the exposure rate from the infection rate and introduced varying levels of quarantine to mitigate the spread of the disease. Although the study resulted in more precise findings compared to prior research, the model's parameters still require further refinement for more consistency.

Mohammad Shanna and Abdallah [8] utilized Net Logo to study virus propagation in a semi-closed environment. To address this problem, they adapted an existing model built by Yang and Wilensky in 2011 and applied WHO's statistics on infection probability and other input parameters for COVID-19. Their proposed system employs a disease-spreading model from the Net Logo library to simulate the spread of the virus across a country, increasing accuracy over the current model.

Researchers J. Hackl and T. Dubernet [9] used MATSim to simulate the spread of contagious illnesses within transportation settings in an urban area. They based their model on actual data from people's daily commute routes during seasonal flu epidemics in Kilchberg. The outcomes of the simulation were compared to the known SIR model, which investigates the complexities of virus epidemics and elements influencing viral transmission. The research findings produced various scenarios of an epidemic in a metropolitan setting, aiding in predicting the occurrence and taking appropriate measures.

2.1 Open Problems in Existing System

The existing model fails to produce the desired results despite utilizing different tools such as SIR, Net Logo, SEIR-PAD, and GIS models. While they attempted to enhance accuracy through distinct clusters, the results were still inaccurate. The k-means clustering method utilized by the existing model does not provide any clear indication of clusters and fails to identify the number of people who are suspected, infected, or recovered. Rather, it only offers a graph representation of the SIR model's recovery rate, infectious rate, and suspicion rate. To achieve the desired results, new techniques should be utilized to overcome the existing model's shortcomings. The existing model's disadvantage is that it calculates the outcome based on the entire population, which results in inaccurate output.

3   Proposed System

The proposed SIR model-based epidemic outbreak prediction system is an upgraded version of an existing model. The earlier system encountered algorithmic and accuracy-related challenges. To overcome these issues, a new approach was adopted by incorporating the DBSCAN Clustering algorithm. This method can efficiently implement the proposed system by generating clusters as required. These clusters can be formed based on specific characteristics such as age, immunity, and resistance, among others. The resulting clusters can distinguish individuals who are suspected, infected, or recovered.

The clustering method output is used as input for the SIR model, which calculates the suspected rate, infectious rate, and recovery rate. The final results are displayed on a web page. This proposed model has several advantages over existing models. To implement the proposed system, follow these steps:

- Understanding the dataset.
- Determining how specific columns are related.
- Preprocessing the data in the dataset.
- Calculating DBSCAN from the data by dividing the data into clusters.
- The output of the cluster will be sent to MODEL parameters to obtain a predictive SIR model.
- The result will be shown on the webpage using the Flask app.

3.1 Project Management Plan

(See Fig. 1).

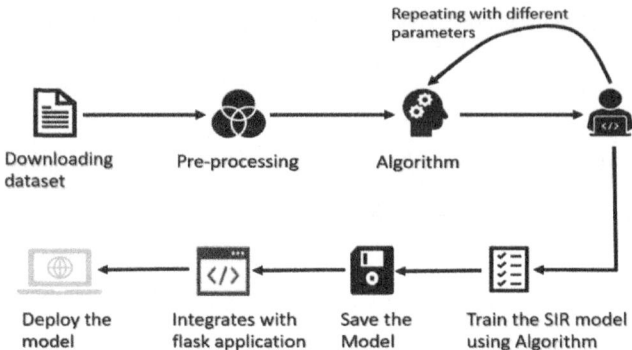

Fig. 1. Flow Diagram

4 Methodology

4.1 Data Collection

A dataset is necessary to achieve good results in any project. The dataset should include Age Immunity, Resistance, and Days to Recover data. Information about the susceptible, infected, and recovered individuals in the population should be collected and tracked over time. This data can be obtained from public health authorities or Kaggle, which is a platform that offers many datasets. The dataset used in this project was downloaded from Kaggle.

4.2 Model Selection

DBSCAN is a popular clustering method that categorizes similar data points in a high-dimensional space based on their proximity to each other. Unlike other clustering algorithms, it determines the size of clusters automatically based on data point density. It is beneficial for datasets with irregular geometries and can identify outliers or noisy points. DBSCAN requires two critical parameters: radius (eps) and the minimum number of points (min samples) needed to generate a compact region. These parameters can be adjusted based on the data being clustered.

5 Implementation

In the SIR model, the total of these three compartments (susceptible, infectious, and recovered) stays steady and equivalent to the underlying number of populations. The fundamental SIR model was introduced, where β is the disease rate or transmission rate or the power of contamination. Furthermore, γ means the recuperation or elimination rate. To settle this arrangement of differential conditions, it needs from beginning qualities for the three-state factors S, I, and R specifically S (t), I (t), and R (t) (Fig. 2).

The SIR model has three compartments (susceptible, infectious, and recovered) that interact dynamically, assuming a homogeneous population. Recovered individuals gain immunity against the sickness, preventing reinfection.

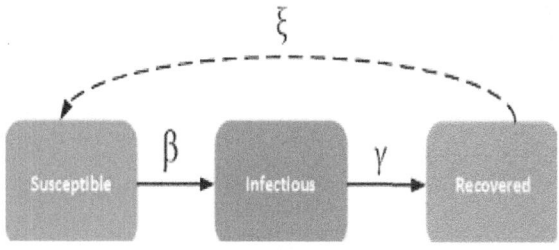

Fig. 2. SIR Model Structure

- S(t): The number of individuals susceptible to the disease at time t.
- I(t): The number of individuals susceptible to the disease at time t.
- R(t): The number of individuals susceptible to the disease at time t.
- N: the total population.
- β: The rate of transmission refers to the speed at which susceptible individuals contract the infection.
- γ: The rate at which infected individuals recuperate and acquire immunity, also known as the recovery rate.

5.1 Equations

Listed below are the differential equations that represent the SIR model:

- $dS/dt = -((\beta SI)/N)$
- $dI/dt = ((\beta SI)/N) - (\gamma I)$
- $dR/dt = \gamma I$
- $dS/dt = (ND * (N - (S))) - ((Beta * S * I) / N)$
- $dI/dt = ((Beta * S * I) / N) - ((Gama + Alpha + ND) * I)$
- $dR/dt = (Gama * I) - (ND * R)$

   where:

- ND is the natural death rate (per day)
- Alpha is the death rate caused by disease (per day)
- Beta is the infection rate (per day)
- Gama is the recovery rate (per day)
- dS/dt It measures the rate of change of susceptible individuals over time.
- dI/dt It measures the rate of change of infectious individuals over time.
- dR/dt It measures the rate of change of infectious recovered over time.

The SIR model demonstrates how the number of susceptible, infectious, and recovered individuals changes over time, based on transmission and recovery rates. This model can be used to predict and investigate the impact of interventions on disease transmission.

5.2 Overall Design of the Proposed System

(See Fig. 3).

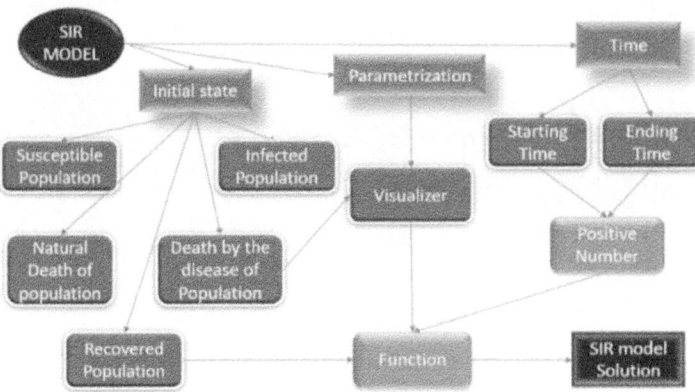

Fig. 3. System Architecture

6 Results and Discussion

DBSCAN clustering method identifies and forms distinct clusters, revealing patterns and relationships between variables. Comparing columns through scatter plots provides a comprehensive understanding of the data, aiding in result interpretation and analysis. This approach results in more accuracy.

6.1 Age vs Resistance

Age and resistance are inversely related in our project. A younger person has stronger immunity than an older person. Therefore, younger individuals affected by COVID-19 tend to recover more quickly, while older individuals may take longer to recover compared to other age groups (Fig. 4).

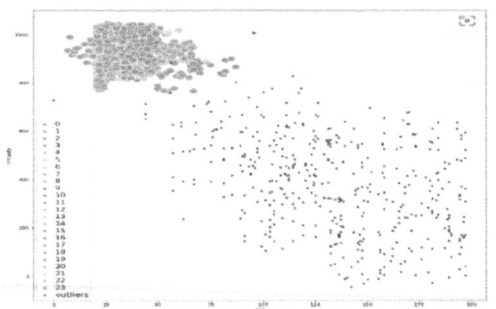

Fig. 4. Age vs Resistance

6.2   Immunity vs Resistance

Immunity and resistance are interrelated as a person's resistance increases with their immunity. Conversely, someone with lower immunity will have lower resistance. The level of immunity and resistance is determined by an individual's age group. However, regardless of age essential for recovery from covid-19 (Fig. 5).

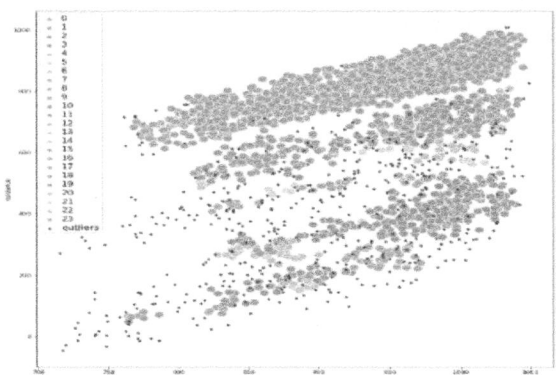

Fig. 5. Immunity vs Resistance

They are not restricted from computing Immunity and Resistance. Based on age, taking other health conditions into account, and assessing Immunity and Resistance. There is no set formula for determining Immunity and Resistance.

6.3   Varying Transmission Rate (β)

The SIR model often uses a plot to analyze the effect of transmission rates on epidemics. This involves adjusting β while keeping γ and initial conditions constant. The plot shows how the number of susceptible, infected, and recovered individuals changes over time for each β value. It helps to demonstrate how changes in transmission rates impact the severity, timing, and overall scale of the epidemic (Fig. 6).

6.4   Varying Recovery Rate (γ)

The SIR model also includes a plot that compares the impact of different recovery rates on the epidemic's progression. The plot adjusts the recovery rate (γ) while.

keeping the transmission rate (β) and initial conditions constant. It displays the count of susceptible, infected, and recovered individuals over time for each value of (γ) (Fig. 7).

6.5   Sensitivity Analysis

A sensitivity analysis plot can show how small changes in model parameters affect the outcome of an epidemic. By varying the parameters around their baseline values, the plot displays how many individuals get infected. This plot can help identify the key parameters that drive the epidemic and inform interventions to prevent disease spread (Fig. 8).

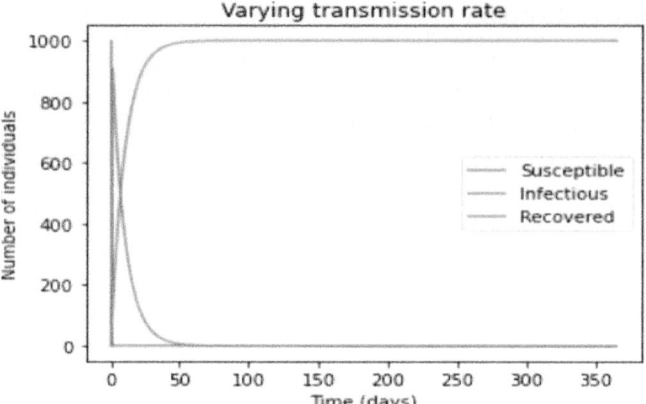

Fig. 6. Transmission Rate (β) Analysis

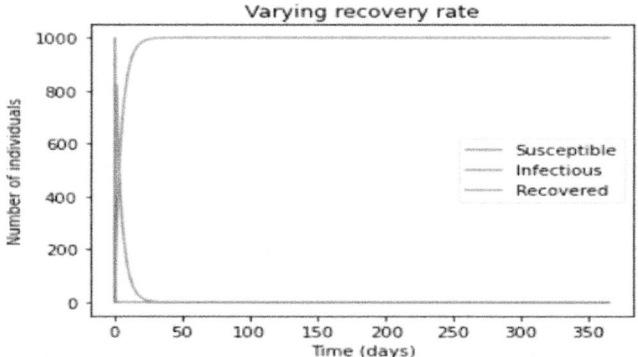

Fig. 7. Recovery Rate (γ) Analysis

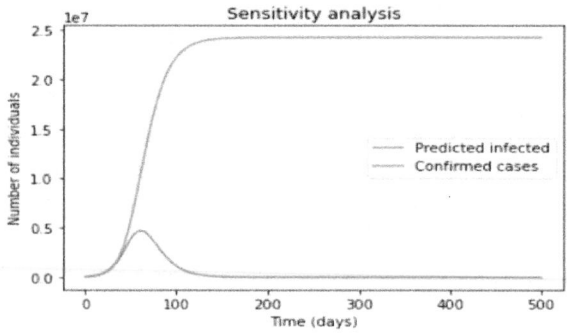

Fig. 8. Sensitivity Analysis

6.6   Basic Reproduction Number (R0)

R0 is a metric used to measure the contagiousness of a disease. It is calculated by dividing the transmission rate by the recovery rate. If R0 is above one, the disease will spread throughout the population less than one, the disease will fade away (Fig. 9).

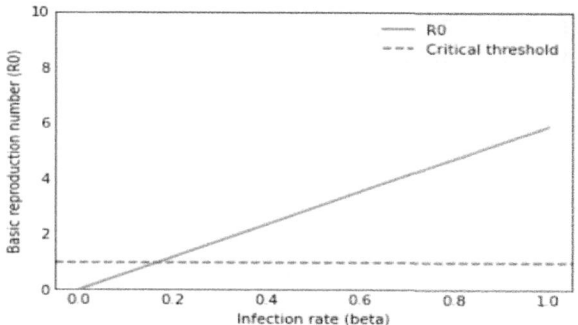

Fig. 9. Reproduction Number (R0) Plot

6.7   Peak Infection

The peak infection rate is crucial in determining the severity and duration of the epidemic and understanding virus transmission. It is influenced by factors like transmission rate and population susceptibility. This parameter is determined by the timing and magnitude of the peak (Fig. 10).

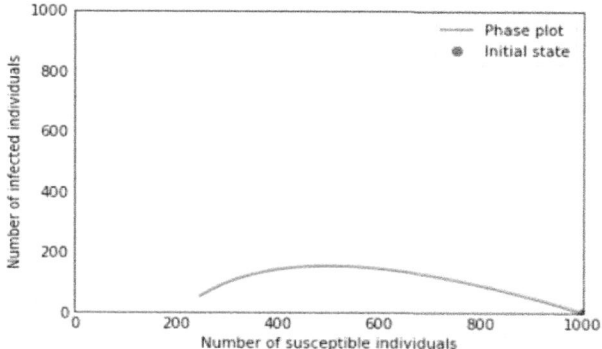

Fig. 10. Peak Infection Plot

Simulating the SIR model with different parameters and initial conditions can reveal the effects of modifying these factors on the epidemic's development and interventions.

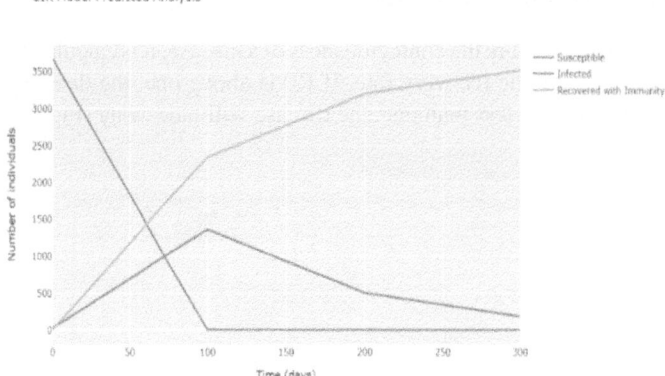

Fig. 11. SIR Prediction

Findings from this model have the potential to impact public health policies and initiatives aimed at controlling infectious diseases (Fig. 11).

The SIR model predicts the number of Susceptible, Infectious, and Recovered persons over time-based on epidemic baseline conditions and model parameters, providing insights into the epidemic path and assessing interventions such as natural death or death rate due to disease. However, the accuracy of the model depends on its assumptions and data inputs and should be complemented with real-world data (Fig. 12).

Fig. 12. User Input Web Interface

The SIR model output was pushed into the web interface using the Flask framework. The Plotly library comes in handy while creating the web frame. The flask program runs on a server after loading the appropriate modules, with the results shown in a web interface. The web interface allows users to input the population size, rate of infection, and a COVID-19 dataset (Fig. 13).

The final output displays the ultimate prediction of the SIR model based on the user-provided input data. It represents the culmination of the model's analysis, considering various factors such as transmission and recovery rates, initial conditions, and other

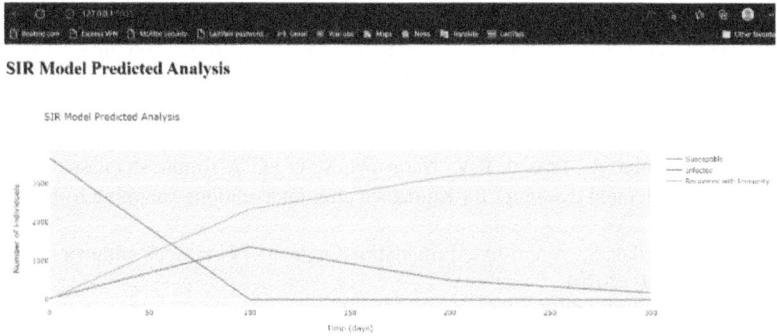

Fig. 13. The analysis of the output is based on the user input parameters.

relevant data points. The SIR model plays a crucial role in predicting and understanding the spread of infectious diseasesThe SIR model is capable of effectively processing real-time data. By leveraging this model, it is possible to develop and implement effective interventions and control strategies that can help minimize the impact of epidemics on society.

7 Conclusion

In this research, the DBSCAN clustering technique was applied to a COVID-19 dataset to find discrete groupings of regions with similar trends in confirmed cases, deaths, and recoveries. The SIR model was then used to anticipate the virus's spread within each group, taking into consideration the initial number of Susceptible, Infected, and Recovered with immunity persons, the natural death (ND) rate, the death rate due to disease (Alpha), as well as the transmission and recovery rates. The SIR model output was shown on a web page using Plotly and Flask app, this allows us to examine how the number of Susceptible, Infectious, and Recovered with immunity persons over time inside each cluster.

Overall, this approach can fill a gap in the existing systems, by providing an intuitive and user-friendly interface that allows users to easily input parameters and generate visualizations of outbreak predictions provides a valuable instrument for forecasting the spread of infectious illnesses and may assist improve public health policies and initiatives targeted at reducing the effect of such outbreaks.

References

1. World Health Organization. WHO Director-General's opening remarks at the media briefing on COVID-19 (2020). Accessed 14 Mar 2020
2. Merriam-Webster Dictionary. Pandemic. https://www.merriam-webster.com/dictionary/pandemic. Accessed 14 Mar 2020
3. World Health Organization, Novel Coronavirus–China., Disease outbreak news: Update 12 January 2020

4. Wikipedia. Timeline of the 2019–20 coronavirus pandemic in November 2019 – January 2020
5. World Health Organization. Director-General's remarks at the media briefing on 2019-nCoV on 11 February 2020 (2020) . Accessed 21 Feb 2020
6. Marina Bagić Babac - Resetting the Initial Conditions for Calculating Epidemic Spread: COVID-19 Outbreak in Italy. IEEE Xplore. Retrieved Jan 2020
7. Ashutosh, S., Simha, A., Prasad, R.V., Narayana, S., et al.: A simple stochastic SIR model for COVID-19 infection dynamics for Karnataka after interventions– learning from European trends (2020)
8. Shanna, M., Abdallah, S.: Agent-based simulation for covid-19 outbreak within a semi-closed environment. IEEE (2020)
9. Hackl, J., Dubernet, T.: Epidemic spreading in urban areas using agent-based transportation models. Futur. Internet **11**(4), 1–15 (2019)

Detecting Fake and Genuine Facebook Users Using Gradient Boosted Trees and Interpretable Machine Learning

Siva Srikanth Reddy Janke[✉], Rahim Shaik, and Linda Joseph

Department of Computer Science and Engineering, Hindustan Institute of Technology and Science, Chennai, India
jankesivasrikanthreddy@gmail.com, rahimshek999@gmail.com, 3lindaj@hindustanuniv.ac.in

Abstract. Fake Facebook accounts can be used to spread misinformation or invade privacy. Traditional methods of identifying fake accounts are time- consuming, inefficient, and unreliable. Black-box AI models can make biased or unfair decisions, further discrediting computing power. Create a system to accurately identify fake and real Facebook users. To identify fake accounts and reduce human intervention, the proposed system must be fast, precise, and extensible. The proposed system aims to objectively and fairly identify fake Facebook accounts, reducing the risk of malicious activity and improving user experience. The proposed system should use explainable AI (XAI) to increase trust in technology and ensure user security by providing transparency and interpretability in decision-making.

Keywords: explainable AI · interpretable AI · machine learning · fake users · Facebook dataset · model evaluation · model transparency · ethical AI · and responsible AI

1 Introduction

Social media platforms are now an essential component of our day-to-day lives, as they enable us to communicate with one another, share information, and enjoy a variety of forms of entertainment in this era of rapid technological advancement. Because it is the most popular social media platform, Facebook, which has over 2.8 billion active users worldwide, is an attractive target for malicious activities such as the creation of fake accounts. The creation of these fake accounts can serve a variety of nefarious purposes, such as the dissemination of false information, the conquest of individuals, or the violation of user privacy. As a consequence of this, there is a pressing need for a solution that is reliable and trustworthy, with the ability to accurately detect fake and genuine users within a dataset obtained by Kaggle.

Manual verification or rule-based systems are examples of traditional approaches to detecting fake accounts; however, these methods are time-consuming, expensive, and

P. Dassan et al. (Eds.): ICICSCNT 2023, CCIS 1970, pp. 131–144, 2024.
https://doi.org/10.1007/978-3-031-75957-4_12

frequently ineffective. Because of the ever-increasing volume of data, manual verification is becoming increasingly impractical, and rule-based systems might not be able to identify the intricate patterns and characteristics of fake accounts. The use of black-box models can lead to decisions that are biased or unfair, which can erode trust in the technology. However, machine learning presents a promising approach to the detection of fake accounts.

This project's goal is to develop an AI solution that can be interpreted and explained in order to differentiate between fake and real Facebook users so that it can help address the challenges described above. Collecting data from users, pre- processing that data, developing features, selecting and training a machine learning model, and continuously evaluating and improving the model's performance—all of these will be required to implement the solution. A system that can accurately classify Facebook users as fake or genuine and provide clear explanations for its decisions will ultimately be the product of this project.

This project was inspired by the requirement for a dependable and trustworthy solution that could identify fake accounts on social media platforms such as Facebook. The conclusion of this project may bring about an improvement in the safety and security of social media platforms and shield users from harmful activities. This project aims to promote accountability and transparency in the deployment of artificial intelligence by developing a solution that is interpretable and explainable. Restoring trust in the technology is also a goal of the project.

The remainder of this document is structured as follows. Section 2 delivers a literature review on Stress Detection through facial and speech analysis. Section 3 is our Problem Statement. Section 4 talks about XAI. Section 5 Methodology. Section 6 Experimental Results and Analysis. Section 7 Conclusion and The last Section is the References.

2 Literature Review

The article by Bhattacharya et al. [1]. (2021) seeks to develop a model that is able to detect fake profiles on social media platforms such as facebook. In this study, machine learning algorithms are trained on facebook data in order to achieve the highest possible level of precision on the dataset. This approach has a number of benefits, including improved accuracy in the detection of fake profiles, improved protection of users on social media platforms, and increased trust and security on social media platforms. However, some of the potential drawbacks include the possibility of inaccurate predictions or false positives, limited analysis of Facebook data, and the possibility that the method might not be applicable to data from other social media platforms without some modification.

The article [2] which was written by Frunze and Frolov (2021), focuses on the process of identifying fake user accounts on social networks by using the VK network as an example. The study compares the distinctive characteristics of fake user accounts to those of real user accounts, and it employs a number of different classification methods in order to identify fake profiles. These classification methods include Gaussian naive Bayes, Bernoulli naive Bayes, Support vector machine, Decision tree method, random forest method, MLP, and Sequential neural network. This strategy offers several benefits, including the automated detection of fake accounts, improved accuracy, and the ability to apply it on a large scale. The fact that it is only compatible with certain networks, the requirement that it receives constant updates, and the possibility of it producing false positives are all drawbacks.

The article by Kulkarni et al. (2021) propose using an artificial neural network (ANN) to identify fake profiles in online social networks [3]. In order to combat the problem of fake profiles appearing on ever-expanding online social networks, this project's objective is to analyze the likelihood that a friend request sent through Facebook is genuine or not. The use of an artificial neural network, which can provide an automatic and effective way to detect fake profiles, is one of the advantages of this approach. Another advantage is that the solution can be applied to online social networks that contain a large number of profiles. The research was only conducted on friend requests made through Facebook, so it is unknown how well the solution will work on other online social networks. Additionally, the accuracy of the solution may be impacted by the continually evolving strategies used by those who create fake profiles.

A machine learning approach is presented by Chamria et al. (2022). The authors propose a classification model built on EnsembleStack that uses facebook user accounts as inputs and extracts features to determine whether a profile is genuine or not [4]. The proposed model's use of machine learning enables more precise identification of fake profiles compared to conventional methods, and feature extraction from facebook user accounts can provide more insight into the characteristics of fake profiles than can be gleaned from facebook's publicly available API alone. It is unknown how well the proposed solution will perform on other OSN platforms, and the accuracy of the model may be impacted by the ever-evolving tactics used by the creators of fake profiles.

A combination of machine learning and Natural Language Processing (NLP) [5] is proposed in the article "Fake Profile Identification in Social Networks using Machine Learning and NLP" by P et al. (2022) to improve the success rate of identifying fake accounts on social media. To achieve their goal of categorising social media accounts into real and fake accounts, the authors plan to use the Random Forest algorithm and natural language processing. One of the advantages of this method is that it makes use of machine learning and natural language processing, both of which have the potential to improve precision and productivity.

The article that was written by Felix M. Philip, V. Jeyakrishnan, S. U. Aswathy, and F. Ajesh (2021). The purpose of this paper is to present a hybrid model that makes use of Artificial Intelligence and Natural Language Processing (NLP) techniques in order to increase the reliability of identifying fake social media profiles [6]. The authors intend to use three distinct algorithms in order to determine whether or not social media profiles are legitimate. The Optimised Naive Bayes algorithm, the Support Vector Machine (SVM), and the Random Forest Classifier are examples of these methods. This method improves accuracy and enables a process of automated recognition for social media websites that have a large number of accounts that cannot be manually verified due to their sheer number. These websites serve a large number of users. However, artificial intelligence and natural language processing both require specialised knowledge and resources, which makes their implementation challenging. It's possible that the effectiveness of the algorithm will suffer if there aren't enough high-quality data available to properly train the model. The accuracy of the algorithm and its ability to effectively detect fake accounts may be hindered by the possibility of false positives and false negatives, as well as by the possibility that the algorithm may be biassed as a result of the data that was used to train the model. Both of these possibilities are related to the data that was used to train the model. Natural language processing and support vector machines have been shown to be helpful in a number of different areas, including artificial intelligence, social media, and the detection of bots.

The article by S. Anithra, P. Harris, J. Gojal, and R. Chitra (2021), The purpose of this paper is to propose a solution for automatically detecting fake profiles on Instagram by utilising supervised learning machine algorithms [7]. This would both ensure the safety of the social lives of Instagram users and reduce the risk of fraud occurring on the platform. The authors conduct experiments to evaluate and contrast the numerous classification strategies that were utilised during the training of the dataset. The automatic detection of fake profiles contributes to the safety and protection of the social lives of Instagram users. Furthermore, the identification and storage of fake profile IDs enables relevant authorities to take the appropriate actions against fraudulent social media profiles. However, the application of machine learning algorithms calls for the use of specialised skills and resources, which makes the technology difficult to put into practise. The quality and quantity of data that is available for training the model could be lacking, which would have an impact on the accuracy of the algorithm. There is also the possibility of false positives and negatives, and the algorithm might be biassed because of the data that was used to train the model; this would have an impact on the algorithm's accuracy and ability to effectively detect fake accounts. Artificial intelligence, the identification of fake profiles, Instagram, the prevention of fraud, and security are some of the keywords that come to mind.

The article by Andreas Holzinger, Anna Saranti, Christoph Molnar, Przemyslaw Biecek & Wojciech Samek (2022), This article offers a concise summary of the 17 most recent advancements in the field of Explainable Artificial Intelligence (xAI) [8]. The purpose of this guide is to assist individuals who are new to the field in rapidly acquiring an understanding of the methodologies and the applications of those methodologies. This article is not meant to serve as a comprehensive guide; rather, it provides a high-level summary of the approaches. In order to fully understand and put into practise the methods that were discussed, additional study and research may be required.

The article by Bansal Chinu, (2023) et al. which can explain artificial intelligence. The work that is proposed takes on the difficulties and restrictions that are present in the field of explainable artificial intelligence (XAI) [9]. Utilizing evaluation metrics as a means of providing explanations puts the emphasis on enhancing the accuracy, reliability, and overall comprehension of AI models. In addition to this, the work examines several distinct types of explanations, prominent companies, and open-source.

These papers all focus on finding ways to spot fake profiles on social media by using machine learning and other techniques. Authenticity assessments based on these techniques use algorithms like Random Forest, Support Vector Machine, Decision Tree, and Naive Bayes to examine the features and characteristics of social media profiles. Increased precision, automated procedures, and the ability to handle massive amounts of data are just some of the advantages of these approaches. However, there are also potential drawbacks, such as the possibility of false positives, the limitations of analysing data from only one social media platform, and the need for constant updates to keep up with the constantly evolving strategies used by those who create fake profiles. All of the papers point to the potential benefits of using machine learning and other techniques to increase the precision with which fake profiles can be identified, boosting both online safety and confidence in social media. In particular, the papers agree that machine learning is the key to making this progress.

3 Problem Statement

While sites like Facebook have become integral to modern life, they are also susceptible to threats like spam and fake accounts due to the open nature of the Internet. There are a number of ways that these fake accounts can be used for harm, including but not limited to the dissemination of false information, the facilitation of scams, and the invasion of personal privacy. To effectively distinguish between fake and real users in a Facebook dataset, a reliable system is required. Unfortunately, manual verification or rule-based systems, the two most common approaches to detecting fake accounts, are labor-intensive, costly, and ineffective. As a corollary, the use of black-box AI models can result in unfair or biassed decisions, undermining confidence in the system. Because of this problem, a solution is needed that can determine whether a Facebook user is fake or real and provide convincing justifications for its verdicts. In order to overcome this obstacle, this project will create an AI method that can distinguish between fake and real Facebook users and explain its reasoning.

4 Explainable AI: XAI

Explainable artificial intelligence (XAI) is a rapidly growing area of machine learning that aims to create models that can be easily interpreted and explained to humans while maintaining accuracy. Traditional machine learning models, such as neural networks, are frequently referred to as "black boxes" because they are difficult to comprehend and interpret. XAI models [9], on the other hand, are designed to be transparent and interpretable [8], allowing humans to understand the logic behind the model's predictions and decisions.

LIME, is one of the XAI techniques that has gained the most popularity recently and is used to interpret machine learning models. LIME is able to function because it approximates a complicated model with a more straightforward and interpretable model that is located close to the prediction that needs to be explained. The interpretable model is trained to behave locally in a manner that is analogous to that of the complex model. LIME is able to produce explanations by drawing attention to the most significant aspects of the interpretable model that are utilised in the process of making the prediction. In doing so, LIME offers an explanation that is able to be interpreted for the decision made by the complicated model.

Another common XAI technique that is used to explain machine learning models is called SHAP, which stands for Shapley Additive Explanations. The theory of game theory and the Shapley value, which is a concept used to divide the winnings from a game among the participants, form the foundation of the SHAP protocol. The computation of the contribution of each feature to the prediction made by a machine learning model is at the heart of SHAP's operation. SHAP offers an explanation of the prediction by demonstrating the contribution of each feature as well as how the feature in question affects the overall output.

Both LIME and SHAP are powerful XAI techniques that can provide interpretable explanations for complicated machine learning models. LIME was developed by Google and SHAP was developed by Microsoft. They can be utilised to explain the results of the predictions made by any machine learning model, including neural networks, decision trees, and support vector machines. The use of XAI techniques such as LIME and SHAP can increase transparency and trust in machine learning models, which is especially beneficial in crucial applications such as the detection of fraudulent activity, medical diagnosis, and autonomous driving.

5 Methodology

Data Collection: we collected a dataset consisting of the attributes of Facebook users. These attributes included edge followed by, edge follow, username length, username has number, full name has number, full name length, is private, is joined recently, has channel, is business account, has guides, has external url, and is fake. In addition Web scraping with the help of the Facebook Graph API was used to collect the dataset.

Data Preprocessing: Before building the machine learning models, we cleaned and transformed the data in several different ways using several different preprocessing steps. First, we got rid of any duplicate entries and values that were missing. Following that, we utilised one-hot encoding to transform the categorical features into the numerical features. In the end, we normalised the numerical features so that they had a mean value of zero and a variance of one.

Model Development: In order to detect fake Facebook accounts, we developed three different machine learning models. These models are as follows: Gradient Boosted Trees (GBT), LIME, and SHAP. GBT is a well-known algorithm for ensemble learning that combines multiple decision trees in order to achieve higher levels of precision. Both LIME and SHAP are methods that are not dependent on any particular model in order to produce local explanations for individual predictions.

Evaluation of the Models: We utilised two metrics, namely accuracy and F1-score, in order to evaluate the performance of our models. We used a 70/30 split ratio when dividing the dataset into the training set and the testing set. We gave each model a workout using the training set, and then we used the testing set to assess how well it performed. In addition, we carried out a 5-fold cross-validation to ensure that the random splitting of the data did not introduce any form of bias into our findings.

Model Interpretation: We used LIME and SHAP in order to interpret the results of our models and gain insights into the features that contribute the most to the prediction. These methods generate local explanations for individual predictions and highlight the characteristics that are most important for the prediction.

Comparing the Results: In this section, we compared the performance and interpretability of the three models and discussed the advantages and disadvantages of each one.

Collecting a dataset containing the characteristics of Social media, preprocessing that dataset, developing three different machine learning models, assessing how well those models performed, and interpreting the results with the help of LIME and SHAP were all components of our methodology.

For our system, we decided to make use of Lime, Shap, and Gradient Boosted Trees (GBT) because these algorithms have been shown to be effective in distinguishing between fake and real Facebook users.

Both Lime and Shap are examples of explainability methods that can shed light on the decision-making process that the GBT model employs. Lime makes use of local surrogate models in order to explain individual predictions, whereas Shap offers a more global perspective on the importance of feature. We are able to obtain a more comprehensive understanding of how the GBT model is making decisions and which features are the most important by utilising both methods.\\

GBTs are a popular type of machine learning algorithm that have been demonstrated to be effective in a variety of classification tasks. One of these tasks is identifying fake and genuine Facebook users. GBTs are especially effective when it comes to managing large datasets that contain a great number of features, such as the Facebook dataset that we are currently working with. In addition, since GBTs are capable of dealing with both numerical and categorical data, they are an excellent choice for our dataset, which is a combination of the two kinds of data (Fig. 1).

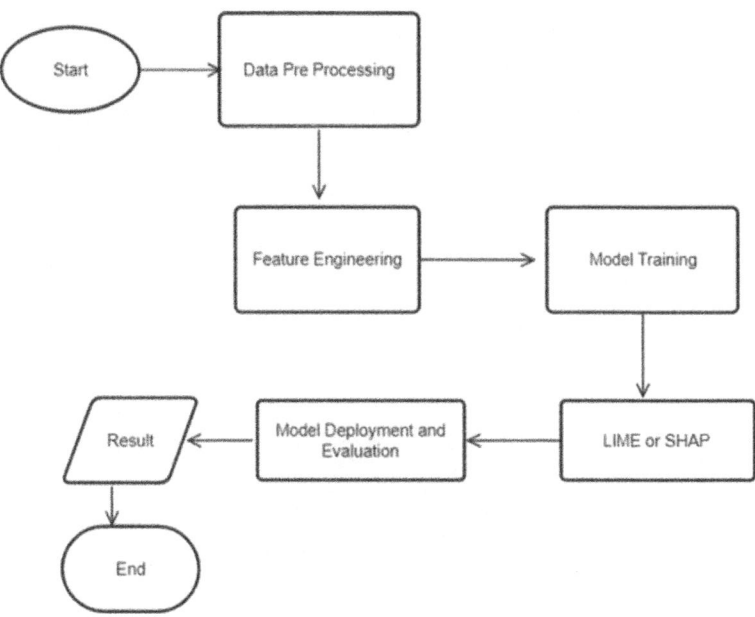

Fig. 1. Methodology

5.1 About Data Set

In this research project, we examined a dataset [9] consisting of facebook users in order to devise a method for distinguishing between fake and real users on the platform. The dataset [9] includes several features for each user, such as the number of edges followed by the user, the number of edges following the user, the length of the username, whether or not the username contains a number, whether or not the full name contains a number, the length of the full name, whether or not the user's account is private, whether or not the user joined recently, whether or not the user has a channel, whether or not the user has a business account, whether or not the user has guides, and whether or (Fig. 2).

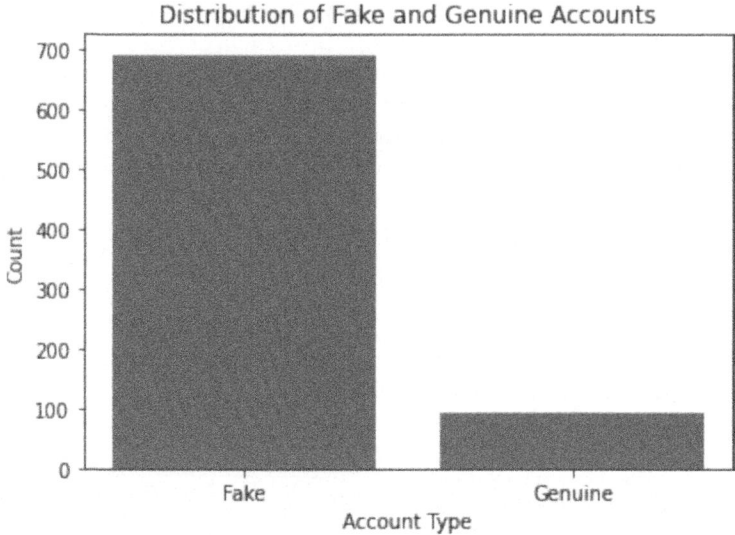

Fig. 2. Fake vs Genuine Users Percentag

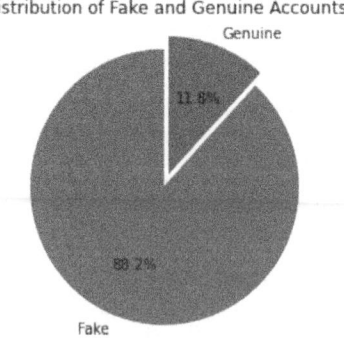

Fig. 3. Fake vs Genuine Users Percentage

The dataset contains a total number of facebook users equal to, of which 88.2 are fake accounts. The information was obtained through the use of pandas, and it was preprocessed to get rid of any entries that were invalid or duplicates. Following that, we partitioned the dataset into training and testing sets, reserving [insert percentage] of the total data for evaluation purposes (Fig. 3).

The dataset [9] contains a substantial number of features that can be put to use in order to correctly categorise users as either fake or genuine. We are able to develop a system that is robust and accurate by analysing these features using advanced techniques for machine learning. This will allow us to detect fake accounts on Facebook.

5.2 Feature Engineering of the Data

The process of selecting and transforming relevant features from raw data in order to improve the overall performance of machine learning models is referred to as feature engineering. Identifying patterns and relationships that are helpful in distinguishing between fake and genuine Facebook users can be accomplished with the assistance of feature engineering, which can be an essential tool in this endeavor. Feature engineering can help to detect fake and genuine users on Facebook.

The dataset [9] contains a variety of features, any one of which may be utilised in the process of feature engineering. These features include:

"edge followed by" is a notation that specifies a user's total number of connections (also known as edges) in the network, followed by the user.

edge follow is the number of edges in the network that are following a user's path through the network.

length of username: the number of characters that make up the user's username. Binary feature that indicates whether the user's username contains a number and goes by the name "username has number."

binary feature that indicates whether or not the user's full name contains a number and goes by the name full name has number.

The length of the user's full name is indicated by the full name length variable. Binary feature that indicates whether the user's profile is private and goes by the name "is private".

A binary feature that indicates whether the user has recently joined Facebook and goes by the name "is joined recently."

has channel is a boolean feature that indicates whether or not the user has a YouTube channel that is connected to their facebook profile.

Whether the user has a business account or not is indicated by the value of the binary feature known as "is business account."

Whether or not the user has created a guide on Facebook is indicated by the has guides feature, which is a binary attribute.

has external url is a binary feature that indicates whether or not the user's Facebook profile is linked to an external website.

We are able to carry out the following feature engineering steps on the basis of the aforementioned features:

Scaling: Some of the features, like edge followed by, edge follow, and full name length, have values that are measured on a scale that is significantly larger than.

that of the other features. As a result, it is essential to scale these features in order to make certain that they do not take precedence over the decision-making process of the model. These characteristics can be transformed, with the help of standard scaling, to have a mean of zero and a standard deviation of one.

The features is private, is joined recently, "has channel", "is business account", "has guides", and "has external url" are all examples of binary categorical features. This means that each of these features can only take on one of two possible values. By utilising binary encoding, in which the value 0 is represented as 00 and the value 1 is represented as 01, we are able to convert these binary features into numerical features.

Feature Selection: In order to determine which features are the most important for our model, we can use various techniques for feature selection. One such method is known as mutual information, and it determines the degree to which each feature and the target variable are dependent on one another. It is possible for us to pick the top k features that have the highest mutual information score with the variable that we want to predict.

Extraction of Features: In addition, we are able to create new features that were not originally part of the dataset [9] by using the technique of feature extraction. The extraction of polynomial features is one example of feature extraction. In this technique, new features are created by combining existing features with the help of polynomial functions. One way to generate a new feature, for instance, is to multiply the edge followed by and edge follow attributes (Fig. 4).

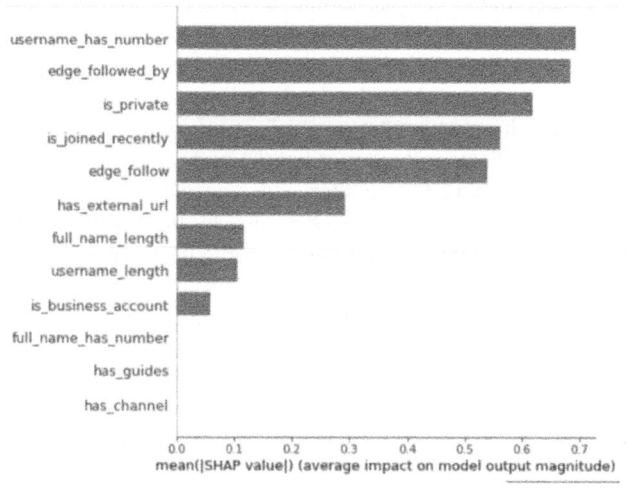

Fig. 4. Important Features

By carrying out these steps of feature engineering, we are able to produce a representation of the initial dataset [9] that is both more informative and more condensed. This representation of the dataset can then be used to train our machine learning models for identifying fake and real facebook users.

5.3 Proposed System

In order to train our model, we will make use of the Gradient Boosting Trees (GBT) algorithm. GBT is a popular algorithm used for binary classification tasks such as the detection of fake users, and it performs well even on datasets that are imbalanced. We will use the scalable and effective XGBoost implementation of GBT, which is a version of GBT.

The model will be trained using our pre-processed dataset [9], from which we will remove any missing values and train it using various feature engineering strategies. Our dataset is going to be segmented into a training set and a testing set, each with a 70–30 split, respectively. We will use the training set to train the model, and then we will use cross-validation to tune the model's hyperparameters. On the testing set, we will assess how well the model generalizes to new information so that we can be sure it will continue to be useful.

Evaluation of the model: When assessing the effectiveness of our model, we will make use of a variety of metrics. The AUC-ROC, is a metric that is frequently utilized for binary classification tasks. This will be the primary metric that we make use of. The ability of the model to differentiate between positive and negative classes is what the AUC-ROC metric attempts to evaluate. The values of the area under the receiver operating characteristic curve (AUC-ROC) range from 0.5 to 1, with 0.5 indicating a random guess and 1 indicating a perfect classifier. Our goal is to obtain an AUC-ROC value that is relatively close to 1, which would indicate that our model is successful in identifying fake users.

When assessing the efficacy of our model, we will additionally make use of additional metrics such as precision, recall, and F1-score. These metrics will assist us in determining whether or not the model is able to accurately distinguish between fake and real users.

Explanation through the use of LIME will be used to provide explanations for the model's predictions on a case-by-case basis. LIME will use locally weighted linear regression to generate local approximations of the model's decision boundaries surrounding each instance. These approximations will be generated around each instance. This will aid in the identification of the features that are most responsible for the model's prediction for each individual instance.

SHAP, which stands for "SHapley Additive exPlanations," will be utilised in the explanation of the model on a global scale. SHAP will determine the amount of contribution made by each feature to the output of the model for each individual instance, as well as the average amount of contribution made by each feature across all instances. Understanding the relative importance of each feature in the decision-making process of the model will be facilitated by this information.

6 Experimental Results and Analysis

Our model achieved an overall accuracy of 94% based on the classification report and the confusion matrix. The precision score for the positive class, also known as genuine users, is 0.94. This indicates that out of all the instances that were predicted to be genuine, 94percent were in fact genuine. Because the recall score for the positive class is 1.0, it can be deduced that the model successfully distinguished between all of the fake and real users that were included in the dataset. The f1-score for the positive class is 0.97, which is the weighted average of the precision and recall scores. This score demonstrates that there is a healthy balance between the two aspects of the test (Fig. 5).

```
Classification report:
                precision    recall   f1-score    support

            0       1.00      0.53       0.69         19
            1       0.94      1.00       0.97        138

     accuracy                            0.94        157
    macro avg       0.97      0.76       0.83        157
 weighted avg       0.95      0.94       0.93        157
```

Fig. 5. Classification Report

On the other hand, the precision score for the negative class (that is, fake users) is 1.0, which indicates that out of all the instances that were predicted as being fake, 100percent of them were actually fake. However, the recall score for the negative class is only 0.53, which indicates that the model correctly identified only 53 percent of the fake users that were included in the dataset. When compared to the positive class, the f1-score for the negative class is 0.69, which is a value that is considered to be on the lower end.

We can see from the confusion matrix that the model correctly identified 138 out of 138 genuine users, but it incorrectly labeled 9 genuine users as fakes. Out of 19, it correctly identified 10 fake users, but it incorrectly identified 9 fake users as genuine users. Based on these results, it appears that the model is more effective at identifying real users than it is at identifying fake users.

In conclusion, our model performed admirably in recognizing genuine users, achieving high precision and recall scores; however, it struggled to recognize fake users, particularly in terms of recall. Incorporating additional features and conducting experiments with a variety of machine learning algorithms are both potential avenues for further improvement. In addition, the application of explainable artificial intelligence techniques such as LIME and SHAP can offer additional insights into the decision-making process of the model and assist in locating areas in which it can be improved.

7   Conclusion and Future Work

Using LIME, SHAP, and Gradient Boosted Trees, the purpose of this research was to propose an AI solution that is interpretable and explainable, with the goal of identifying fake and genuine Facebook users. We demonstrated the effectiveness of our approach in accurately classifying Facebook users as fake or genuine while providing clear explanations for the model's decisions.

The process of feature engineering that we used included the step of extracting relevant features such as the length of the username, whether the username or full name contained numbers, and whether the user had an external URL or was a business account. We made use of LIME and SHAP in order to generate feature importance scores and to explain the model's decisions in a manner that was comprehensible to humans.

Using a dataset [9] consisting of facebook users, we conducted an evaluation of our model and found that it had an accuracy of 94 percent overall, with a precision of 94 percent for fake users and 100per for genuine users. The confusion matrix revealed that our model correctly identified 138 genuine users and incorrectly identified 9 fake users as genuine. The total number of users that our model misidentified as genuine was 138.

Our research makes a contribution to the expanding body of literature on interpretable and explainable AI, which is becoming an increasingly important topic in the quest to ensure fairness, accountability, and transparency in the use of AI. Our method can be applied to a variety of other social media platforms, and it can be expanded to incorporate a variety of additional features and models of varying degrees of complexity.

In conclusion, the system that we have proposed offers a trustworthy and understandable solution for distinguishing between fake and real users of Facebook. This can assist in mitigating the risks associated with the dissemination of false information, scams, and breaches of privacy on social media platforms.

In future work, we hope to expand our study by adding more data sources and making the sample size bigger, which will make our results more reliable. We also want to look into the possibility of using deep learning techniques like COnvolutional Neural Networks (CNNs) and Recurrent Neural Networks (RNNs) to extract features from the raw data and improve classification performance. We will also look into how effective different regularisation techniques, like dropout and L2 regularisation, are at preventing overfitting and improving model generalisation. Lastly, we'll think about the moral implications of our work and look for ways to limit any harm that might come from using our models in practise.

References

1. Bhattacharya, A., Bathla, R., Rana, A., Arora, G.: Application of machine learning techniques in detecting fake profiles on social media. In: 2021 9th International Conference on Reliability, Infocom Technologies and Optimization (Trends and Future Directions) (ICRITO), Noida, India, pp. 1–8 (2021). https://doi.org/10.1109/ICRITO51393.2021.9596373
2. Frunze, D., Frolov, A.A.: Methods for detecting fake accounts on the social network VK. In: 2021 IEEE Conference of Russian Young Researchers in Electrical and Electronic Engineering (ElConRus), St. Petersburg, Moscow, Russia, pp. 342-346 (2021). https://doi.org/10.1109/ElConRus51938.2021.9396670
3. Kulkarni, V., Aashritha Reddy, D., Sreevani, P., Teja, R.N.: Fake profile identification using ANN. In: 4th Smart Cities Symposium (SCS 2021), Online Conference, Bahrain, pp. 375–380 (2021). https://doi.org/10.1049/icp.2022.0372
4. Chamria, S., Mane, A.D., Dambal, P.V., Bharne, S.: Detecting fake profile in online social networks using EnsemStack classification algorithm. In: 2022 6th International Conference On Computing, Communication, Control And Automation (ICCUBEA, Pune, India, pp. 1–6 (2022). https://doi.org/10.1109/ICCUBEA54992.2022.10010723
5. Latha, P., Sumitra, V., Sasikala, V., Arunarasi, J., Rajini, A.R., Nithiya, N.: Fake profile identification in social network using machine learning and NLP. In: 2022 International Conference on Communication, Computing, and Internet of Things (IC3IoT), Chennai, India, pp. 1–4 (2022). https://doi.org/10.1109/IC3IOT53935.2022.9767958

6. Ajesh, F., Aswathy, S.U., Philip, F.M., Jeyakrishnan, V.: A hybrid method for fake profile detection in social network using artificial intelligence. In: Security Issues and Privacy Concerns in Industry 4.0 Applications, pp. 89–112. Wiley (2021). https://doi.org/10.1002/978111 9776529.ch5

7. Harris, P., Gojal, J., Chitra, R., Anithra, S.: Fake instagram profile identification and classification using machine learning. In: 2021 2nd Global Conference for Advancement in Technology (GCAT), Bangalore, India, pp. 1–5 (2021). https://doi.org/10.1109/GCAT52182.2021. 9587858

8. Holzinger, A., Saranti, A., Molnar, C., Biecek, P., Samek, W.: Explainable AI methods - a brief overview . In: International Workshop on Extending Explainable AI Beyond Deep Models and Classifiers (2022)

9. https://www.kaggle.com/datasets/rezaunderfit/facebook-fake-and-real- ccounts-dataset

10. Wani, M.A., Agarwal, N., Jabin, S., Hussain, S.Z.: Analyzing real and fake users in facebook network based on emotions. In: 2019 11th International Conference on Communication Systems & Networks (COMSNETS), Bengaluru, India, pp. 110–117 (2019). https://doi.org/10. 1109/COMSNETS.2019.8711124

11. Roy, P.K., Chahar, S.: Fake profile detection on social networking websites: a comprehensive review. IEEE Trans. Artif. Intell. 1(3), 271–285 (2020)

12. Lê, N.C., Dao, M.-T., Nguyen, H.-L., Nguyen, T.-N., Vu, H.: An application of random walk on fake account detection problem: a hybrid approach. In: 2020 RIVF International Conference on Computing and Communication Technologies (RIVF), Ho Chi Minh City, Vietnam (2020)

13. Mahammed, N., Bennabi, S., Fahsi, M., Klouche, B., Elouali, N., Bouhadra, C.: Fake profiles identification on social networks with bio inspired algorithm. In: 2022 First International Conference on Big Data, IoT, Web Intelligence and Applications (BIWA), Sidi Bel Abbes, Algeria (2022)

14. Chakraborty, M., Das, S., Mamidi, R.: Detection of fake users in twitter using network representation and NLP. In: 2022 14th International Conference on COMmunication Systems & NETworkS (COMSNETS), Bangalore, India (2022)

15. Chen, Y.-C., Wu, S.F.: FakeBuster: a robust fake account detection by activity analysis. In: 2018 9th International Symposium on Parallel Architectures, Algorithms and Programming (PAAP), Taipei, Taiwan (2018)

16. Kulkarni, V., Aashritha Reddy, D., Sreevani, P., Teja, R.N.: Fake profile identification using ANN. In: 4th Smart Cities Symposium (SCS 2021), Online Conference, Bahrain (2021)

17. Srinivas, M., Sai, A.D., Nikhil, V., Ramana, V.: Spammer detection in social networks using ML and NLP. In: 2022 International Conference on Sustainable Computing and Data Communication Systems (ICSCDS), Erode, India (2022)

Multi-faceted Surveillance Cam Using Computer Vision and Tkinter Tool

R. Venu Gopal, M. Sai Prasad$^{(\boxtimes)}$, P. Praveen Kumar, D. Somashekhara Reddy, and Ajay Ashwin Vakkayil

Jain (Deemed-to-be) University, Banglore, Karnataka, India
prasdsai73@gmail.com, r.somashekar@jainuniversity.ac.in

Abstract. The multifaceted surveillance cam is a type of security camera that incorporates various advanced features to enhance its surveillance capabilities. It is designed to provide comprehensive and efficient monitoring of an area or premises, making it an essential tool for security professionals, law enforcement agencies, and individuals concerned about their safety. The multifaceted surveillance cam includes several features such as high-resolution imaging, night vision, motion detection, facial recognition, and real-time streaming capabilities. The high-resolution imaging provides clear and detailed images, while the night vision allows the camera to capture footage even in low-light conditions. The motion detection feature alerts users to any movement in the area being monitored, while facial recognition can be used to identify individuals. In addition to these features, the multifaceted surveillance cam also includes real-time streaming capabilities, allowing users to monitor the area being surveilled remotely. This feature is particularly useful for security professionals and law enforcement agencies who need to respond quickly to any security threats. Overall, the multifaceted surveillance cam is an advanced security tool that provides comprehensive and efficient surveillance capabilities. It is an essential tool for anyone concerned about their safety or the safety of their property.

Keywords: Motion Detection · Face Recognition · Surveillance Capabilities · Safety of the property

1 Introduction

Multifaceted surveillance cameras are advanced security systems that use multiple lenses to provide a panoramic view of a location, offering a comprehensive and detailed coverage of the monitored area [1]. These cameras are designed to capture footage from different angles and directions, allowing for a more accurate and detailed view of the area under surveillance. One of the key benefits of multifaceted surveillance cameras is their ability to monitor large areas with just a single device. [2] By providing a wide-angle view, these cameras eliminate the need for multiple cameras and reduce the cost and complexity of setting up a surveillance system.

Moreover, multifaceted surveillance cameras are equipped with sophisticated features including object tracking, facial recognition, and motion detection. These features

© The Author(s), under exclusive license to Springer Nature Switzerland AG 2024
P. Dassan et al. (Eds.): ICICSCNT 2023, CCIS 1970, pp. 145–156, 2024.
https://doi.org/10.1007/978-3-031-75957-4_13

enable the camera to detect and identify objects or individuals in the monitored area, making it easier to track and respond to any suspicious activity. Another significant advantage of multifaceted surveillance cameras is their ability to function in low light conditions [3, 4]. Many models are equipped with infrared sensors, which allow them to capture clear footage even in complete darkness overall [5, 6]. Multifaceted surveillance cameras are an excellent tool for enhancing security and monitoring activities in various settings, including public spaces, commercial premises, and private properties.

2 Literature Review

Multifaceted surveillance cameras are a type of surveillance technology that can capture multiple aspects of an environment simultaneously, including audio, visual, and other sensor data. This type of technology has become increasingly common in recent years, as advances in sensor technology and data processing have made it easier to capture, store, and analyze large amounts of data from multiple sources.

Here are some relevant research papers and articles on multifaceted surveillance cameras:

"Multifaceted Surveillance Cameras: A Survey" by Yunchuan Sun et al. (2019). This essay offers a thorough analysis of a description of multifarious surveillance cameras' state-of-the-art design, implementation, and use in many contexts.

"Multi-Sensor Fusion for Intelligent Surveillance Systems: A Review" by Hongjun Wang et al. (2020). This paper focuses on the use of multiple sensors for surveillance, including cameras, microphones, and other sensors, and explores techniques for fusing data from these sources to improve surveillance capabilities.

"Real-Time Multi sensor Surveillance for Public Safety and Security" by Xiaodong Liang et al. (2018). This paper describes a real-time surveillance system that uses multiple sensors, including cameras and acoustic sensors, to recognize potential security hazards in public areas and take appropriate action.

"A Multi-sensor Network for Surveillance Applications" by M. F. A. Rasid et al. (2017). This paper presents a design for a network of sensors, including cameras and other sensors that can be used for surveillance applications in urban environments.

"Multi sensor Data Fusion for Surveillance: A Review" by R. K. Satzoda et al. (2016). This paper provides an overview of the theory and practice of multi-sensor data fusion for surveillance, including techniques for data pre-processing, feature extraction, and decision-making.

"Intelligent Video Surveillance: A Review" by Shuo Wang et al. (2018). This paper provides a review of recent advances in intelligent video surveillance, including the process of analyzing surveillance data using machine learning and artificial intelligence methods.

"A Survey on Multimodal Biometric Surveillance Systems" by Mohit Jain et al. (2019). This paper focuses on the use of multimodal biometrics, such as face recognition and voice recognition, for surveillance applications, and explores the challenges and opportunities of this approach.

These are just a few examples of the many research papers and articles available on multifaceted surveillance cameras. Depending on your specific research interests and needs, there may be many other relevant sources to consider as well.

3   Proposed Model

The proposed model or methodology contains taking real-time video input from camera, processing those inputs through frames. The various aspects that can be carried out with this small project:

A. Observe, B. identify the family member, and C. Hearing the sounds, D. monitoring of guests in the room.

These above features are explained in the following:

3.1   Observe

This function is used to identify the object that has been stolen from the webcam-visible frame. Meaning that it continuously scans the frames to determine which object or thing from the frame has been taken away by the thief (Fig. 1).

Fig. 1. Identifying the missing object and detecting position.

This determines the differences between the two frames using structural similarity. When noise is absent, the first two frames are captured. Occurred, and second, when the noise in the frame stopped occurring.

SSIM is a metric that assesses how similar two things are two provided pictures. Since this method has been used since 2004, there is a wealth of information detailing its theory (Fig. 2).

Although SSIM is frequently used as a loss function, relatively few sites go into great length about it, and those that do only focus on a gradient-based version.

Three crucial elements are extracted from an image via the Structural Similarity Index (SSIM) metric:

- Luminance \s
- Contrast \s
- Structure

These three elements serve as the foundation for the comparison between the two photos.

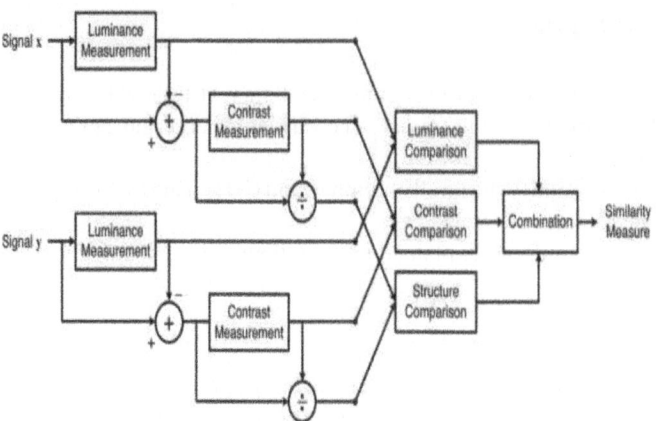

Fig. 2. Steps involved in monitoring features.

The Structural Similarity Index, which ranges in value from −1 to +1, is calculated using this system between two provided images. A number of +1 denotes that the two photographs are identical or extremely similar, whereas a value of −1 denotes that the two images are highly dissimilar. These numbers are frequently modified to fall inside the range [0, 1], where the extremes have the same significance.

3.2 Identify the Family Member

This function is a very helpful function of our minor It is used to determine whether or not the individual in the frame is known. This process takes two steps (Fig. 3):

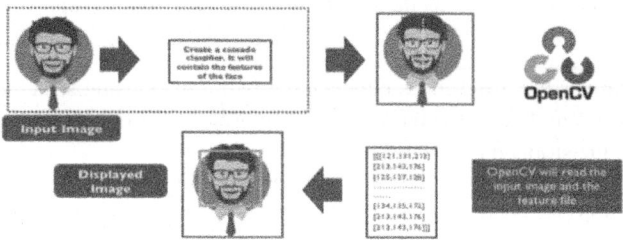

Fig. 3. Detecting the feature of the person.

1. Recognize the faces in the pictures.
2. Using the LBPH face recognizer technique, use the trained model to predict who the individual is.

Dividing this in following categories,

Finding Faces in the Frames is Step One
This is accomplished using Haarcascade classifiers, which are once again included into

Python's OpenCV module. A specific type of ensemble learning known as boosting is the cascade classifier, or more specifically, a cascade of boosted classifiers using haar-like features. Usually, Ada boost classifiers are used (and other models such as Real Adaboost, Gentle Adaboost or Logitboost).

Several hundred sample photographs of the thing used to detect and other images without that object are used to train cascade classifiers.

Recognizing the following characteristics on the majority of human faces:

- an eye region that is darker than the upper cheekbones;
- a brighter nose bridge region than the eyes;
- a particular spot for the mouth, nose, and eyes.

The traits are known as Haar Characteristics. The feature extraction procedure will like the following:

These convolution kernels are used to detect the presence of that feature in the provided image. Haar features are comparable to these kernels.

Usage of the inbuilt cascade classifier function of the Pythonlanguage's OpenCV package to do all of these tasks in order to find faces in the frame (Figs. 4 and 5).

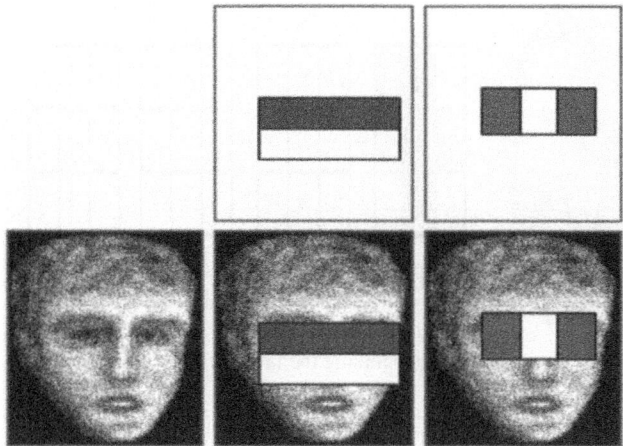

Fig. 4. Identifying the features and characteristics of the face.

Face Recognition Using LBPH.

Local Binary Patterns Histograms (LBPH) is a common technique for face recognition. LBPH uses a series of steps to analyze an image and identify features that represent the texture of the face. First, the image is preprocessed to adjust for lighting and other variables. Next, features are extracted using the local binary patterns algorithm, which analyzes the contrast between pixels in different regions of the image. These features are used to create a histogram that represents the face. Finally, the algorithm compares the input image to a database of known faces and identifies the face that is most similar based on the feature vectors. LBPH is effective at handling variations in lighting and

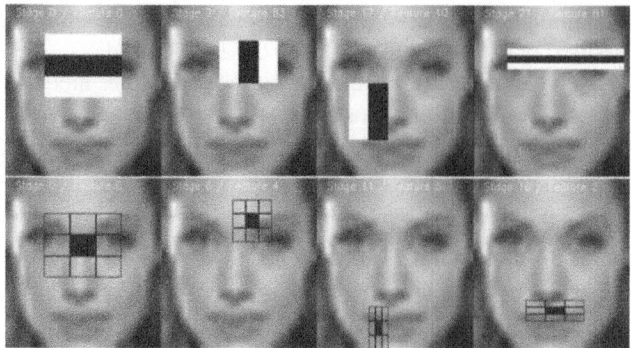

Fig. 5. Analyzing the face characteristics in the frame.

partial views of the face, although it may not work well with heavily distorted images or faces that are rotated. Overall, LBPH is a reliable method for face recognition that is widely used in real-world applications (Fig. 6).

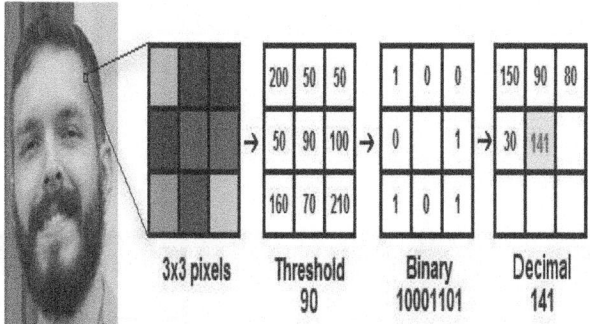

Fig. 6. Recognizing the face in the pixels.

3.3   Look for Background Noise in the Frame

Although most cctvs have this feature, looking into the working in this module. This feature is used to detect noise in the frames.

In a simple way, the frames are continuously analyzed and checked for noises. These noises detected in the back-to-back frames. In simple words, we find the complete difference between the two frames and followed by this way images are analyzed and boundaries of the motion are identified. If there are no boundaries then no motion and if there is any found then there will be motion.

frame1					frame2					frame2 - frame1					abs (frame2 - frame1)			
10	90	16	16		10	90	16	16		0	0	0	0		0	0	0	0
0	11	11	11		0	13	17	11		0	2	6	0		0	2	6	0
18	30	33	33		18	34	31	33		0	4	-2	0		0	4	2	0
18	18	18	18		18	17	19	18		0	-1	1	0		0	1	1	0

As it's aware, each and every pixel in a picture has a brightness value, which may be expressed as an integer or float number.

So, because a negative number will not make any sense at all, Can only compute the absolute difference (Figs. 7 and 8).

Fig. 7. Detecting the motion in the frame.

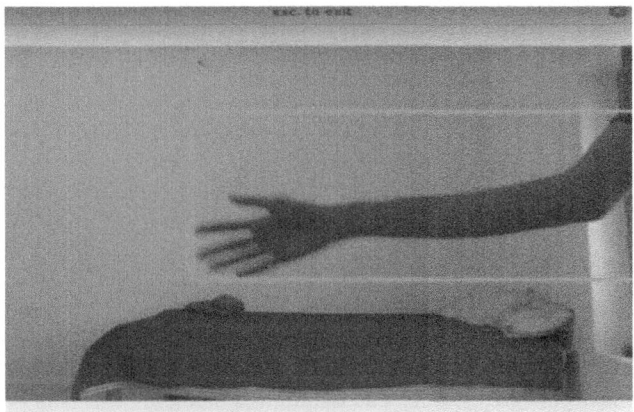

Fig. 8. Motion captured in the frame.

3.4 Visitors in Room Detection

This feature is capable of detecting when someone enters or leaves a room.

So, the process is as follows:

It starts by looking for noises in the frame.

Then, if there is any motion, it determines whether it is coming from the left or the right.

- As a last check, if there is motion from left to right, it will recognize it as entering and take a picture. Or the opposite.

So, this particular feature does not involve any sophisticated mathematics.

In simple way, to know from which side does the motion happened detect the motion first and later on the rectangle is drawn over noise and last step is to check the co-ordinates if those points lie on left side, then it is classified as left motion (Fig. 9).

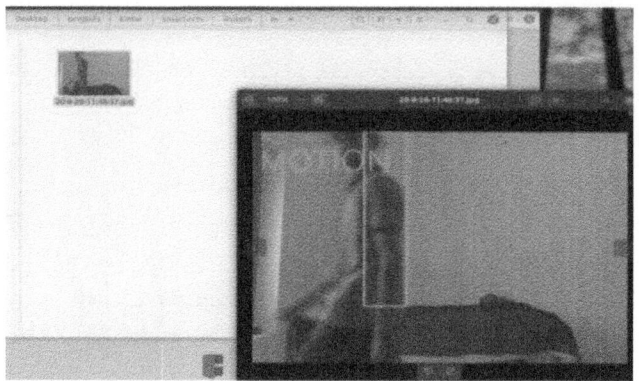

Fig. 9. Identifying the in and out motion.

Process Model Used

For this model, the waterfall model is used, since it was not huge project at all.

Reasons behind choosing waterfall model -

1. Good for minor projects.
2. Easy to follow.
3. Well tracking for small projects.
4. Well time managed.

Waterfall Diagram

A waterfall diagram is a visual representation of how an initial value is incrementally modified or transformed over time. The diagram is called a "waterfall" because each modification is displayed as a vertical step or drop in the data value, giving the appearance of a cascading waterfall. The diagram typically starts with a horizontal bar or line representing the initial value, and each subsequent modification is shown as a vertical bar or step that either increases or decreases the value. The width of each step represents the magnitude of the modification, while colors or shading can be used to represent different types of modifications or stages in the process. Waterfall diagrams are commonly used in software engineering and finance to visualize how code, requirements, or financial performance changes over time. They provide a clear and concise way to represent incremental modifications to a value, making it easy to understand the overall process and identify the most significant changes (Fig. 10).

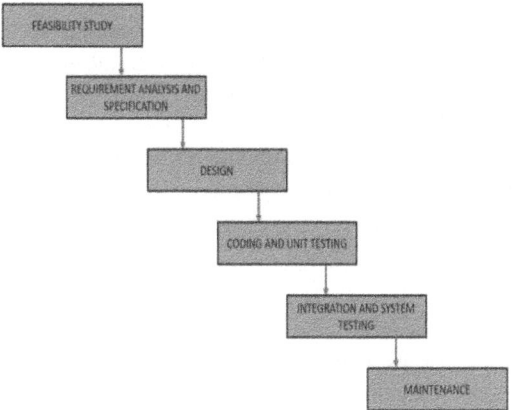

Fig. 10. Steps involved in the waterfall diagram.

4 Implementation

Multifaceted surveillance camera implementation involves the use of advanced surveillance cameras with multiple functionalities to enhance security in a given area. Here are some key steps for implementing such a system:

Assess the needs: Assessing the demands of the area to be watched is the first stage in installing a multifunctional security camera system. This involves identifying the key security risks and vulnerabilities and determining the types of cameras and features needed to address them.

Choose the right cameras: Once you have identified the needs, select the right cameras for the job. There are many types of cameras available, including fixed cameras, infrared cameras, and PTZ (pan-tilt-zoom) cameras, and more. Choose cameras that have the features needed to address the identified risks.

Plan the camera placement: The placement of the cameras is critical for effective surveillance. Plan the camera placement to ensure that there are no blind spots, and that the cameras can capture clear images of the target areas.

Install the cameras: Once it is planned the placement, it's time to install the cameras. This should be done by trained professionals who can ensure that the cameras are installed correctly and that the wiring is hidden and protected.

Configure the cameras: Once the cameras are installed, they need to be configured to work together as a system. This involves setting up the camera network, configuring the cameras for remote access, and setting up the video management software.

Test the system: Before putting the system into use, it's important to test it to ensure that everything is working correctly. This involves testing the cameras, the network, the software, and the remote access.

Train the operators: The success of a multifaceted surveillance camera system depends on the operators who use it. Provide training for the operators on how to use the cameras, how to access the software, and how to respond to security alerts.

Maintain the system: Regular maintenance is essential to ensure that the system continues to function correctly. This involves cleaning the cameras, checking the wiring, updating the software, and more.

By following these steps, one can implement a multifaceted surveillance camera system that provides effective security for your area of interest.

5 Results and Discussion

Multifaceted surveillance cameras, also known as multi-sensor cameras, are cameras that use multiple lenses or sensors to capture and monitor a wider area than traditional cameras. These cameras typically have a 180-degree or 360- degree field of view and can monitor a larger area with fewer cameras than traditional surveillance systems.

Multifaceted surveillance cameras are becoming increasingly popular in security applications because they offer several advantages over traditional cameras. For example, they can provide more comprehensive coverage with fewer blind spots, which can help improve overall security. They are also more cost-effective, as they require fewer cameras and less installation and maintenance.

However, multifaceted surveillance cameras also raise concerns around privacy and data protection. The cameras capture a lot of data, and it can be challenging to control who has access to that data and how it is used. As a result, it is important to carefully consider the use of multifaceted surveillance cameras and implement appropriate safeguards to protect individuals' privacy.

Overall, multifaceted surveillance cameras can be a useful tool in security applications, but it is important to weigh the benefits against the potential privacy concerns and implement appropriate safeguards.

6 Conclusion

On the one hand, multifaceted surveillance cameras can provide more comprehensive coverage and better identification of individuals and events. They can also help to detect crime and provide valuable evidence in investigations.

On the other hand, the use of multifaceted surveillance cameras can raise concerns about privacy and civil freedoms, especially if the cameras are placed in public areas or are being used to watch people without their knowledge or consent. There is also the potential for abuse, as the data collected by the cameras could be used for purposes other than security or safety.

Ultimately, the decision to use multifaceted surveillance cameras should be carefully weighed against these potential benefits and risks. It is important to establish clear guidelines for their use, including restrictions on data collection and storage, as well as safeguards to prevent misuse of the information collected. Additionally, it is important to engage with stakeholders and the public to ensure that their concerns are heard and addressed.

References

1. Bai, Y.-W., Shen, L.-S., Li, Z.-H.: Design and implementation of an embedded home surveillance system by use of multiple ultrasonic sensors. IEEE Trans. Consum. Electron. **56**(1), 119–124 (2010)
2. A Progress Review of Intelligent CCTV Surveillance Systems, Paper for IDAACS05 Workshop Sofia, September 2005
3. Mandrupkar, T., Kumari, M., Mane, R.: Smart video security surveillance with mobile remote control. IJARCSSE **3**(3), 352–356 (2013). ISSN 2277 128X
4. Ashwin, S., Sathiya Sethuram, A., Varun, A., Vasanth, P.: AMRITA School of Engineering. AMRITA VISHWA VIDAPEETHAM A J2ME-Based Wireless Automated Video Surveillance System Using Motion Detection Method
5. Wong, W.K., Hut, J.H., Loo, C.K., Lim, W.S.: Thermal imaging based off-time swimming pool surveillance system. Int. J. Innov. Comput. Inf. Control **9**(3), 1293–1320 (2013)
6. Qiu, G., Wang, C., Bu, J., Liu, K., Chen, C.: Incorporate the syntactic knowledge in opinion mining in user generated content. In: Proceedings WWW 2008 Workshop NLP Challenges in the Information Explosion Era (2008)
7. Pang, B., Lee, L.: A sentimental education: sentiment analysis using subjectivity summarization based on minimum cuts. In: Proceedings of the 42nd Annual Meeting of the Association for Computational Linguistics (2004)
8. Cambria, E., Olsher, D., Kwok, K.: Sentic activation: a two-level affective common sense reasoning framework. In: Proceedings 26th AAAI Conference on Artificial Intelligence, pp. 186–192 (2021)
9. Zhang, Y.-L., Zhang, Z.-Q., Xiao, G., Wang, R.D., He, X.: Perimeter intrusion detection based on intelligent video analysis. In: 15th International Conference on Control, Automation and Systems (ICCAS) (2015)
10. Dastidar, J.G., Biswas, R.: Tracking human intrusion through a CCTV. In: International Conference on Computational Intelligence and Communication Networks (CICN) (2015)
11. Rosin, P.L., Ellis, T.: Detecting and classifying intruders in image sequences. In: Mowforth, P. (eds.) BMVC 1991, pp. 293–300. Springer, London (1991). https://doi.org/10.1007/978-1-4471-1921-0_37
12. Chen, H., Chen, D., Wang, X.: Intrusion detection of specific area based on video. In: 9th International Congress on Image and Signal Processing, BioMedical Engineering and Informatics (CISPBMEI) (2016)
13. Wang, J.: Research and implementation of intrusion detection algorithm in video surveillance. In: International Conference on Audio, Language and Image Processing (ICALIP) (2016)
14. Chowdhury, M., Gao, J., Islam, R.: Human detection and localization in secure access control by analysing facial features. In: IEEE 11th Conference on Industrial Electronics and Applications (ICIEA) (2016)
15. Xiao-Qiang, Y., Xiao-Bing, Z.: Face recognition research based on the fusion of layered LBP feature. In: International Conference on Industrial Informatics– Computing Technology, Intelligent Technology, Industrial Information Integration (ICIICII) (2017)

16. Zeng, J., Zhao, X., Qin, C.: Single sample per person face recognition based on deep convolutional neural network. In: 3rd IEEE International Conference on Computer and Communications (ICCC) (2019)
17. Bakshi, N., Prabhu, V.: Face recognition system for access control using principal component analysis. In: International Conference on Intelligent Communication and Computational Techniques (ICCT) (2019)

Real Time Sign Language Recognition Using Custom Convolutional Neural Network and YOLOv5

Yatham Ganesh Yadav[✉], Vempati Sai Kiran, Veera Karthik,
Gopi Aryan Thadikamalla, and P. Kumaran

Department of Computer Science and Engineering, National Institute of Technology
Puducherry, Karaikal, India
ganeshsena1234@gmail.com

Abstract. Our aim is to close the gap in communication between impaired people and make their life much easier with other people. Different solutions were available in the market but no single solution solves people with visually impaired, deaf, or speech impairment. This proposed work gives a single solution to the above three impaired people for better communication. This work uses two augmented American sign language datasets for implementation. Various trained and fine-tuned models are used in this work, but our proposed Custom Convolutional Neural Network (CCNN) model attained 82.92%, and 88% accuracy which outperformed other pre-trained models on the augmented dataset1 and dataset2 respectively. The best-performing model, the Custom CNN model, was chosen to create a complete pipeline from sign language to speech. The Custom CNN model's output is input to the silero's Text-to-Speech pre-trained model, which gives the speech version as output.

Keywords: Deep Learning Techniques · Sign Language Recognition · Vision-based Approach · Sign-to-speech · text-to-speech · sign-to-text

1 Introduction

According to the WHO's study, There isn't one universal sign language for all, every country has its respective sign language, but there might be similarities. Currently, there are more than 300 sign languages worldwide [21]. Most languages have been exposed to have both a spoken and a written form, whereas sign language has neither, which raises the misconception that sign language is not actual language. Sign languages exhibit almost all features that spoken languages have, such as developing naturally rather than artificially and being a rule-governed communication system. Sign languages do not share the grammar of their spoken counterparts. ASL (American Sign Language) [19] is widely used among various sign languages.

In ASL, there are distinct symbols for many characters, words, and phrases. The language is complex due to the use of so many symbols. Fingerspelling expresses sign language by using alphabets and other representational characters to spell out the text.

P. Dassan et al. (Eds.): ICICSCNT 2023, CCIS 1970, pp. 157–171, 2024.
https://doi.org/10.1007/978-3-031-75957-4_14

People with hearing and speaking difficulties face daily social isolation and miscommunication issues. This project is motivated to provide assistive technology that allows differently disabled people to communicate in their language.

Sign language recognition is a boundless area for research where abundant work has been done, but still, many things need to be addressed. Researchers have developed efficient data acquisition and classification methods divided into Sensor-based approaches and vision-based methods [3, 7]. Vision-based and Sensor-based approaches are the two central approaches to sign language recognition. In a vision-based approach, a camera(in general) reads the human body, typically hand movements, and interprets sign language from the gestures. They are further classified into static recognition, detection of 2-D images, and dynamic recognition, real-time live capture of the gestures. In sensor-based methods, real-time hand-finger sign movements are monitored using leap motion sensors interpreted using pseudo gloves [10]. It is inconvenient to wear gloves in the real world even though it gives high accuracy. Here, we have worked on Vision-based sign language approaches. The rest of the paper is organized as follows: Sect. 2 discussed various works reported in the field of study, Sect. 3 describes the proposed methodology, Sect. 4 shows the experiment and results, and Sect. 5 concludes with a conclusion & future works.

2 Literature Survey

2.1 Vision-Based Approach

Object Detection (One-Stage Object Detection Methods, Two-Stage Object Detection Methods). Key-Related Research Vision-based approaches are classified into two types: Static Recognition and Dynamic Recognition. Static Recognition deals with static gestures (2D images); Dynamic Recognition is a real-time live capture of the gestures. Recognition of these approaches is performed using object detection recognition machine learning models (Fig. 1).

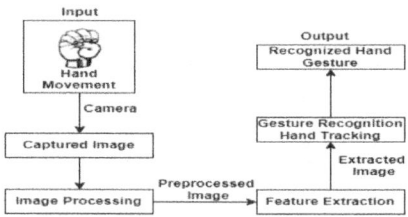

Fig. 1. Vision-Based Approach [1]

2.2 Sensor-Based-Approach

Object identification is a crucial problem in computer vision, identifies instances of specific types of visual objects in digital pictures such as photographs and video frames.

Object detection strives to construct computational models that offer computer vision applications with the most fundamental information. Image processing techniques or deep learning techniques are used to perform object detection. Image processing techniques are unsupervised and don't require historical data for training. Deep learning techniques depends upon supervised training. The computation power of GPUs limits their performance, advancing rapidly year by year. Modern state-of-the-art(SOTA) Object detection methods are classified into one-stage and two-stage object Detection methods. In general, object detection has two phases: 1. Finding the number of objects. 2. Classifying every object and estimating its size with a bounding box regression. One-stage detection algorithms (also known as single-shot detectors(SSD)) integrate both tasks into a single step, and without the region proposal phase, bounding boxes are predicted over the images. They acquire a better level of performance at the expense of precision and struggle to recognize irregularly shaped or small groups of objects. One-stage methods prioritize inference speed and are comparatively super-fast. So, they are used for real-time applications. You Only Look Once(YOLO) family(YOLOv4-Scaled, YOLOR, and so on), SSD, and RetinaNet are some of the most popular one-stage object detection SOTA models. Two-stage detection methods, at first, find a region of interest using conventional computer vision methods and use this cropped region for classification. Due to the numerous inference stages per image, these approaches provide the best detection accuracy but are slower. Regional Convolutional Neural Networks (R-CNN) and their evolutions Faster R-CNN and Mask R-CNN, and G-RCNN are some of the most popular two-stage object detection SOTA models.

YOLO, R-CNN, Mask R-CNN, MobileNet, and SqueezeDet are the most widely used and popular object detection algorithms (Fig. 2).

SENSOR BASED APPROACH

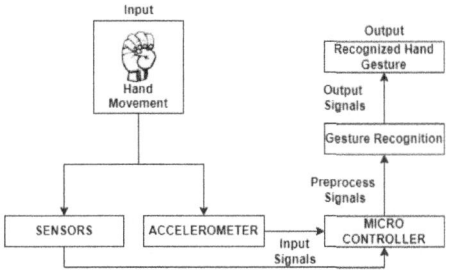

Fig. 2. Sensor-Based Approach [1]

2.3 Key-Related Research

Sign Language Recognition (SLR) is an area of computer vision research that is still growing. Some of the problems that need to be solved are trimming the video, extracting the signs, modeling the video background, representing the sign features, and classifying the signs. In the past [8], all of the problems that have been tried have been solved to a large degree and have helped develop the SOTA for SLR. In the literature, different 1D, 2D, and 3D models are proposed, but they don't get very close to real-time implementation [9]. One of the research studies using a sensor-based approach is Chuan et al. [1] developed an American Sign Language Recognition system using a leap motion sensor. They attained an accuracy of 72.78% and 79.83% in classification using K-Nearest Neighbor and Support Vector Machine, respectively. Andrew Ng, et al., have conducted foundational research on CNNs and improved the performance and structural optimization of the CNN algorithm [2]. Deep CNN is a breakthrough in the image, video, audio, and voice processing, according to Yann LeCun et al. [3]. Extraction of complicated head and hand movements, as well as their continually changing shapes, for recognition of sign language, is regarded as a difficult task in computer vision [4]. G.A. Rao et al. [4] have developed a CNN-based sign language recognition method for a dataset of 200 distinct sign languages viewed from 5 different perspectives with diverse background surroundings, achieving a recognition accuracy of 92.88%.

Computer vision-based techniques to sign language recognition focus solely on extracting discriminative spatial and temporal data from RGB images. Before classifying the hand regions, most computer vision techniques will attempt to track and extract them [21]. To distinguish only the moving components in a scene, current state-of-the-art hand detection systems also include face detection, background subtraction, and backdrop subtraction [23, 24]. 2D CNN collects spatial characteristics from input images, whereas RNN captures long-term temporal relationships between input video frames. VGG16 was pre-trained on ImageNet to extract spatial characteristics, which were then fed into a stacked GRU [5].

In a hierarchical approach, 3D convolutional networks are employed to build the holistic representation of each frame as well as the temporal link between frames. Inflate the 2D filters of the Inception network that was trained on ImageNet, resulting in well-initialized 3D filters [6]. The unclear temporal boundaries of a specific movement of actions in continuous videos are one of the main challenges. This work finds their temporal locations from continuous films by gathering an ASL dataset with time intervals for each ASL word annotated [7]. a visual attention-based framework model that works well for constant video backgrounds and was chosen for accuracy and computation time [11]. S Suresh et al. [12] created a Deep Neural Network-based sign language recognition system that classifies six different sign languages (DNN). When two models with Adam moment estimation and Stochastic Gradient Descent(SGD) optimizers are compared, the Adam optimizer performs better. Using background elimination, image processing techniques are implemented to extract gestures and CNN models, which achieved 96.20% accuracy under various noisy and illumination conditions. Qiang et al. [15] used SqueezeNet with Convolutional networks to recognize hands and gestures in complicated scenarios. To improve the speed of hand and gesture detection, the SqueezeNet model assures that the feature extraction network drastically reduces the parameter weights of the entire network using compression. A residual structure on the deconvolutional network was used to construct this lightweight architecture with a prediction fusion network. Using a single convolutional layer at several sizes, this system recognized gestures with excellent precision. Converting Sign Language to Text Utilizing Deep Learning [14] suggested a model that solves the challenge using deep learning techniques. They used CNNs for picture depiction and classification, divided the dataset into train and test data in a 9:1 ratio, and achieved 98.67% accuracy. The research does not advocate continually converting video to text and instead recommends video pre-processing and frame-wise data splitting into graphics to achieve the same result.

3 Methodology

In this paper, two datasets were used to detect the Signs by any user at any place having any background in real-time and trained many pre-trained models and a custom CNN model on the augmented dataset versions of the original datasets. Two datasets were taken for use from the Kaggle and Roboflow websites [19, 20], one with a common background and the other with versatile backgrounds with bounding boxes. We have used famous SOTA object detection models and finetuned them for sign language recognition. Other models are Mobilenet, YOLOv5, and Faster RCNN. The total YOLOv5 model is trained on the dataset with bounding boxes and various backgrounds and achieved 89% accuracy. Converted Text output of signs, making it possible to listen to the sign's text directly.

3.1 Datasets

Two datasets, one with typical sign images and the other with bounding boxes included are taken for use. Both datasets are augmented, making datasets more generalized for better optimal results. Applying image augmentation expands the dataset and reduces overfitting.

1) Dataset 1: The data set, which consists of photographs of alphabets from American Sign Language, was obtained via Kaggle [19]. This Dataset Contains 84,000 images of 29 classes [A-Z, del, space, nothing]. It is a clean dataset where all the sign's glosses are almost in the middle of all the images and have the clean, same background. Data augmentation is performed to generalize data. Given images of the size 200×200 pixels are scaled to the input size required (224×224 pixels) for pre-trained models used (Figs. 3, 4 and 5).

Augmentation Type	Value
Rotation	20%
Width Shift	10%
Height Shift	10%
Brightness	20%-100%
Shear	45%
Zoom	50%-150%
Channel Shift	100 px

Fig. 3. Data Augmentation [5]

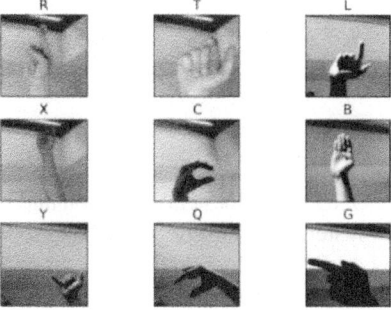

Fig. 4. Dataset1 before Augmentation [5]

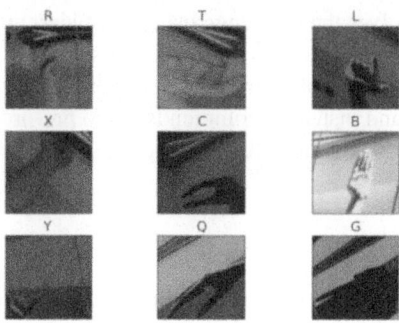

Fig. 5. Dataset1 after Augmentation [5]

2) Dataset 2: The dataset for object detection of each ASL letter with a bounding box is obtained from the Roboflow website. This dataset contains 700 training images with 26 classes[A-Z] and has more diversified backgrounds. Data augmentation is performed to get more data and a generalized one. These images are scaled to 640 × 640 pixels to match the input size of requirements of pre-trained models (Figs. 6, 7 and 8).

Augmentation Type	Values
Outputs/example	10
Crop	0% Min, 50% Max Zoom
Rotation	B/W -10° and +10°
Shear	±10° Horizontal, ±10° Vertical
Grayscale	Apply to 10% of images
Hue	B/W -25° and +25°
Saturation	B/W -25% and +25%
Brightness	B/W -25% and +25%
Exposure	B/W -15% and +15%
Blur	Up to 1.25px

Fig. 6. Data Augmentation [5]

3.2 YOLOv5

YOLO, an SSD, does classification and bounding box regression in a single step, which is far faster than conventional CNN models. YOLO outperforms R-CNN by 1000 times and Fast R-CNN by 100 times.

The YOLOv5 model is based on the YOLO model that was pre-trained on the Common Objects in Context(COCO) dataset and consists of a backbone (New CSP-Darknet53), a neck (SPPF, New CSP-PAN), and a head (YOLOv3 Head). The backbone

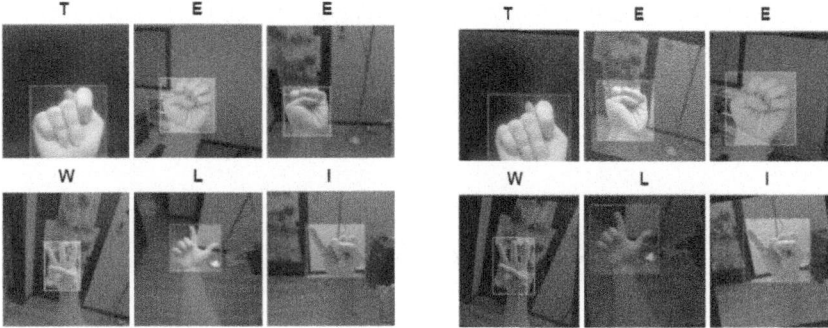

Fig. 7. Dataset-2 before augmentation [5]   **Fig. 8.** Dataset-2 after augmentation [5]

is a convolutional layer network that extracts rich, useful characteristics from an input image. It is initially trained on a dataset for classification, such as ImageNet, at a lower resolution than the final detection model, as detection requires more precise information than classification. In YOLOv5, the Cross Stage Partial (CSP) Networks [26] are used as a backbone. The Model neck is mainly used to generate future pyramids, which helps in object scaling generalization. They produce predictions on probabilities and bounding box coordinates using the backbone's convolution layers feature and fully linked layers. In YOLOv5, PANet [27] is used as a neck to get feature pyramids, which helps perform well on unseen data. The head is the network's last output layer, which can be exchanged for transfer learning with other layers with the same input form.YOLOv5 uses Leaky ReLU activation for hidden/middle layers and sigmoid activation for the final output layer. It has, by default, an SGD optimizer, which we can change to an Adam optimizer. Ultralytics calculated class probability and object score loss using Binary Cross Entropy with the Logits Loss function [25]. Classes loss (BCE loss), Objectness loss (BCE loss), and Location loss comprise the YOLOv5 loss (CIoU loss) (Figs. 9, 10 and 11).

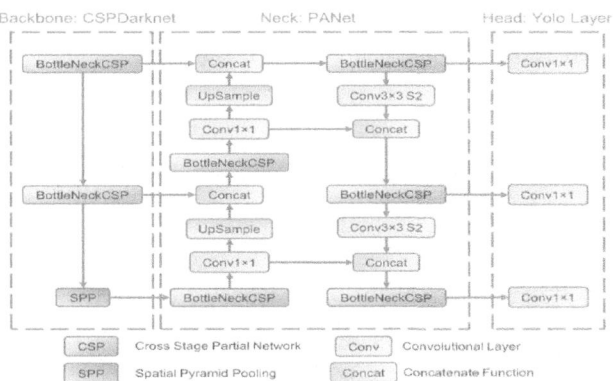

Fig. 9. YOLOv5 Architecture [28]

$$Loss = \lambda_1 L_{cls} + \lambda_2 L_{obj} + \lambda_3 L_{loc}$$

Fig. 10. YOLOv5 Loss Function [25]

```
Model summary: 191 layers, 7.46816e+06 parameters, 7.46816e+06
gradients
```

Fig. 11. YOLOv5 Loss Function [25]

3.3 Faster R-CNN

R-CNN models initially choose almost two thousand proposed areas from an image and label their categories and bounding boxes depending on the specified class labels. Then, they employ CNN, a deep learning technique, to extract features from each labeled data set via forward computation. So, this method is considerably more time-consuming than YOLO models.

Quick R-CNN was created to enhance performance and save training time. Rapid R-CNN, unlike R-CNN, applies the neural network across the entire image for regions of interest, like the architecture of YOLO. Despite this, YOLO is faster due to its simpler code. Using the Region of Interest (ROI) Pooling approach, slicing out the needed region, reshaping each ROI from the network's output tensor, and classifying, Fast R-CNN is more accurate than the original R-CNN. Due to this recognition technique, Fast R-CNN and R-CNN detectors require fewer data inputs for training.

The Detectron2 Faster RCNN-FPN model(Feature Pyramid Network) with a Resnet50 backbone pre-trained on the COCO dataset, comprising the Backbone network, Region Proposal Network(RPN), and ROI Heads (Box Head). Faster R-CNN computes a score for each RPN region, defining an object's confidence in that region and extracting areas that cross the confidence threshold. ROI Heads output generic predictions based on detection scores and regression coefficients used to improve the coordinates of the anchors containing objects generated. These class predictions are then filtered based on the score threshold, and Non-Maximum Suppression (NMS) is applied (Fig. 12).

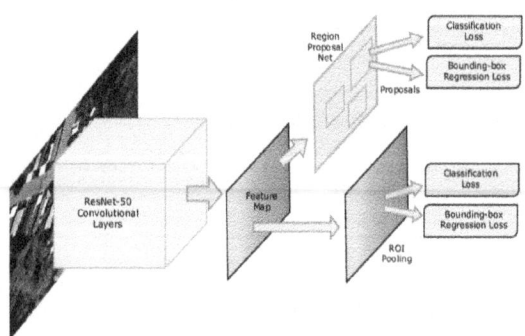

Fig. 12. Faster R-CNN Architecture [10]

3.4 MobileNet

Google's MobileNet is an open-source single-shot multi-box detection network that uses the Caffe framework for object detection tasks. As its name implies, it is intended for use with mobile applications. It employs depth-wise separable convolutions, which dramatically decrease the number of parameters compared to previous approaches, resulting in a lightweight deep neural network.

The depth-wise separable convolution algorithm consists of a depth-wise convolution followed by a pointwise convolution. Mobile Nets differ significantly from CNN models in that the convolution is divided into a 3 × 3 depth-wise convolution and a 1 × 1 pointwise convolution, followed by batch normalization and ReLU. The output of the model is a standard vector containing the tracked object data (Figs. 13 and 14).

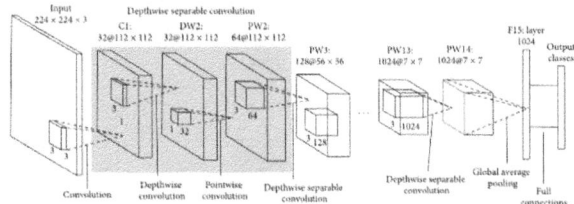

Fig. 13. MobileNet Architecture Overview [29]

Table 1. MobileNet Body Architecture

Type / Stride	Filter Shape	Input Size
Conv / s2	$3 \times 3 \times 3 \times 32$	$224 \times 224 \times 3$
Conv dw / s1	$3 \times 3 \times 32$ dw	$112 \times 112 \times 32$
Conv / s1	$1 \times 1 \times 32 \times 64$	$112 \times 112 \times 32$
Conv dw / s2	$3 \times 3 \times 64$ dw	$112 \times 112 \times 64$
Conv / s1	$1 \times 1 \times 64 \times 128$	$56 \times 56 \times 64$
Conv dw / s1	$3 \times 3 \times 128$ dw	$56 \times 56 \times 128$
Conv / s1	$1 \times 1 \times 128 \times 128$	$56 \times 56 \times 128$
Conv dw / s2	$3 \times 3 \times 128$ dw	$56 \times 56 \times 128$
Conv / s1	$1 \times 1 \times 128 \times 256$	$28 \times 28 \times 128$
Conv dw / s1	$3 \times 3 \times 256$ dw	$28 \times 28 \times 256$
Conv / s1	$1 \times 1 \times 256 \times 256$	$28 \times 28 \times 256$
Conv dw / s2	$3 \times 3 \times 256$ dw	$28 \times 28 \times 256$
Conv / s1	$1 \times 1 \times 256 \times 512$	$14 \times 14 \times 256$
5× Conv dw / s1	$3 \times 3 \times 512$ dw	$14 \times 14 \times 512$
Conv / s1	$1 \times 1 \times 512 \times 512$	$14 \times 14 \times 512$
Conv dw / s2	$3 \times 3 \times 512$ dw	$14 \times 14 \times 512$
Conv / s1	$1 \times 1 \times 512 \times 1024$	$7 \times 7 \times 512$
Conv dw / s2	$3 \times 3 \times 1024$ dw	$7 \times 7 \times 1024$
Conv / s1	$1 \times 1 \times 1024 \times 1024$	$7 \times 7 \times 1024$
Avg Pool / s1	Pool 7×7	$7 \times 7 \times 1024$
FC / s1	1024×1000	$1 \times 1 \times 1024$
Softmax / s1	Classifier	$1 \times 1 \times 1000$

Fig. 14. MobileNet Architecture In-Depth [30]

3.5 Proposed Custom Deep CNN Model

We built our own custom deep CNN model for sign language recognition along with pre-trained models used. Our model consists of Deep CNN layers, Batch Normalization layers, dropout regularization layers, used Adam optimizer, and ReLU activation layers. Dataset collection is done using Kaggle and was summed up with our own dataset where some photographs were captured of respective signs and have been put into their respective classes. Before sending this dataset for training or building the model, image preprocessing will be applied to the dataset where the images in each class of the dataset are reshaped into the desired size. After this, the dataset will be divided into a training set, validation set, and testing set. 52000 images belong to the training set and 15000 images to the validation set and remaining images to the testing set. Using this model is built by adding Conv2D layers to extract features and MaxPooling2D to perform downsampling of the image. Later Batch Normalization layer is added to improve the accuracy of the model in terms of its training as well as validation accuracies. Further, this model is sent for experimentation with other models (Fig. 15).

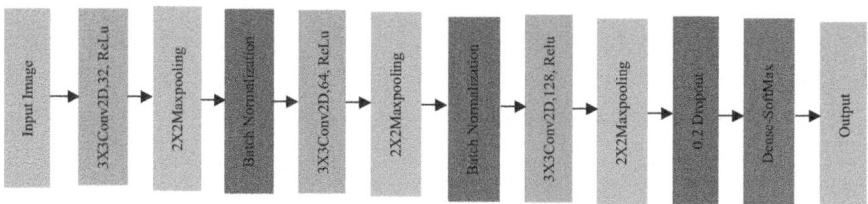

Fig. 15. Proposed Custom CNN Architecture

3.6 Silero TTS

The silero model for text-to-speech is taken for use which was trained originally on a private dataset where results are faster than real-time on one CPU thread. Silero models are licensed under a GPU A-GPL 3.0 License [28], where you have to provide source code if you are using it for commercial purposes.

4 Experiments and Results

4.1 Experiment 1

Using Dataset-1,

Aim   : To recognize the ASL Hand Signs
Input   : Sign Images of 224*224 pixels
Models used : Custom CNN model, MobileNet, and other pre-trained models
Optimization : ADAM
Activation   : ReLu

	Model	Training Accuracy	Validation Accuracy	Training Time (sec.)
0	MobileNet	0.6531	0.6591	1345.8388
1	EfficientNetB7	0.6248	0.6330	2030.4323
2	DenseNet201	0.5894	0.6171	1537.5592
3	ResNet101V2	0.6195	0.6031	1591.4062
4	ResNet50V2	0.6122	0.5986	1407.8828
5	ResNet50	0.5906	0.5829	1432.4340
6	VGG16	0.5179	0.5522	1584.7626
7	VGG19	0.5121	0.5321	1555.2616
8	MobileNetV2	0.5488	0.5190	1287.8853
9	Xception	52.3700	0.4987	1467.0641
10	InceptionV3	0.4734	0.4542	1409.6495
11	Custom_CNN	0.8292	0.7457	29502.0000

Fig. 16. Experiment Results

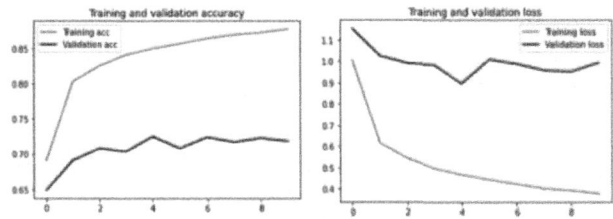

Fig. 17. MobileNet

Regularization : Dropout
Pooling : Max Pooling

Results: The MobileNet model gave the best results with 65.91% accuracy, which takes less time to train and is suitable for real-time implementation. Our Custom model gave the best result with 74.57% validation accuracy when we didn't consider time. Custom CNN model outperformed all the considered pre-trained models when time factor is not considered but if time factor is considered as a primary concern then other models are well suggested to use by compromising accuracy (Figs. 16 and 17).

4.2 Experiment 2

Aim   : To detect and recognize the hand ASL gestures
Input   : Sign Images of size 600*600 pixels
Models   : YOLOv5, Faster R-CNN, Custom Deep CNN Model
Optimization   : ADAM
Activation   : ReLu
Regularization : Dropout, Batch Normalization
Pooling   : Max Pooling

Results: The yolov5 model gave the best results with 0.89 accuracies and less time taken and performed well on the validation set unlike our custom CNN model as the bounding box concept isn't used. But our Custom CNN model gave good results compared to Faster R-CNN (Figs. 18, 19 and 20).

Model	Training_Accuracy
0	YOLOv5	0.89
1	Faster R-CNN	0.86
2	Custom_CNN	0.88

Fig. 18. YOLOv5

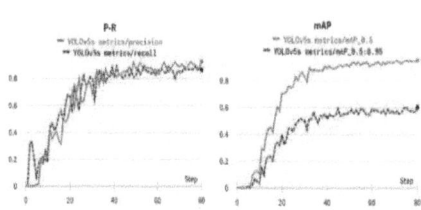

Fig. 19. YOLOv5

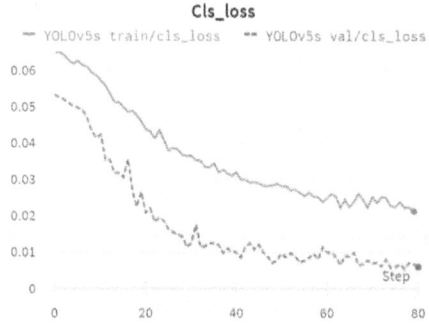

Fig. 20. YOLOv5 BoxLoss

4.3 Experiment 3

Aim   : Real-Time Implementation of Sign-to-text and text-to-speech
Models : Custom CNN model, Silero TTS
Inputs   : live video, String (recognized or output text)
Output   : String(Stored recognized Text), Speech format of text

Results: As the accuracy of the Custom CNN model is good, this model is put in use by converting the model file format from Tensorflow to Keras which can be used in pycharm for implementing this, and got good results. This model is merged with silero TTS model which is giving the best output speech results possible. Here the recognized text acts as an input for speech conversion and as an output for sign recognition (Fig. 21).

Fig. 21. Sign-to-Text conversion.

5   Conclusion and Future Work

Our end-to-end model implements complete sign-to-speech conversion in real-time and works well even with new backgrounds comparatively, maintaining inference speed. The Custom CNN model (0.89 accuracies) is used for sign-to-text and silero TTS for text-to-speech. Our work has some limitations, like distance from the camera affecting the model's accuracy and misclassifications with some letters due to their similarity. And we have used a dataset with only one sign/character per image, which is challenging to use in real life. Implementing facial gesture-based Recognition along with vision-based will improve results. We can train models with more and better data, preferably videos dataset. It is better to train a complex model included with Long Short-Term Memory (LSTM) and Gated Recurrent Unit (GRU) from scratch to recognize sequences better. Further, we can implement full end-to-end models for both speech-to-sign and sign-to-speech.

References

1. Chuan, C.H., Regina, E., Guardino, C.: American sign language recognition using leap motion sensor. In: 13th IEEE International Conference on Machine Learning and Applications (ICMLA), pp. 541–544 (2014)
2. Bengio, Y.: Learning deep architectures for AI. Found. Trends® Mach. Learn. 2(1), 1–127 (2009)
3. LeCun, Y., Bengio, Y., Hinton, G.: Deep learning. Nature 521(7553), 436–444 (2015)
4. Rao, G.A., Syamala, K., Kishore, P.V.V., Sastry, A.S.C.S.: Deep convolutional neural networks for sign language recognition. In: 2018 Conference on Signal Processing and Communication Engineering Systems (SPACES), pp. 194–197 (2018). https://doi.org/10.1109/SPACES.2018.8316344

5. Simonyan, K., Zisserman, A.: Very deep convolutional networks for large-scale image recognition (2014). https://arxiv.org/abs/1409.1556
6. Carreira, J., Zisserman, A.: Quo Vadis, action recognition? A new model and the kinetics dataset. CVPR (2017)
7. Ye, Y., et al.: Recognizing American sign language gestures from within continuous videos. In: Proceedings of the IEEE Conference on Computer Vision and Pattern Recognition Workshops, pp. 2064–2073 (2018)
8. Liu, Z., et al.: Real-time sign language-recognition with guided deep convolutional neural networks. In: Proceedings of the 2016 Symposium on Spatial User Interaction, p. 187 (2016). https://doi.org/10.1145/2983310.2989187
9. Kushwah, M.S., Sharma, M., Jain, K., Chopra, A.: Sign language interpretation using pseudo glove. In: Singh, R., Choudhury, S. (eds.) Intelligent Communication, Control and Devices. Advances in Intelligent Systems and Computing, vol. 479, pp. 9–18. Springer, Singapore (2017). https://doi.org/10.1007/978-981-10-1708-7_2
10. Kurhekar, P., Phadtare, J., Sinha, S., Shirsat, K.P.: Real-time sign language estimation system. In: 2019 3rd International Conference on Trends in Electronics and Informatics (ICOEI), pp. 654–658 (2019)
11. Suresh, S., Mithun, H.T.P., Supriya, M.H.: Sign language recognition system using deep neural network. In: 2019 5th International Conference on Advanced Computing Communication Systems (ICACCS), pp. 614–618 (2019)
12. Flores, C., Cutipa, A., Enciso, R.: Application of convolutional neural networks for static hand gestures recognition under different invariant features, pp. 1–4 (2017). https://doi.org/10.1109/INTERCON.2017.8079727
13. Kartik, P.V.S.M.S., et al.: Sign language to text conversion using deep learning. In: Ranganathan, G., Chen, J., Rocha, Á. (eds.) ICICCT 2020, pp. 219–227. Springer, Singapore (2021). https://doi.org/10.1007/978-981-15-7345-3_18
14. Ranganathan, G., Chen, J., Rocha, Á.: Inventive Communication and Computational Technologies. Diss. Department of Computer Science, City University of Hong Kong (2021)
15. Qiang, B., et al.: SqueezeNet and fusion network-based accurate fast fully convolutional network for hand detection and gesture recognition. IEEE Access **9**, 77661–77674 (2021)
16. Sridhar, S., Sanagavarapu, S.: SqueezeCapsNet – transfer learning-based ASL interpretation using SqueezeNet with multi-lane capsules. In: 2021 IEEE 12th Annual Ubiquitous Computing, Electronics Mobile Communication Conference (UEMCON), pp. 0001–0007 (2021)
17. Adaloglou, N., et al.: A comprehensive study on sign language recognition methods. IEEE Trans. Multimedia (2019)
18. Dadashzadeh, A., Targhi, T., Tahmasbi, A.: HGR-Net: A Two-stage Convolutional Neural Network for Hand Gesture Segmentation and Recognition (2018)
19. https://www.kaggle.com/grassknoted/asl-alphabet
20. American Sign Language Letters - v2 aug (roboflow.com)
21. Konstantinidis, D., Dimitropoulos, K., Daras, P.: Sign language recognition based on hand and body skeletal data. 3DTV-Conference (2018)
22. Zheng, L., Liang, B., Jiang, A.: Recent advances of deep learning for sign language recognition. In: DICTA 2017 - 2017 International Conference on Digital Image Computing: Techniques and Applications, pp. 1–7 (2017)
23. Lim, K.M., Tan, A.W.C., Tan, S.C.: A feature covariance matrix with serial particle filter for isolated sign language recognition. Expert Syst. Appl. **54**, 208–218 (2016)
24. Lim, K.M., Tan, A.W.C., Tan, S.C.: Block-based histogram of optical flow for isolated sign language recognition. J. Vis. Commun. Image Represent. **40**, 538–545 (2016)
25. [1911.11929] SPNet: A New Backbone that can Enhance Learning Capability of CNN (arxiv.org). https://pytorch.org/docs/master/generated/torch.nnBCEWithLogits Loss.html

26. https://arxiv.org/abs/1803.01534
27. https://github.com/snakers4/silero-models/blob/master/LICENSE
28. https://www.researchgate.net/figure/The-network-architecture-of-Yolov5-It-consists-of-three-parts-1-Backbone-CSPDarknet_fig1_349299852
29. https://www.hindawi.com/journals/misy/2020/7602384/fig1/
30. https://www.researchgate.net/figure/layers-of-MobileNet-architecture4_fig2_331675538

Modernized Speed Regulation Technique of DC Motor Through Arduino

R. Dhanasekar[1]([✉]), L. Vijayaraja[1], L. Sakthi Vel Murugan[1], S. V. Santhosh Balaji[1], R. Srigiridharan[1], and P. Subha[2]

[1] Department of Electrical and Electronics Engineering, Sri Sairam Institute of Technology, Chennai, India
{dhanasekar.eee,vijayaraja.eee}@sairamit.edu.in, {sit20ee034, sit20ee020,sit20ee035}@sairamtap.edu.in
[2] Department of Information Technology, Sri Sairam Institute of Technology, Chennai, India
subha.it@sairamit.edu.in

Abstract. In prevalent DC motor is extensively used in many applications, industrial functions and etc., Speed manipulate of DC motor is complicated nowadays. So, the speed management of direct motor (DC) for a number of purposes is very necessary to make it easier. Speed of DC motor can be managed in many ways. The utilized voltage is controlled via Arduino. So that we can manipulate the pace of the motor. The PC makes use of software program application to manipulate the velocity of the motor. Arduino is used to acquire the information like revolution per minutes from computer. Arduino is at once related to PC thru the USB cable and command is given by way of the PC to the Arduino to manipulate the pace and path of the DC motor. From there the command like the velocity of the motor at which it is desires rotate, at which course the motor needs to rotate and the revolution per minutes are given to the motor by way of the Arduino UNO.

1 Introduction

The pace manipulate DC motor is managed by means of the approach based totally on Adaptive Neuro-Fuzzy Inference System (ANFIS). The DC motor is excited by using Neuro-Fuzzy controller. This approach offers higher overall performance when in contrast to PI controller technique [1]. Speed manipulate of DC motor is additionally finished with the aid of various armature voltage. It is in contrast with the older technic of controlling the usage of traditional controllers and determined that it offers an 50% overload disbursed potential [2]. The DC motor is the most extensively used converter in modern-machine equipment and robots, it is wanted to be manipulate of personal parameters. For this controlling the cutting-edge simulation software program MATLAB/Simulink is used [3]. The pace manipulate of DC motor is additionally executed the use of traditional converters like (PID, IMC). A comparative evaluation of all controllers have been completed [4]. Motor velocity is additionally managed by way of the use of IGBT and the triggering is carried out by way of PWM converters beneath a number of loading circumstance and various armature voltage and subject voltage [5]. To keep away from the double closed-loop gadget, Beetle Antennae Search (BAS) is

P. Dassan et al. (Eds.): ICICSCNT 2023, CCIS 1970, pp. 172–182, 2024.
https://doi.org/10.1007/978-3-031-75957-4_15

introduced to clear up the parameters. BAS-PID controllers are mixed with self-tuning feature [6]. The Internal Model Control technique to manipulate (IMC) velocity of DC motor offers a mild overall performance version when in contrast to the Proportional-Integral-Derivative (PID) approach [7]. Fuzzy based PID controller gives better pace response however traditional controllers presenting higher velocity response with the aid of altering load at the value of lengthy settling time [8]. Controlling the velocity of a DC motor based totally on armature voltage control, armature resistance manipulate and area excitation manipulate with regular flux motors shunt and collection field. This idea permits enhancing the experimental end result [9]. Tuning of FOPID parameters the use of Ant Colony Optimization Technique (ACO) is carried out and excited by way of exterior methods [10]. The genetic algorithm (GA) offers higher robustness in evolutionary optimization [11]. Tuning of FOPID parameters the usage of Ant Colony Optimization Technique to manipulate the pace of the motor [12]. Voltage manage is completed with the aid of pulse width modulation (PWM) and path is managed with the aid of algorithmic code [13]. The transmitter is Android cell and the Bluetooth is receiver which controls the velocity if the DC motor [14]. The obligation cycle of the pulse width modulation is additionally diverse to manipulate the velocity the usage of Android mobile phones [15]. Among the range of voltage manipulate methods, pulse width modulation is the fine technique for velocity manage and path is managed by means of microcontrollers. Speed of a DC motor varies proportional to the fixed voltage with a constant grant voltage [16–19]. In the manage of pace o DC motor, whilst the usage of armature manipulate technique the armature voltage will increase and the pace of the motor decreases, whilst in discipline resistance technique pace will increase when the subject resistance will increase [20].

2 Proposed Methodology

Figure 1 shows the components which are all presents in the proposed prototype. Based on the connection and working the block diagram has two section. One is charging section, another one is controlling section. From the name itself the combination of solar panel, DPST switch, Battery are called charging circuit. Whenever the battery is under dry condition, it requires some external source to recharge the battery. Hence the DPST switch is need to connect or disconnect the solar panel for charging the battery. The combination of NPN - transistor, IC7432, switches are called controlling circuit. The Arduino UNO is connected to the PC or Laptop and to the switch and IC 7432. Figure 2 shows the components are connected together to perform the task which is mentioned in the abstract.

Arduino is the primary part of this circuit, because it controls signals (HIGH, LOW) for the components presented. The Arduino gives the source to all elements through digital pins (1,2,3,8,9). The NPN-transistor has three terminals (Emitter, Base, Collector). The collector has to collect the battery constant voltage. Emitter is the component which emits the regulated voltage. In between these two terminals, the base terminal is there which is used to conduct the voltage from collector to emitter. In this case NPN needs some base signal to turn ON the transistor. The base signal is provided by the Arduino (3rd PIN). The regulated voltage act as the source of IC7432 and the control signals

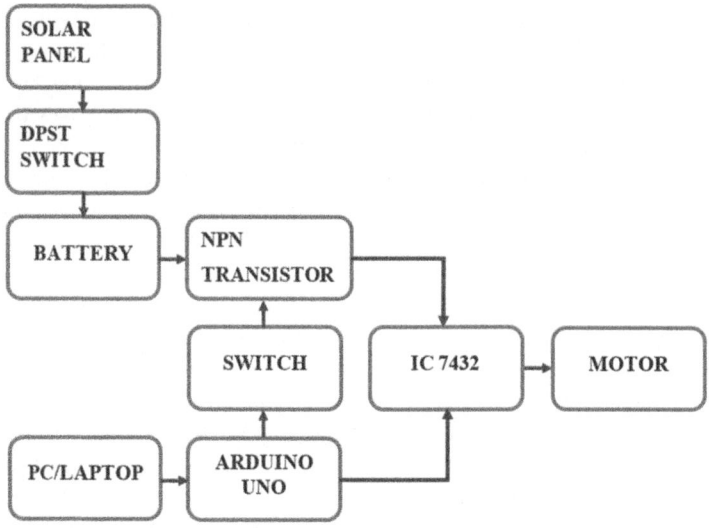

Fig. 1. Block Diagram of Proposed Methodology

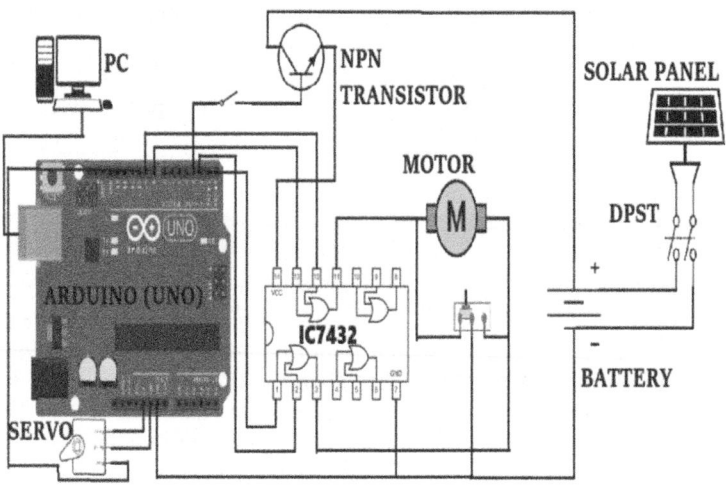

Fig. 2. Circuit Diagram of Proposed Methodology

of IC7432 provided by Arduino (1,2,8,9 PIN's). The servo mechanism is used for the application purpose which is connected to Arduino (Vin, GND, A0).

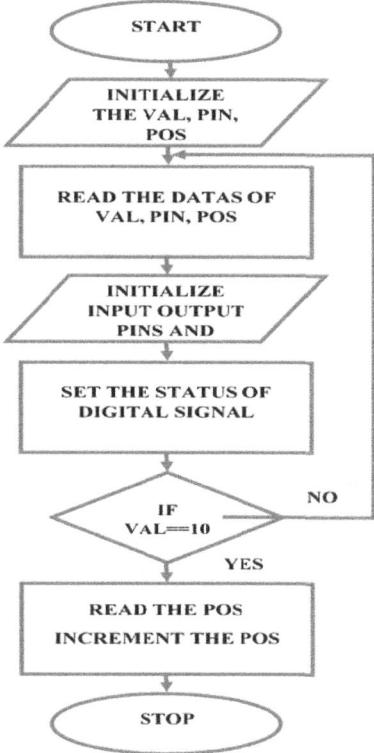

Fig. 3. Flowchart of Proposed model

3 Simulation Work for the Proposed System

3.1 Working Methodology

Figure 3 shows the flow diagram of software program. The program code that can be control the overall actions of the hardware. The flow chart follows the TOP-DOWN process. The single loop can need the low memory allocation and it reduce the time taken for compilation. The loop repeats until the motor speed is reached to the decided rated value. If the motor reaches the rated value the DC Servo motor started automatically due to the control signal. The program can be written by using C-Language. For servo mechanism the suitable library #include <Servo.h> is selected. Initially three integers will be initialized. The input/output pins were selected by using void setup () function. In this case the output control signal pins (1,2,8,9) was selected and servo added pin also selected. The Void loop function was declared for run the program again and again. In this loop function the control signals (High, Low) are selected, for NPN transistor the base signal also selected as the constant [Analog write (pin, Val)]. Servo motor rotate position was controlled by using for loop condition. The motor speed is controlled by the technique voltage regulation. The battery fixed voltage is change under regulated voltage the regulated voltage is can be varied by using the program constant VAL. According to the constant VAL variation is same as the motor speed.

3.2 Simulation Results

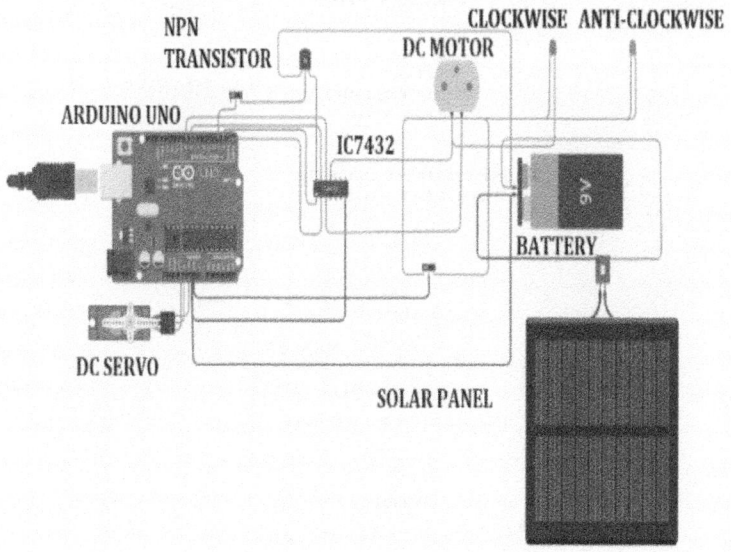

Fig. 4. Simulation of proposed model

Figure 4 shows the hardware simulation. The interconnection and setup pins are chosen by uploading programming code to the Arduino microcontroller (ATmega328P). The NPN base signal values are varied by using the coding. The two LED (Red, Blue) are presented in the simulation diagram. When the motor rotates clockwise the Red colour indication will blowing, when the motor rotates Anti-clockwise the Blue colour indication will glow. The servo motor gets the supply from the Arduino and it's done the rotation of clockwise (0–180°) and Anti-clockwise (180–0°). The clock pulse is provided by Arduino.

3.3 Direction Control

3.3.1 Clockwise Direction

Figure 5 shows the simulation of the hardware when it is rotates clockwise direction. In this case the motor terminals positive and negative are connected to the gate in which the gate has one end as higher potential and another end has lower potential respectively. As the result the motor start rotates clockwise. At the time the blue indication will starts glow.

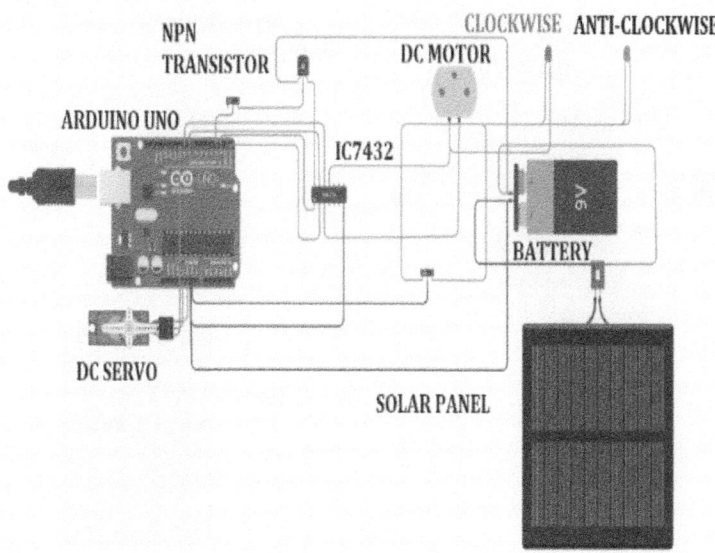

Fig. 5. Clockwise rotation

Table 1. Representation of control signal vs speed for simulation work

Base value of control signal	Speed in RPM
0	0
10	150
20	300
35	590
40	670
50	840

3.3.2 Anti-clockwise Direction

Figure 6 shows the simulation of hardware when it is rotates anticlockwise direction. In this case the motor terminals negative and positive are connected to the gate in which the gate has one end as higher potential and another end has lower potential respectively. As the result the motor start rotates anti-clockwise. At the same time the red indication will starts glow.

Table 1 shows the parameters which is presented in the simulation work. The table consists the two parameter one is control signal value and another is motor speed in rpm. Figure 7 shows the graphical representation of control signal value and the motor speed. The graph was plotted by using the Table 1. The tabulation was made by using the simulation results. When the NPN transistor base value will increase the speed of the motor also increase.

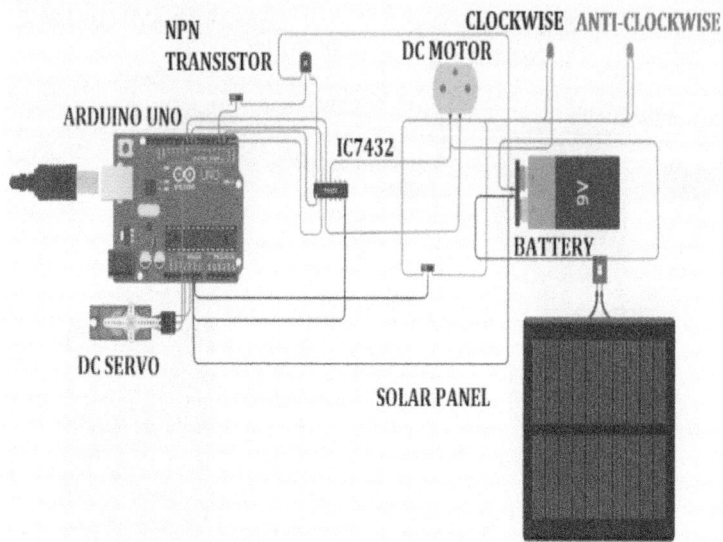

Fig. 6. Anti clockwise rotation

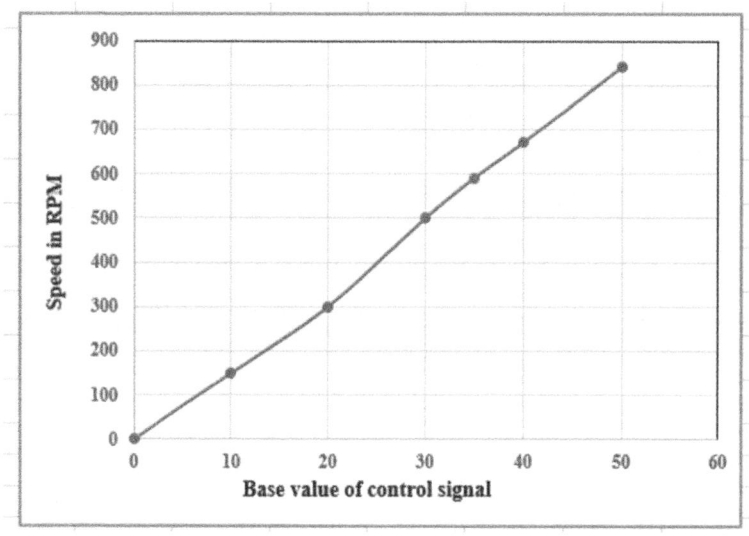

Fig. 7. Graphical representation of control signal vs speed for simulation work

4 Hardware Implementation

Figure 8 shows the hardware setup of the proposed system. The NPN, IC7432, indicators are located in breadboard. The Arduino was located near to bread board. The experimental result was similar to the simulation result, when the base signal value was increases the motor speed also increased. When base signal value was decreases the motor speed also decreased. Based on the the signal of IC7432 the motor rotates i the direction clock and anti-clock wise. When motor rotates Anti-clockwise the Blue LED was blowing. Figure 9 shows the prototype of the designed model speed control of DC motor using Arduino UNO connected to the laptop. From the Laptop the Arduino program is compiled and the instructions such as revolution per minute and the direction at which the motor want to be rotated will be initialized.

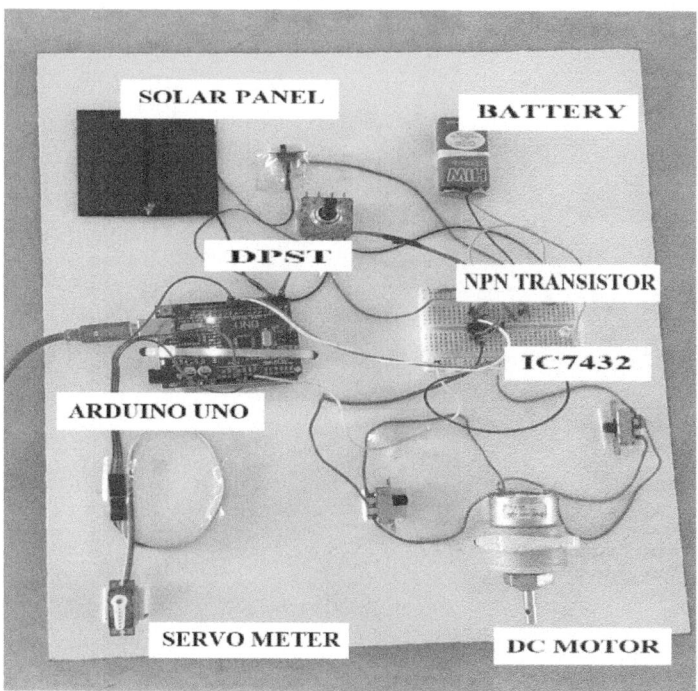

Fig. 8. Prototype of Hardware

Table 2. Representation of control signal vs speed for hardware implementation

Control signal	Speed in RPM
0	0
10	150
20	290
30	480
40	590
50	700

Table 2 shows the parameters which is presented in the testing of hardware. The table consists the two parameters, one control signal value and another motor speed in rpm. The motor speed was noted for each control signal value.

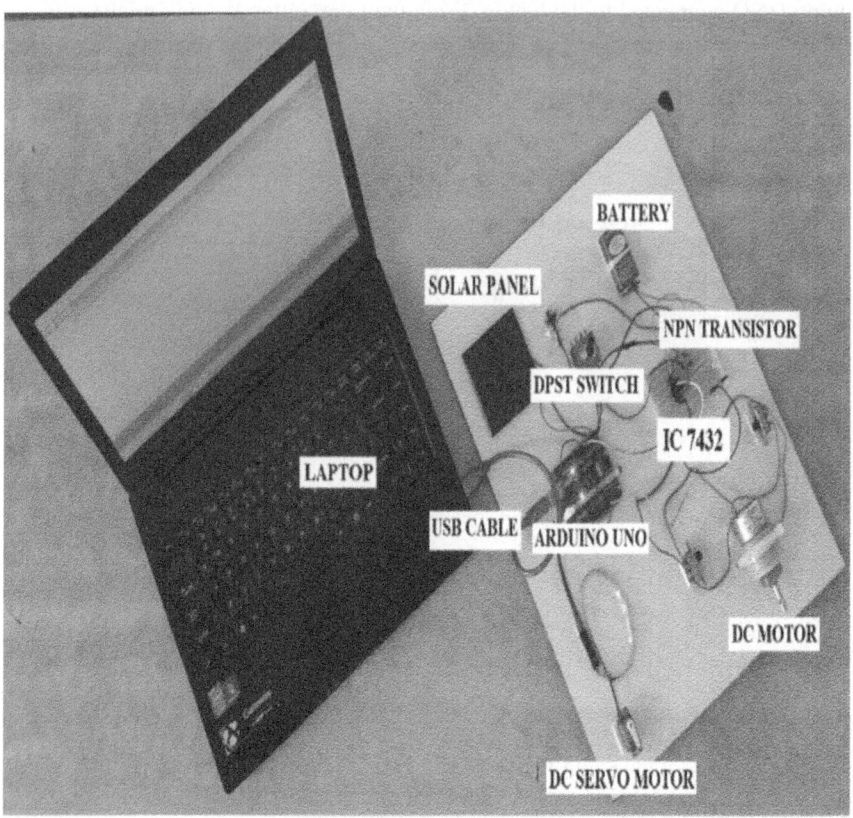

Fig. 9. Prototype of Hardware Connected with Laptop

Figure 10 shows the graphical representation of base value of control signal and the speed of the motor. The graph was plotted by using the Table 2. The tabulation was made by using the hardware results. The base value of NPN transistor is directly proportional to the speed of the motor. When the base value increases the motor speed also increased.

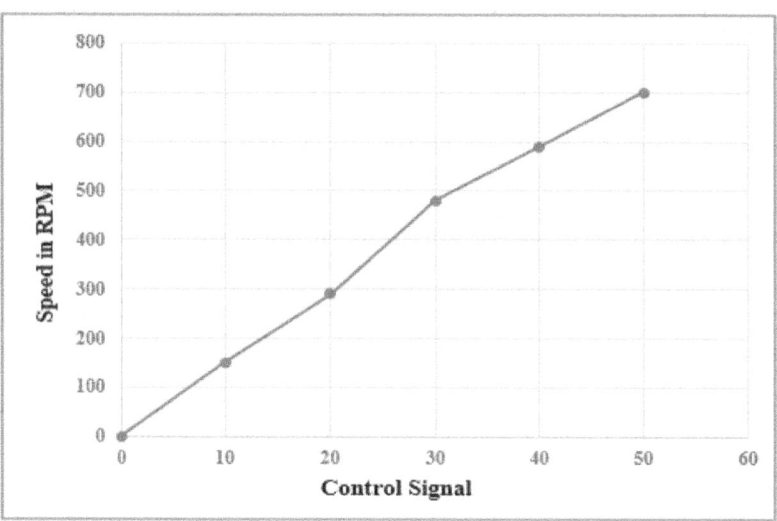

Fig. 10. Graphical representation of control signal vs speed for hardware implementation

5 Conclusion and Future Scope

The motor speed was controlled by the voltage regulation method, with the base signal (VAL) proportional to the speed of the motor. The direction of rotation of the motor was successfully controlled by IC7432, and proper indication was maintained. In the future, the proposed model will be modified to incorporate voice recognition technology, allowing the user to control the motor speed (in RPM) through voice commands.

References

1. Pavan Kumar, S., Krishna Veni, S., Venugopal, Y.B., Kishore Babu, Y.S.: A neuro-fuzzy based speed control of separately excited DC motor. In: IEEE Transactions on Computational Intelligence and Communication Networks, pp. 93–98 (2010)
2. George, M.: Speed control of separately excited DC motor. Am. J. Appl. Sci. (2008). ISSN 1546-9239
3. Mohammed, J.A.: Modeling, analysis and speed control design methods of a DC motor. Eng. Tech. J. **29**(1), 141–155 (2011)
4. Ahmad, M.A., Kishor, K., Rai, P.: Speed control of a DC motor using controllers. Autom. Control Intell. Syst. **2**(6-1), 1–9 (2014)
5. Kumar, A., Munsh, S.P., Memon, M.R., Mishra, S.: Speed control of DC motor using IGBT. National Institute of Technology Rourkela (2007)
6. Lin, X., Liu, Y., Wang, Y.: Design and research of DC motor speed control system based on improved BAS. In: 2018 Chinese Automation Congress (CAC) (2018)
7. Ahmed, U.O., Patrick, A.A., Kwembe, B.A.: DC motor speed control using internal model controller: industrial transformation strategy. Int. J. Eng. Adv. Technol. (IJEAT) **9**(5), 300–306 (2020)
8. Singh, V., Garg, V.K.: A comparative study on speed control of D.C. motor using intelligence techniques. Int. J. Electron. Electr. Eng. **7**(4) (2014)

9. Frayyeh, H.F., Mukhlif, M.A., Abbood, A.M., Keream, S.S.: Speed control of direct current motor using mechanical characteristics. J. Southwest Jiatong Univ. (2019)
10. Yadav, J., Kumar, P., Gupta, D.N., Kaur, M.M.: Speed control of DC motor using ACO based FOPID controller. Int. J. Innov. Res. Comput. Sci. Technol. (IJIRCST) 8(4) (2020)
11. Ahuja, A., Aggarwal, S.K.: Design of fractional order PID controller for DC motor using evolutionary optimization techniques. WSEAS Trans. Syst. Control 9 (2014)
12. Idir, A., Kidouche, M., Bensafia, Y., Khettab, K., Tadjer, S.A.: Speed control of DC motor using PID and FOPID controllers based on differential evolution and PSO. Int. J. Intell. Eng. Syst. (2018)
13. Singh, R., Kumar, S., Kumar, R., Kumar, V.: Synchronisation, speed and direction control of DC motor. Int. J. Electron. Electr. Eng. 7(9), 943–952 (2014)
14. Ganesh, C.V., Suryaprakash Reddy, G., Rajashree, Akash Kumar, G.: DC motor and direction control using android. IJCRT 9(6) (2021). ISSN 2320-2882
15. Bhavithra, M., Gengagowri, B., Valarmathi, S., Josephine, J.: Speed and direction control of DC motor by current and voltage controlled device by android application. Int. J. Adv. Res. Sci. Eng. 7(02) (2018)
16. Balakrishna, N., Vijay Kumar, L., Sandeep, S., Durga Prasad, D., Srikanth, B.: Speed and direction control of DC motor using android application. Int. J. Res. Dev. (IJRD) 7(6) (2022)
17. Krishna, M., Aditya Reddy, G., Shiva Sai, R., Bhanu Prasad, P.: Speed and direction of DC motor using bluetooth, vol .11, no. 11 (2020). ISSN 0377-9254
18. Saini, H.K., Firoz, S., Pandey, A.: Arduino based speed control. J. Electr. Electron. Eng. 3(4) (2017)
19. Mane Devanand, B., Koli Rohan, B., Bhosale Sushilraje, S., Jogdand Vishakha, V.: Arduino based speed control of DC motor by using LabVIEW. Int. J. Innov. Res. Sci. Eng. Technol. 6(5) (2017)
20. Al-Sagar, Z.S., Saleh, M.S., Mohammed, K.G., Sameen, A.Z.: Modelling and simulation speed control of DC motor using PSIM. IOP Conf. Ser. Mater. Sci. Eng. (2020)

Design and Performance Analysis of Proposed Dual Channel Redundant LoRaWAN for Industrial Control and Safety Shutdown Systems

Sobanbabu Nandhagopal[✉] and A. Mohan Babu

SRM Institute of Science and Technology, Ramapuram, Chennai, India
{sn2271,mohanbaa}@srmist.edu.in

Abstract. As industry 4.0 is currently being implemented in industries, there is a huge demand for highly reliable and Hard real time wireless data communication between sensor/actuator end nodes and Control System and to integrate the same as part of IIoT. High availability and faster data transfer are the most critical requirement for exchanging feedback/control signal between end nodes and control System of a closed loop control and safety shutdown system to avoid any downtime in a plant due to false positive or false negative values. In this work, Redundant LoRaWAN is designed which transmit/receive the same data using dual channels and the data is picked from a healthy channel using an analog Multiplexer in case of failure of any one channel. An Experimental test bed is setup to establish communication and analyse the performance. The test results confirm that this Redundant LoRaWAN reduces the probability of failure of communication to less than 1% in an Industrial control and shutdown application.

Keywords: Industry 4.0 · IIoT · LoRaWAN

1 Introduction

Internet of Things (IoT) is evolving and expanding with billions of devices with various wireless communication technologies from short range to long range in sensor level to Cloud Server level and mainly serves in the area of Smart city, Agriculture, Healthcare, Warehouse and Transportation etc., which are mostly soft real time system in nature and mostly used for Monitoring purpose. ZigBee, BLE, ZWAVE, WIFI [8] are used for the short ranges and LoRaWAN, NB-IoT, Sigfox are used for the long range communication in non critical monitroing application.

As far as Industry 4.0 is concerned, Industrial Internet of Things (IIoT) connects locally or remotely located Sensors/actuator nodes on a Wireless sensor and actuators network using either conventional short range Industrial wireless communication such as Wireless HART, ISA100.11a [11] or Long range LPWAN communication technologies [5] such as LoRaWAN, NB-IoT, Sigfox via gateways. Also data from Control system and Machines are integrated with IIoT through existing conventional M2M communication

P. Dassan et al. (Eds.): ICICSCNT 2023, CCIS 1970, pp. 183–197, 2024.
https://doi.org/10.1007/978-3-031-75957-4_16

[6] network such as Industrial Ethernet, Modbus, Profibus, CAN bus etc., using protocol converters.

In IIoT, The data collected from end nodes through wireless network is used for monitoring and Non critical supervisory control. Currently, data exchange among time critical closed loop & safety shutdown Systems are realised only on Industrial Automation OT network with up/down link to IioT. Industrial enviroment is harsh due to unpredictable variations of temperature, pressure, humidity, noise, Electro magnetic interference and high signal attenuation and hardware fault which are threat to failure of the data communication and this prevents using IIoT for critical applications. This demands to design and develop robust wireless communication technologies to meet the various performance requirements suh as high availability, reliability and Low latency etc., of Hard real time Closed loop and shutdown systems in order to seamlessly integrate with IIoT.

IIoT implementation using LPWAN helps to realise the industrial automation solutions where large number of battery powered end nodes are geographically distributed in hazardous and remote locations which needs larger coverage area with low power consumption such as offshore pipeline management, Power grid monitoring and control, Tank level control etc.,

In this work, the authors propose to design a novel dual channel redundant LoRaWAN wireless network with Class C communication to make LoRaWAN suitable for Machine to Machine and Hard real time communication in closed loop and shutdown system. This ensures availability and interruption free communication in case of hadware fault which is not provided in data repetition redundancy [7] or frame replication [4]. Availability is achieved by integrating a Signal splitter in the sender nodes, Analog Multiplexer based switching technique to select pockets from healthy receiver nodes on the physical and Link layer. Adaptive intentional time delay is introduced between pockets during switching to ensure synchronised data transfer as per the feedback sampling time and scan time of controllers of a closed loop control system. Various performance metrics are analysed and reported during both normal and redundancy change over due to hardware fault condition.

This paper discusses about Sect. 2 Requirements of Real Time data communication for Industrial Control and Safety Shutdown System, Sect. 3 Overview of LoRaWAN Protocol and performance metrics, Sect. 4 Implementation of the proposed Redundant LoRaWAN, Sect. 5 Performance analysis of test bed setup Finally, Sect. 6 contains our conclusions and possible future developments.

2   Requirements of Real Time Data Communication for Industrial Control and Safety Shutdown System

The following are the most essential requirements of real time communication.

2.1   Highly Deterministic Hard and Firm Real Time Communication

Based on the operational requirements defined by the International Society of Automation, delivery of all the data transmitted must be guaranteed exactly in real-time with the lowest latency and immediate action on an event is required for Safety Shutdown systems to prevent any disaster in the plant due to crossing of safety critical limits, Communication must be of highly deterministic hard real time in nature within few ms. Periodical or event based closed loop regulatory Control with feedback loops may or may not require faster communication which is classified under firm real time system. Communication failure will upset the process if no action is taken within a second or few seconds. Existing IIoT communication protocols are to be improved to meet this requirement. Polling time of data as per the cycle time of various systems shown in Table 1 are to be ensured for the proper functioning of each system.

2.2 Soft Real Time Communication

Supervisory Closed loop Control systems with slow feedback signals and event-based control action, Open loop control systems with Control action initiated by an operator based on historical data collected from sensors, Event based Messaging or alerting systems and Data logging systems are also real time systems but communication failure does not cause any major accidents or operational consequence in the plant. Currently available IIoT meet these requirements.

Table 1. Communication range and Cycle time requirement for different applications [13]

Application	Communication Range	Cycle Time
Safety Shutdown	300 m	10–250 ms
Power System Protection	>1 km	50 ms
Factory Automation	300 m	100 ms
Process Control	200 m	10–1000 ms
Monitoring and supervision	500 m	>1000 ms
Building Automation	100 m	100–1000 ms

2.3 Increased Availability and Scalability

Wireless communication in Industries with harsh environment face number of challenges. All the hardware systems should be highly reliable and fault tolerant with a low probability of hardware failure which also allows hot-swapping of faulty modules without any communication drop. Network should be easily scalable [13] as per the future expansion of the plant.

2.4 Operating in Unlicensed Bands

Communication should not require any licensed spectrum which increases the cost due to subscription charges and communication infrastructure setup by third-party Communication service providers.

2.5 Power Consumption

Remotely located Devices that are used to set up communication should consume less power [13] with perfect trade-off between power consumption and data update rate and energy harvesting feature to prolong life span as most of the nodes are battery powered in remote location. A higher data rate can be achieved with Class C communication which consumes high power for Machine to machine or a closed loop control system where a fixed power supply is possible.

3   Overview of LoRaWAN Protocol and Performance Metrics

3.1 LoRaWAN Protocol

LoRa is a physical layer Communication technology developed by Semtech for long range wireless applications. It works based on a proprietary Chirp Spread Spectrum (CSS) [3] technique and modulation scheme for communication and spreads the RF signal over a wide bandwidth which is robust and less susceptible to noise and interference [17].

LoRaWAN is a MAC layer protocol which is on top of LoRa physical layer and it represents layer 2 and 3 in the OSI Model as indicated in Fig. 1. Desired Performance enhancements in LoRaWAN can be achieved by adjusting various parameters. It is one of the LPWAN technologies which connect sensor, actuator and Machine nodes to the Cloud via gateways and backhaul LAN/Internet. Data rate for the LoRaWAN is from 300 bps to 5 Kbps [10] with a uplink channel band width of 125 kHz or 500 kHz and downlink channel of 500 kHz [10].

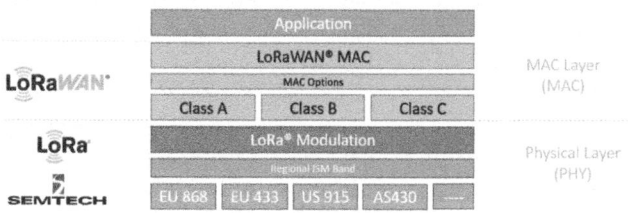

Fig. 1. LoRaWAN OSI Layers [17]

3.2 LoRaWAN Network Architecture

LoRaWAN network used as a part of IIoT consists of End nodes usually sensors or actuators or Machines that communicate using the LoRa transceivers via LoRa Gateways. Gateway modules collect messages from end nodes and relays to the Network Server using a reliable IP connection as shown in Fig. 2.

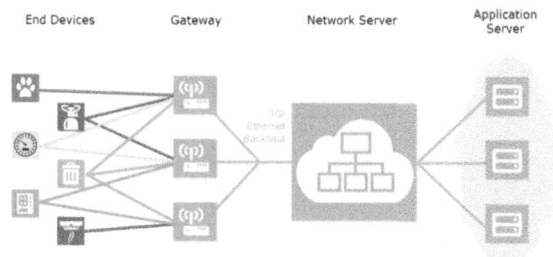

Fig. 2. LoRaWAN IIOT Architecture [17]. Source: http://www.semtech.com/wireless-rf/internet-of-things/what-is-lora/

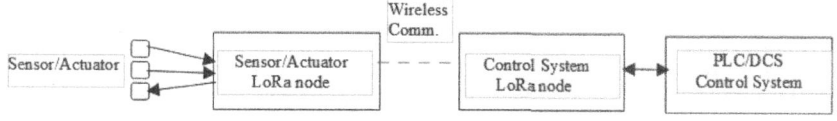

Fig. 3. LoRaWAN M2M Architecture

3.3 LoRaWAN Architecture for Machine to Machine Communication

Direct LoRaWAN wireless link between Control system LoRa node and Sensor/Actuator nodes without any gateway module to establish a peer-to-peer communication is classified as LoRaWAN M2M network as shown in Fig. 3.

3.4 LoRaWAN Communication Classes

As per the specification of LoRaWAN, there are three different communication profiles in between devices and applications [17].

Class A ("Aloha"), Class B "Beaconing", and Class C. (Continuous Reception).

These Communication classes are adopted as per the optimum requirements in each application. Latency and Power consumption are the major difference in these three classes. End devices can always send data via uplink when needed, but receiving of downlink signal is determined by the class of operation. Class C is the most suitable for Actuators and real time applications.

3.5 Unlicensed ISM Frequency Band Used in LoRaWAN

LoRaWAN operates in an unlicensed ISM sub-GHz frequency band that is region dependent [8] as shown in Table 2 and recently the launched 2.4 GHz band which is universally used. 865.0625 MHz, 865.4025 MHz and 865.985 MHz are the mandatory joint request channels in India.

3.6 Network Parameter Adjustment

Maximum battery life and Communication performance of end nodes in LoRaWAN is achieved by using Adaptive Data Rate (ADR) [2] mechanism for optimizing data rates, air time, and power consumption [17].

Table 2. Frequency bands used in differnt regions

Frequency Bands	Regions
902–928 MHz	USA
863–870 MHz	Europe
865–867 MHz	India
430–432 MH	Asia/China
915–928 MHz	Australia
920.6–928.0 MHz	Japan

Following are the various transmission parameters on end devices which are controlled by ADR [2].

Spreading factor [1] determines the speed of data transmission. Lower spreading factors mean a higher data transmission rate and vice versa. N = 2 pow of SF. The SFs in LoRa are orthogonal Concurrent transmissions with different SF on the same channel do not interfere with each other and can be successfully decoded [17] Also it determines the number of raw bits a symbol carries ranging from 6 to 12. For example, if SF 7 is used, one symbol in frame maps to 7 bit. The symbol duration Ts in a frame can be computed as per the Eq. 1.

$$Ts = 2^{SF}/BW \tag{1}$$

If spreading factor is increased, the Signal to Noise Ratio (SNR), receiver sensitivity and the signal range are increased. At the same time bit rate is decreased and the communication time in increased. Coding rate (CR) is another important parameter which is responsible for the Forward Error Correction (FEC) applied on the message in order to protect against burst interference. CR ranges between 1 and 4.

Data rate of LoRaWAN signal is determined by Eq. 2.

$$\text{Data rate} = SF * 4/(4 + CR)/2SF/BW \tag{2}$$

Bandwidth is the amount of data that can be transmitted from one point to another within the network [15].

Transmission power is the energy that the end-device transmitter produces at its output [15]. Also referred to as EIRP Equivalent Isotropic Radiated Power, which is the radiated output power refer to an isotopic antenna radiating power equally in all directions and whose gain is expressed in dBi. + 30 dBm is Max EIRP (default) - TXPower 0 in indiaThe power and the gain of the antenna must be selected accordingly to avoid exceeding the EIRP limitation [8] set by the country. Received Power by an antenna is computed as per the Eq. 3.

$$\text{Recieved Power } Pr = (Pt \ GtGrC^2)/(4piRf)^2 \tag{3}$$

Pt: Transmitted power,
Gt: Gain of the transmitting antenna,

Gr: Gain of the receiving antenna,
C: Speed of the light,
R: Range between the antennas and
f: Operating frequency

Time on Air (TOA) The Time on Air [12] is the most important parameter for LoRa transmissions which specifies the required duration to transmit a single LoRa message. It is adjusted by the available bandwidth (BW), the spreading factor (SF), and the total LoRa message length.

Spreading factors and its corresponding Sensitivity, Data rate, Range and Time on Air are shown in Table 3 at a bandwidth of 125 kHz depends on propagation conditions.

Table 3. Spreading factor, Sensitivity, Data rate, Range and Time on-air

Spreading Factor	Sensitivity	Data Rate	Range	Time on Air
SF7	123.0 dBm	5470 bps	2 km	56 ms
SF8	126.0 dBm	3125 bps	4 km	100 ms
SF9	129.0 dBm	1760 bps	6 km	200 ms
SF10	132.0 dBm	980 bps	8 km	370 ms
SF11	134.5 dBm	440 bps	11 km	40 ms
SF12	137.0 dBm	290 bps	14 km	1400 ms

In this work, 865 MHz (IN865) frequency band is used with channel width of 125 kHz.

3.7 Performance Metrics

Received Signal Strength Indication (RSSI) is used to measure signal coverage in a location and it is a relative measurement that helps you determine if the received signal is strong enough to get a good wireless connection from the transmitter [14] RSSI is measured in dBm and its value is a negative form [14] The closer the RSSI value is to zero, the received signal is stronger [14] RSSI [17] is influenced by the output power of the transmitter, Path loss and antenna gain.

Signal-to-Noise Ratio is a ratio of received signal power to noise power which is computed as per Eq. 4.

$$SNR\ (dB) = P\ received\_signal\ (dBm) - P\ Noise\ (dBm) \qquad (4)$$

Link Budget less Link Budget increases data rate with less spreading factor for the short range.

The relationship between data rate and rage is shown in Fig. 4.

Pocket Delivery Ratio (PDR) Signal coverage in a location is determined by calculating the packet delivery ratio (PDR) as defined by Eq. 5.

$$PDR = Pr/Ps \qquad (5)$$

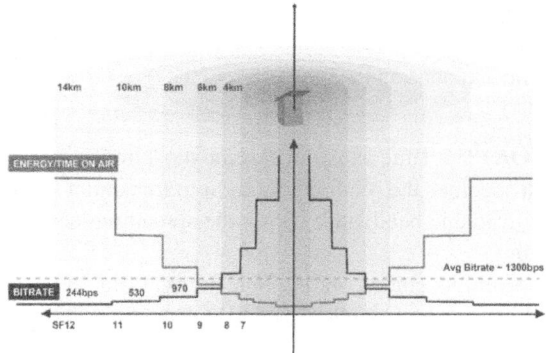

Fig. 4. LoRaWAN data rate vs Range [17]

Pr: No of pockets successfully received by the receiver node and at the network server
Ps: No. of pockets sent by the sending end node

Maximum packet size in LoRaWAN is 255 bytes and in which 13 bytes are used as the header of the pocket. So, the maximum payload possible is 242 bytes and this is applicable only when spreading factor is 7 & 8 [14].

4 Implementation of the Proposed Dual Channel Redundant LoRaWAN

4.1 System Architecture

The Proposed dual channel redundant LoRaWAN is set up between a custom-built sensor and an actuator end node as per the architecture shown in Fig. 5 which is part of Open/Closed Loop Control System. End nodes consist of individual LoRa transceiver modules for each channels connected to microcontrollers for interfacing Sensor/Actuators. Both the channels are integrated in Fault tolerant redundant configuration which simultaneously send/receive the same data from sensor/actuator. Failure of any one channel is detected in the software level and switching of channels is done in Physical layer instantaneously to tap the data from the healthy channel. This experimental test bed is setup in SRM Institute of Science and Technology Ramapuram Campus, Chennai, India.

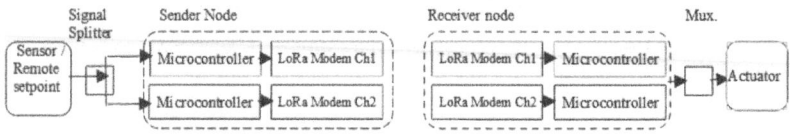

Fig. 5. System Architecture of proposed dual channel LoRaWAN for connecting two end nodes

4.2 SX1276 LoRa Transceiver Module

LoRa Transceiver Module developed using Semtech SX1276 LoRa Chip which supports Frequency band 865–867 MHz and RSSI down to −148 dBm which is readily available in the market is used as LoRa modem in this experimental test bed which send/receive data from/to ESP32 through SPI interface. Pin details of the module is shown in Fig. 6.

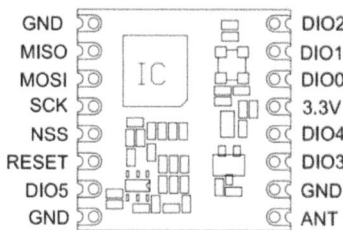

Fig. 6. LoRa SX1276 Transceiver Module Pin details.

4.3 ESP32 Based Micro Controller System

ESP32 is a 3.3 V, 240 MHz, 32 bit SOC Microcontroller System developed by tensilica and manufactured by Espressif Systems with inbuilt Analog to digital, Digital to Analog converter, Wifi, SPI, and I2C interfaces which can be easily deployed by programming using Arduino IDE software. ESP32 is used in this project to process the Analog and digital I/O from sensor and transmits as data through LoRa modem and all the configuration parameters to LoRa modem are performed in ESP32 and downloaded to LoRa modem through SPI.

Sender Node integrated with both ESP32 and LoRa module with an Antenna on a single board made by TTGO is as shown in Fig. 7 is used in this experimental test bed.

Fig. 7. Sender Node

The Receiver Node is integrated as shown in Fig. 8 with LoRa Module and ESP32 board.

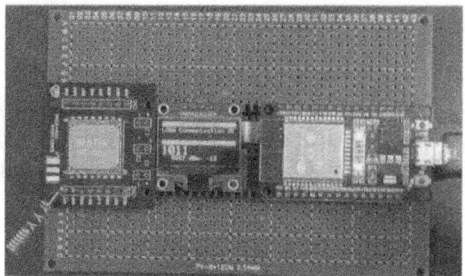

Fig. 8. Receiver Node.

4.4 Sensor/Actuator Integration

Analog measurements from the sensor is converted into the current signal using 0 to 5 V to 4 to 20 mA V to I converter and the same signal is connected to 2 numbers of 4 to 20 mA to 0–3.3 V I to V converter which are in parallel to split the same analog into 2 signals and then connected to Analog input channel of ESP32. This is done to transfer the same analog value simultaneously to the two LoRaWAN channels as shown in Fig. 9.

Fig. 9. Analog Signal Splitter.

4.5 Fault Tolerant Switching of Channels

In the actuator end node, Analog Signals from Analog output channel of ESP32 is connected to the analog multiplexer module to select same analog value transmitted through dual channels upon failure of any one channel as shown in Fig. 10.

4.6 Software Implementation

LoRa Frequency 865.0625 MHz, spreading factor 7 to 12 for adjusting the range of the LoRa network and the Analog scaling functions, Communication pocket transfer and watchdog function for detecting the communication delay and communication failure are configured in the coding part using Arduino IDE software and downloaded to ESP32 microcontrollers.

Fig. 10. Fault tolerant switching using Analog Multiplexer.

5 Performance Analysis of Test Bed Setup

Test bed setup as shown in Fig. 11 was setup and the performance test was carriedout to prove all the relevant performance metrics are well with in the tolerance limit before deploying the system for fault tolerant redudancy test.

Fig. 11. Test bed setup for dual channel redudant LoRa wireless sensor network.

5.1 Communication Range and RSSI with Spread Factor for Highest Data Rate

Highest data rate is achieved with lowest spreading factor, by configuring spreading factor 7, the data pockets are transmitted from sending node and corresponding RSSI values are measured in the receiver node and tabulated in Table 4. Spreading factors were increased upto 12 and corresponding Signal strength noted was plotted as in Fig. 12.

5.2 Pocket Delivery Ratio at Different Ranges

Pockets received in receiving end node are verified for each range and pocket delivery ratio is computed and the result is tabulated in Table 5 by increasing the data payload. The effect on %PDR by payload was found well within the tolerance limit.

Table 4. LoRaWAN Communication range vs Received signal strength

Range	RSSI-dbm
10 m	−20
20 m	−23
50 m	−40
100 m	−60
150 m	−66
200 m	−82
250 m	−87
300 m	−91
350 m	−102
400 m	−110
500 m	−121

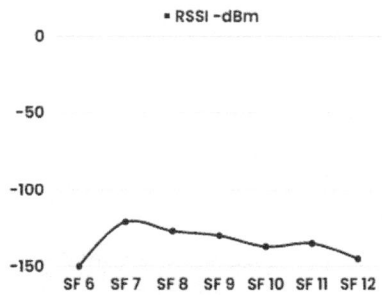

Fig. 12. Received signal strength vs Spreading factor

5.3 Measurement of Signal-to-Noise-Ratio

Signal-to-noise-ratio for the spreading factor 6 with highest payload 240 bytes were measured and the result is as shown in Fig. 13.

5.4 Latency Measurement on Dual Channels

Communication latency on each channel between Sent time Ts and Recieved time Tr are measured on Sending and receiving nodes and the readings are tabulated in Table 6. Laptop Computer loaded with the plotter is Time synched with NTP Time server is used for time measurements.

Table 5. LoRaWAN Communication range vs Pocekt deliver ratio at various Payload

Range	%PDR Payload 60bytes	%PDR Payload 120bytes	%PDR Payload 180bytes	%PDR Payload 240bytes
10 m	99	98	98	97.5
20 m	99	98	98	97.5
50 m	99	98	98	97
100 m	99	98	98	97
150 m	99	98	98	97
200 m	99	98	98	97
250 m	99	98	98	97
300 m	99	98	98	97
350 m	98	97	97	97
400 m	98	97	97	96
500 m	98	97	97	96

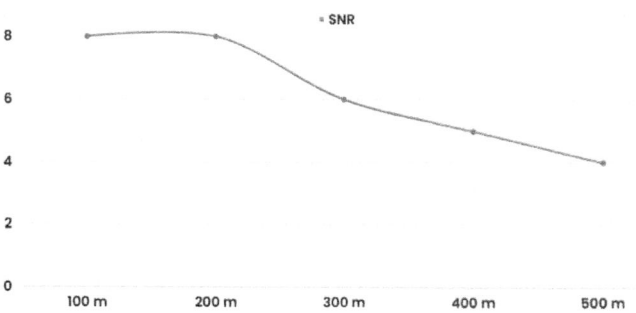

Fig. 13. Signal-to-noise-ratio vs Range in m

5.5 Fault Tolerant Testing

Fault tolerant feature of the sending end node is tested by turning OFF the power to both the ESP32 controllers one by one and confirmed that there is no interruption in LoRa transmission. Similarly receiveing end node is tested.

Table 6. LoRaWAN Communication latency.

Range	Ts in ms Ch. 1 & 2 ON	Tr in ms	Ts in ms Ch. 1 OFF & Ch 2 ON	Tr in ms
10 m	0	58 ms	0	59 ms
20 m	0	58 ms	0	59 ms
50 m	0	58 ms	0	59 ms
100 m	0	58 ms	0	59 ms
150 m	0	58 ms	0	59 ms
200 m	0	58 ms	0	59 ms
250 m	0	58 ms	0	59 ms
300 m	0	59 ms	0	60 ms
350 m	0	59 ms	0	60 ms
400 m	0	60 ms	0	61 ms
500 m	0	60 ms	0	61 ms

6 Conclusion

Dual Channel Redundant LoRaWAN wireless link between two nodes as a peer-to-peer data communication at a maximum data rate of 5 Kbps was set up successfully and the performance test was carried out for the maximum range of 500 m. The purpose of this testing was mainly to prove the redundancy without any data loss during channel switchover due to hardware fault and power failure. As per the recorded results, RSSI signal strength, Pocket delivery ratio and the signal latency with both Channel ON, with one Channel OFF and another channel ON found promising and the Communication failure rate was less than 1% which is most critical for Industrial control and safety system functions. Also, it provides a path for the sucessful implementation of redundancy to eliminate availability issue in LoRaWAN link duing unexpected hardware failures. Based on this work, further testing for the extended range as per the required use case can be carried out.

References

1. Fawaz, H., Khawam, K., Lahoud, S., Martin, S., El Helou, M.: Cooperation for spreading factor assignment in a multioperator lorawan deployment. IEEE Internet Things J. **8**(7), 5544–5557 (2020)
2. de Jesus, G.G.M., Souza, R.D., Montez, C., Hoeller, A.: LoRaWAN adaptive data rate with flexible link margin. IEEE Internet Things J. **8**(7), 6053–6061 (2020)
3. Qadir, Q.M.: Analysis of the reliability of LoRa. IEEE Commun. Lett. **25**(3), 1037–1040 (2020)
4. Fernandes Carvalho, D., Ferrari, P., Sisinni, E., Flammini, A.: Improving redundancy in LoRaWAN for mixed-criticality scenarios. IEEE Syst. J. **15**(3), 3682–3691 (2020)

5. Ainsworth, T., Brake, J., Gonzalez, P., Toma, D., Browne, A.F.: A comprehensive survey of industry 4.0, IIoT and areas of implementation. IEEE SoutheastCon (2021)
6. Gündoğan, C., et al.: The impact of networking protocols on massive M2M communication in the industrial IoT. IEEE Trans. Netw. Serv. Manag. 18(4), 4814–4828 (2021)
7. Sisinni, E., Carvalho, D.F., Ferrari, P., Flammini, A., Gidlund, M.: Adding redundancy to LoRaWAN for emergency communications at the factory floor. IEEE Trans. Ind. Inform. 18(10) (2022)
8. Zourmand, A., Hing, A.L.K., Hung, C.W., AbdulRehman, M.: Internet of things (IoT) using LoRa technology. In: 2019 IEEE International Conference on Automatic Control and Intelligent Systems (I2CACIS 2019), 29 June 2019, Selangor, Malaysia (2019)
9. Cena, G., Scanzio, S., Valenzano, A.: Experimental evaluation of seamless redundancy applied to industrial wi-fi networks. IEEE Trans. Ind. Informat. 13(2), 856–865 (2017)
10. Leonardi, L., Battaglia, F., Bello, L.L.: RT-LoRa: a medium access strategy to support real-time flows over lora-based networks for industrial IoT applications. IEEE Internet Things J. 6(6), 10812–10823 (2019)
11. Adriano, J.D., do Rosario, E.C., Rodrigues, J.J.P.C.: Wireless sensor networks in industry 4.0: WirelessHART and ISA100.11a. In: 2018 13th IEEE International Conference on Industry Applications (2018)
12. Loh, F., Raffeck, S., Metzger, F., Hoßfeld, T.: Improving LoRaWAN's successful information transmission rate with redundancy. In: IEEE 17thh International Conference on Wireless and Mobile Computing, Networking and Communications (2021)
13. Seferagić, A., Famaey, J., De Poorter, E., Hoebeke, J.: Survey on wireless technology trade-offs for the industrial internet of things. Sensors 20(2), 488 (2020)
14. https://Thingsnetwork.org
15. https://docs.arduino.cc
16. TS001-1.0.4 LoRaWAN® L2 1.0.4 Specification. lora-alliance.org
17. Semtech Corporation. http://www.semtech.com
18. https://www.espressif.com/en/support/download/documents

IoT-Based Crop Recommendation System Using SVM and Decision Tree Algorithm

Charumathy, Jaspin[(✉)], and Andrea Joe Lorett

St. Joseph's Institute of Technology, Chennai 600119, Tamil Nadu, India
jaspink@stjosephstechnology.ac.in

Abstract. A high yield with excellent quality is one of the most important concerns in agriculture. The soil significantly improves productivity and yields. A soil assessment assists in identifying all of the soil's features, as well as challenges and requirements related to its development. Nutrient imbalances and deficiencies can have an impact on soil production and well-being; therefore, it is critical to conduct soil evaluations to identify them. This enables farmers to take critical steps to address the problem and sufficiently repair the soil, resulting in increased yields and productivity. Sensors are intelligent enough to provide near-real-time results with minimal effort. The primary objective of this device is to be able to measure soil characteristics, including temperature, pH, water content, moisture, and nutrient content. Better crop recommendations are produced by the proposed IoT-based system using SVM and decision tree algorithms than by the current literature.

Keywords: IoT · soil analyzer · crop recommendation · pH · sensors · soil image classification

1 Introduction

The Indian economy, which relies primarily on agriculture, is unable to use its land resources in the most profitable, and sustainable way. Lack of understanding of agriculture and inefficient use of resources are the main causes. Soil analysis is a vital tool for farmers since it may identify the essential inputs for productive and profitable farming. To identify the ideal quantity of fertilizer to apply to a crop, thorough soil testing takes into account the nutrients already present in the soil. Each state provided between nine and 10 million samples of soil. This lowers the amount of soil samples that can be analyzed in the field, making it more difficult to determine nutrient levels in agricultural regions. [6] Accurate soil analysis helps provide fertilizer samples that meet crop needs while using the soil's current nutrients. Conventional soil assessment includes expensive and time-consuming field sampling and chemical laboratory analysis. Effective decision making requires an understanding of the variability of many soil qualities throughout a field. Many soil sensors can now accurately scan the chemical and physical features of soil. Thus, farmers should have access to soil analysis. Monitoring soil nutrient levels is important for effective fertilizer use and mitigating the environmental impact of

P. Dassan et al. (Eds.): ICICSCNT 2023, CCIS 1970, pp. 198–214, 2024.
https://doi.org/10.1007/978-3-031-75957-4_17

inadequate fertilization. With minimal effort and near-accurate results, new sensors are able to replace in the chemical laboratory tests in real time. [2, 4] Using these sensors, crop recommendations based on soil characteristics are more efficient for farmers, thus improving soil quality and crop yield.

2 Related Work

Hamuda et al. [1] proposed a fresh technique based on color features, as well as morphological deterioration and dilatation, are proposed. This technique isolates the cauliflower harvest area from the weeds and dirt in a shot taken in natural light (cloudy, partly cloudy, and sunny). Using the Hue Saturation Value (HSV) color space, the proposed method recognizes crops, weeds, and soils. Each HSV channel is filtered between particular values to establish an area of interest (ROI) (minimum and maximum thresholds). Morphological erosion and dilatation are then used to further hone the areas. Frames may be broken down into moments to better analyze movement and object density within a movie. Algorithms' efficacy was measured by contrasting their final outputs with those produced by the ground truth method (manual annotation). It was possible to get a sensitivity of 98.91% and an accuracy of 99.04%.

A.P Chakraborty et al. [2] proposed that the growth and development of plants are affected by environmental elements such as precipitation, temperature, and the location of soil conditions [10, 12], and the soil qualities such as soil type, pH value, and nutrient content, are taken into account when proposing produce the appropriate culture for the user. In addition, the farmer selects the appropriate crop and then they will also receive a yield prediction. Machine learning [13], a branch of AI that lets computers learn and improve themselves without human intervention, is the backbone of the proposed system in programming and has better accuracy without human intervention.

Narayani Patil et al. [3] state that a Deep Convolutional Neural Network (CNN) model has been implemented for disease detection using algorithms like Sequential and VGG-16. It detects if crops are infected and identifies other specific diseases on a crop-by-crop basis. The highest achievable accuracy for VGG16 H is 97.53%. In a recommendation system on plant, the dataset is run through a content-based filtering algorithm that has been trained, and many parameters are employed to select plants. A model trained on the acquired dataset delivers crop suggestions depending on the location and growing season of the user.

Sonal Jain et al. [4] presented a unique method for harvesting crops depending on anticipated weather and soil conditions. RNNs are used for seasonal weather forecast. In addition, the outcomes of the predictions are compared to those of the standard ANN, demonstrating greater accuracy. Using a random forest classifier which have an added threshold in the parameter, many crops appropriate to a country are retrieved.

In addition, the approach suggests the ideal planting window for each crop depending on projected weather conditions. Taking into consideration soil parameters, such as the N-P-K ratio and soil temperature, may increase the accuracy of the plant selection technique.

Navod Neranjan Thilakarathne, et al. [5] proposed a new farmers may get help deciding which crops to choose for harvest by using a machine learning-powered cloud-based crop suggestion application. In order to construct a cloud-based recommendation tool, they also analyzed five unique Machine Learning algorithms. It aims to provide a free, open-source precision agriculture solution to contribute to the sector's long-term growth and adoption of precision agriculture for long-term farming [18] solutions. Bock et al., (2011) [15] were the designers of a new methodology for the "Detection and Measurement of Plant Disease Symptoms Using visible wavelength photography and image analysis". This was done using thresholding technique.

Akshata Wani, et al. [6] a well-known way for machines to learn Random Forest uses both classification and regression applications of supervised learning. The IoT [19, 20] business also uses sensors like digital temperature and humidity sensors, light-dependent resistors, and rain sensors modules, and photovoltaic panels are used as hardware and IoT device is made. The accuracy of the system is also improved by comparing several machine learning techniques.

Singh, G., Sharma, et al. [7] presented the use of machine learning methods to improve irrigation water [16] use by forecasting soil moisture using an irrigation frame-work in IoT-driven. It utilizes field data acquired by deployed sensors (temperature, humidity, soil moisture, soil temperature, and solar radiation) and Internet-sourced weather prediction data to estimate moisture content. When assessing several ML [19] approaches, Gradient Boosted Regression Trees (GBRT) may be employed to attain the required outcomes. Comparing the R-squared values and mean squared errors of several machine learning methods reveals that GBRT is the most precise technique.

Tianjun Wu, et al. [8] using convolutional neural network-based learning approaches to map soil properties derived from high-resolution remote sensing photos utilizing geographical objects as the fundamental units. Many sources of geographical data are given as environmental variables for each spatial item, and strong tree-based machine learning methods are used to determine the relationships between soil quality and environmental parameters. The effectiveness of the recommended technique is assessed using a dataset including soil sample locations and geographic data from many sources.

Vijayan Tiruvengadam et al. [9] IoT infrastructures enable the generation of data for broad and remote agricultural areas, which may be used by this machine learning technique for crop selection. The recommendations are dependent on nitrogen, phosphorus, potassium, pH, temperature, relative humidity, and precipitation. These traits define the culture selected. The machine learning methods used by WEKA give the optimum model. The major purpose of this case study is to create a prediction model for high-yield crops and precision agriculture. The suggested system's model contains the required measures for developing technologies, the Internet of Things, and agriculture.

Reshma et al. [11] It was proposed that soil type, that groundwater table, that local population, that daily and seasonal needs of the people, that farm labor force, that same 15 plantations, that same quantity of arable land, etc. as a factor this is considerable. On the basis of these criteria, a study is conducted to determine whether crops [14] are suited for a given type of soil. Make use of decision trees and SVMs [17] in data analysis. Web applications are made to display this data in response to user interaction. Soil moisture sensors, soil nutrient sensor (NPK) probes, pH, humidity, temperature sensors, and microcontrollers, microprocessors with Wi-Fi and cloud storage are all examples of the types of sensors employed by IoT systems. When the sensor is installed, relevant characteristics are measured and the live, time-stamped data is sent to a cloud server.

3 Proposed System

Even though conventional techniques like crop rotation with cover crops and soil amendment with the organic matter have increased agricultural yields in this industry, they fall short of fully utilizing the domain knowledge that is already available. The ability to estimate soil fertility does not guarantee us a higher yield, but it does show us how much the soil has improved thanks to our adopted practices. The forecast of soil fertility will aid in condensing the challenges faced by farmers and serve as a platform for informing agriculturalists of the effective evidence needed to obtain a greater output. This problem can be solved using analyzing soil making changes according to the results. The proposed approach is utilized to calculate the various soil parameters, including moisture, temperature, humidity, pH level, and micronutrient levels. Accurately calculating each soil's individual qualities is difficult. We can identify the sort of soil that is ideal for growing the crop in this project by using the SVM algorithm.

In general, there are many different types of soil available. In our area, the most common forms of soil include alluvial, red, black, laterite, silt, sand, and gravel. Each type of soil has unique qualities. But to measure the characteristics of soil, we need a tool or instrument that will use the Internet of Things to provide accurate readings.

The objectives of the proposed system are:

- Image-based soil type classification by SVM.
- Design and development of a microcontroller-based interface for soil parameter reading,
- Record device results.
- Based on knowledge of soil types and soil properties, crops suitable for production are recommended.

4 System Architecture

The system architecture represented in Fig. 1 of our project provides a detailed description of the datasets, pre-processing, extraction, and training of data using a supervised learning algorithm for classification. The datasets we used in this study are open-source and from various sources. Image capture, preprocessing, segmentation, classification, analysis, and crop recommendation are all steps in soil analysis.

Classification, Analysis, and Recommendation are the three main processes that make up the architecture.

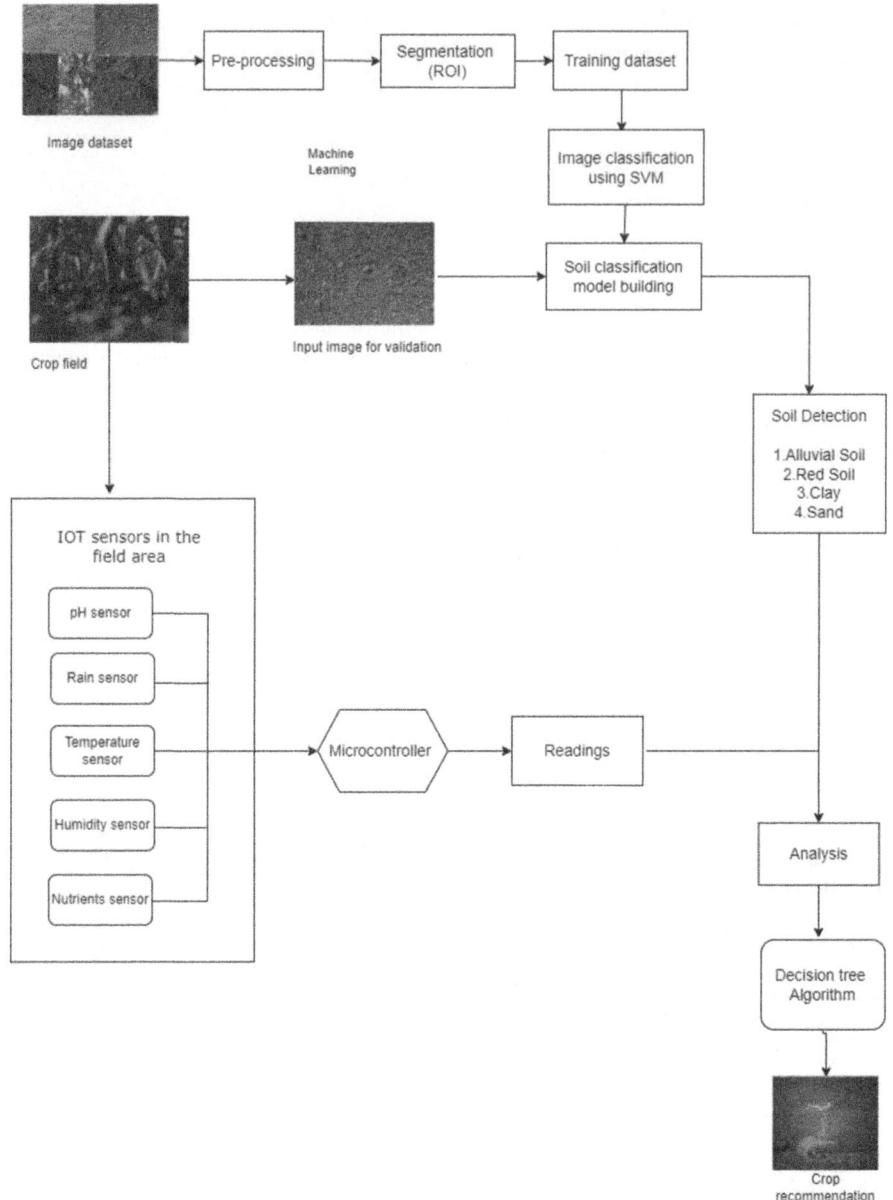

Fig. 1. Proposed System Architecture

Algorithm 1: Image Classification using SVM

INPUT: Soil Images, Label l

OUTPUT: Classified Image

1: Read the Soil Images

Preprocessing:

2: Gray Scaling of the Original Image

    2.1: I=rgb2gray(a)

3: Improve the image contrast

Segmentation:

For the preprocessed image:

4: Divide the image based on contours into small parts

5: Using ACM,

    5.1 Estimate the mean of the Energy function

        upts = find(phi<=0); //interior points

        vpts = find(phi>0); //exterior points

        u = sum(I(upts))/(length(upts)+eps); //interior mean

        v = sum(I(vpts))/(length(vpts)+eps); //exterior mean

    5.2 Minimizing the curves energy

        dphidt = F./max(abs(F)) + alpha*curvature;

    5.3 Evolve the curve boundaries

        phi(idx) = phi(idx) + dt. *dphidt;

End for

Classification:

For the segmented image:

    6: For the training set S,

    6.1 : Train the SVM Classifier

    6.2 : Mark the hyperplane

    6.3 : Return the Classified Image

End for

7: Return Soil Type based on features

Analysis:

For each type of sensor

    8: Record the readings for each sensor

    (Rain, pH, moisture, NPK, temperature)

    9: Display the readings

End for

Algorithm 2: Crop Recommendation using Decision Tree Algorithm

INPUT: Crop recommendation dataset

OUTPUT: Suitable crop for the soil type

Begin

    1: Read the dataset (CSV file)

    2: Retrieve the soil type from the classification.

    3: With the readings from the device,

    For each sensor

      Compare the readings accordingly

    End for

    4: Based on the range of values

    a. Using the Decision tree Algorithm,

    b. Plot the root and leaf nodes

    c. Mark the root node as Rain and other readings will be leaf nodes

    d. With the compared data, recommend the crop according to the soil type.

End

Image processing with the SVM method is used for this. In the analysis step, readings of qualities like pH level, rainfall content, humidity, nutrient content, and temperature are kept track of. Based on the readings and the classification of the soil, crop recommendations are made, which aids the process.

5 Working

In the first step, Machine learning is used to classify images using Supervised Learning algorithms such as Support Vector Machine for image classification and Decision Tree Algorithm for Crop Recommendation. Also, the input for the crop recommendation i.e., soil parameters are calculated from the IOT device (Fig. 3). All the sensors which read the various soil parameters such as pH, moisture, humidity, temperature and nutrient level are embedded in the device.

The type of soil is displayed as shown in Fig. 1, with the use of ACM that helps in the segmentation process of the acquired image and SVM helps in the prediction of the type of soil. All the process and steps with detailed explanation is as follows:

Support vector machine (SVM) is considered a possibility because to its superior generalization performance independent of a priori information, especially in high-dimensional input fields. A SVM may create an N-dimensional hyperplane for classification purposes. The term support vector machines are used to describe a group of algorithms that aim to optimize a mathematical function for a specific set of data.

Separating hyperplane, maximum margin hyperplane, soft margin, and kernel function are the four pillars upon which SVM rests.

Separating Hyperplane:
In two dimensions, a plane can split space in half, but in three dimensions, no other shape will do. The separating hyperplane is the line that divides the different dataset samples.

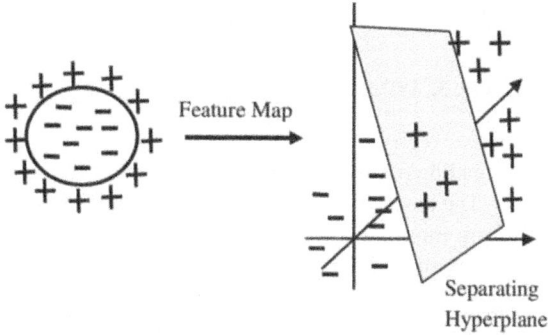

Fig. 2. Hyperplane Separation

A. Hyperplane of Maximum Margin.

To locate the hyperplane with the maximum possible margin, a line through the center must be chosen. In other words, one would choose a line that divides the two groups while maintaining the greatest feasible distance from each of the individual expression profiles (Fig. 2).

B. Margin Soft.

It is possible that SVM can handle data inaccuracies by placing incorrect expression profiles on the "wrong side" of the separation hyperplane. Figure 2 demonstrates the necessary modifications to the SVM algorithm to include a "soft margin" for handling such circumstances. This means that some information may fall on either side of the boundary between the two categories without significantly altering the final tally. Both the B1 and B2 hyperplanes are labelled in Fig. 2. As B1's margin is greater than that of B2, it offers more potential advantages. Vapnik's idea, which he put out there, suppose we have N data sets to train on, $\{(xi, yi)\}$ Ni = 1, with data input xi and labels in binary class yi ¼ $\{-1, +1\}$, the SVM classifier satisfies the conditions.

C. Function of Kernel.

The kernel function is the mathematic method by which the SVM may classify a set of data that was initially conceived as one-dimensional as two-dimensional. Data is often transformed from a lower-dimensional space to a higher-dimensional one using a kernel function.

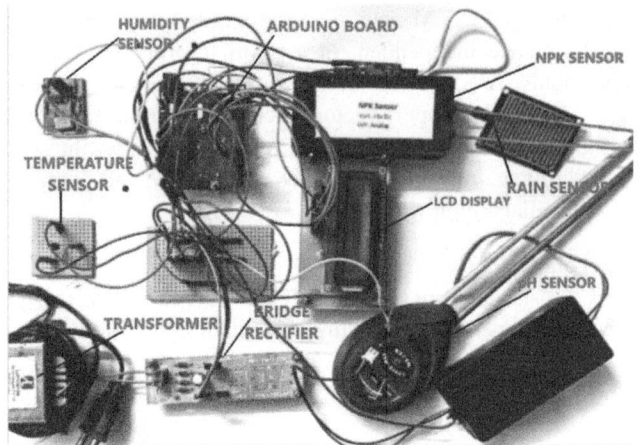

Fig. 3. IoT device for soil properties analysis

In the next step, the readings of soil properties are taken with the help of the IoT device as constructed in Fig. 5. The device has five sensors namely pH, NPK, temperature, rain, and humidity. Sensors are inter-connected to Arduino Uno with a bridge rectifier and the data are read in LCD display.

pH sensor – This sensor monitors the soil's salt content. It depicts the quantity of copper obtained from the soil. The amount of hydrogen ions in a solution, which can be measured with a pH metre, can show how acidic or alkaline the solution is. The pH metre keeps track of the difference in voltage between the electrode that shows the pH and the electrode that measures the pH. A combination electrode, a glass electrode, and a calomel reference electrode are all parts that are used often. In addition to measuring the pH of liquids, a superior probe can also measure the pH of semi-solids.

NPK sensor – It is used to determine the soil's fertility. Potassium, phosphorus, and nitrogen constitute a substantial component of soil fertiliser. Understanding the concentration of nutrients in the soil may provide information about the nutritional abundance or deficit of plant-growing soils. The sensor needs no chemical reagent. It can be used with almost any microcontroller because it has a fast response time, is easy to swap out, and gives accurate measurements. The long life of the probe is due to its high quality, resistance to rust and electrolysis, and resistance to corrosion from salt and alkali. So, it can be used on any kind of soil. Some types of soil that can be found are alkaline, acidic, substrate, seedling, and coconut bran.

Humidity sensor – Electronic devices known as humidity sensors are able to monitor, as well as measure and record, the relative humidity and air temperature in their immediate surroundings. Whether put in the air, soil, or restricted places, humidity sensors detect and report the surrounding environment's humidity and air temperature. Humidity measurements reveal the amount of water vapour in the air.

Temperature sensor – Tools called temperature sensors can read environmental conditions like heat and cold and may be found in a wide variety of household appliances and tools. The geotechnical monitoring sector is one of the many that may benefit from temperature sensors. A temperature sensor may also be thought of as a simple gadget that measures and converts the degree of cold or heat present. To detect the temperature of boreholes, soil, massive concrete dams, or structures, specialized temperature sensors are employed.

Rain sensor – This sensor detects the rain in the agriculture field & generates an alarm whenever there is rain so that we can take some proper actions to conserve the water as well as crops. Consequently, we can enhance the underground water levels through a recharge technique used underwater. The sensor detects the rain and gives alert to concerned persons in different fields like irrigation, automobile communication, home automation, etc.

Arduino Uno – Arduino is a system that includes both a microcontroller (a pro-grammable circuit board) and the IDE (Integrated Development Environment) software needed to write programmes for the board. So, basically it acts as a microcontroller and its functions include controlling, sequencing, monitoring, and displaying data. An analog-to-digital converter (ADC) is included inside the controller. It takes information from the user and displays it on an LCD screen.

6 Methodology

The work is composed of three parts:

In the Classification part, the datasets of diverse soil types, including alluvial soil, red soil, clay soil, gravel, and sand are gathered from a variety of sources and are trained. GUI design is used in MATLAB for an efficient display of the classification process. Pre-processing of the picture for several features, such as taking into consideration the present image's HSV, autocorrelation, density, energy, etc., and enhanced contrast of the image is displayed as shown in Fig. 4.

In order to segment the picture and explicitly remove the boundaries and process only the soil portion, the Active Contour Model (ACM) or the snake's model is established. They are computer-generated curves that move within images to find the soil boundaries by cutting out the unwanted objects in the picture. The picture is then categorized using SVM a supervised approach for learning that splits the hyperplane into upper bound and lower bounds and using certain values the kind of image is determined. With 500 iterations, the accuracy between the taught and tested images is attained. The output display of the image classification is shown in Fig. 4.

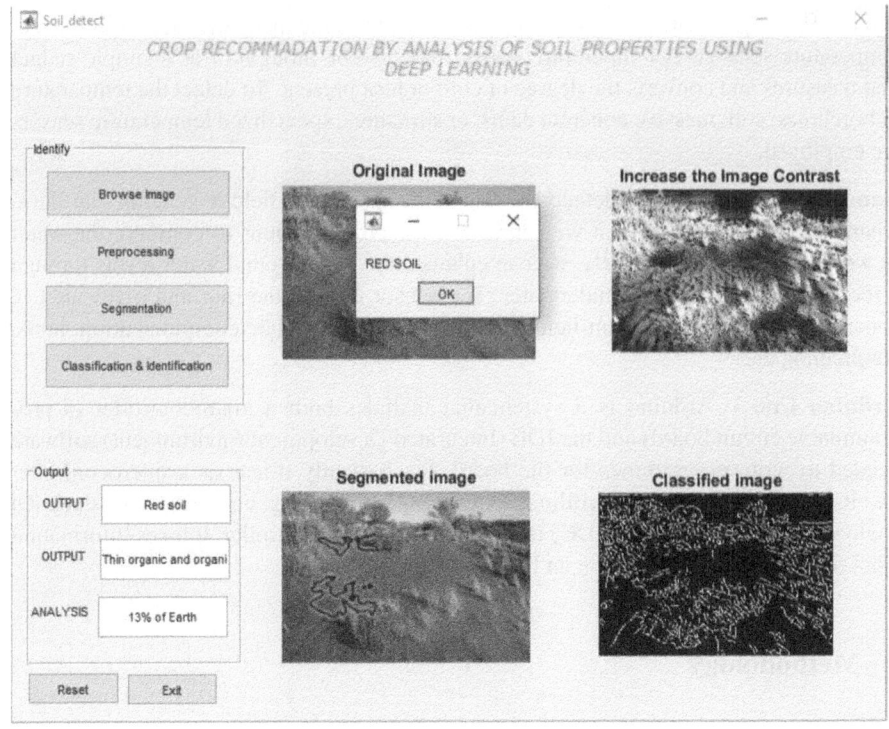

Fig. 4. Image Classification using SVM

In the Analysis part, the characteristics of the soil that are crucial for production are examined. The device is shown in Fig. 3 comprises 5 sensors that monitor temperature, humidity, nutrient level, rainfall, and soil pH. An Arduino Uno board contains the sensors, and the readings are shown on the LCD screen. The software is loaded into Arduino's programmable IC, and the sensors are identified by the ports to which they are connected. On the controller's left side are the analog sensors, while on the right is the digital LCD display. The transformer takes the AC current coming out of the outlet and changes it into a current (between 7 and 12 V) that the board can handle. In order to uniformly distribute the current supply across the whole device, the bridge rectifier converts the incoming AC current to DC using four diodes, a capacitor, and a transistor.

In the Recommendation part, the input is provided together with the soil type and readings gathered from the IoT device (Fig. 6).

For each of the sensor readings as well as the kind of soil, predefined live datasets comprising about 18,412 possibilities are gathered for 5 types of crops such as tomato, onion, rice, coffee and potato. (Table 1) These datasets are carefully trained and given labels in accordance with their contents.

```
ph:  0.00
temp:  28.42
rain:  339.00
humidity:  11.00
N:       0
p:       0
k:       0
ph:  0.18
temp:  28.42
rain:  340.00
humidity:  12.00
N:       0
p:       0
k:       0
ph:  0.16
temp:  28.42
rain:  337.00
humidity:  12.00
```

Fig. 5. Sensors value from IoT device

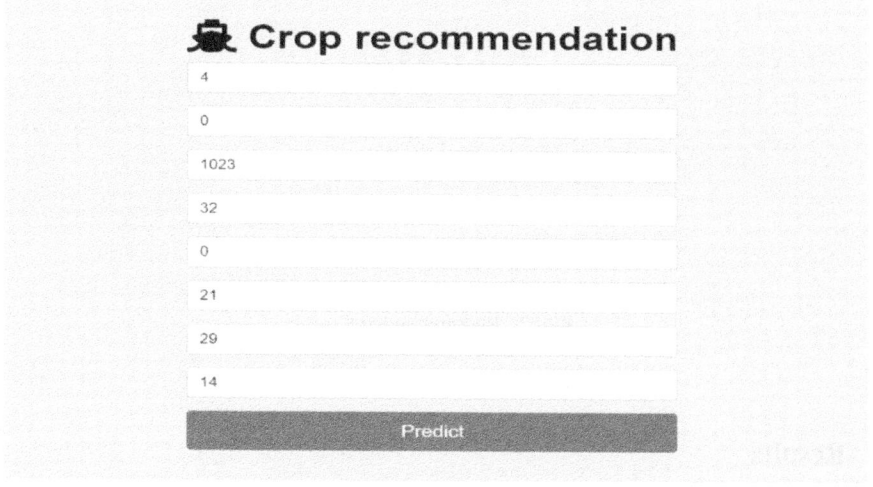

Fig. 6. Crop Recommendation GUI

We employ labelled datasets, thus we use the Decision Tree algorithm, for the recommendation phase. Here, user-defined constraints are employed where the readings of the Rain sensor are defined as root nodes as the value varies in large numbers compared to the other values in the analysis of soil and decisions can be performed easily. With the root node and other recordings along with the soil type as a leaf node, and each of

Table 1. Sample Dataset

CROP	RAIN	TEMP	pH	HUMI	N	P	K
TOMATO	1023	31.85	0	4	0	20	31
ONION	1023	31.36	6	4	25	19	35
RICE	1023	31.85	0.1	3	0	19	31
COFFEE	1023	30.38	0.02	2	0	29	20
POTATO	1023	31.36	0	5	0	19	25

the values of the properties of the soil, Crop recommendation is done accordingly and accuracy is calculated.

Thus, by feeding the inputs to the final system by specifically analysing the soil properties using an IoT device and the type of soil obtained from the image classification process, the crop is recommended as shown in Fig. 6 and Fig. 7.

Fig. 7. Recommended Crop

7 Results

The proposed methodology of the crop recommendation system includes the use of machine learning algorithms, specifically Support Vector Machine (SVM) and decision trees, to predict the most suitable crop for a given set of environmental and soil parameters.

To evaluate the performance of these algorithms, various metrics are examined. One such metric is accuracy, which measures the percentage of correct predictions made by the algorithm. Other metrics may include precision, recall, and F1 score, among others (Fig. 8).

```
DecisionTrees's Accuracy is:  99.97284822155851
                  precision    recall  f1-score   support

       COFFEE        1.00       1.00      1.00       1240
        ONION        1.00       1.00      1.00       1014
       POTATO        1.00       1.00      1.00        212
         RICE        1.00       1.00      1.00        399
       TOMATO        1.00       1.00      1.00        818

     accuracy                             1.00       3683
    macro avg        1.00       1.00      1.00       3683
 weighted avg        1.00       1.00      1.00       3683
```

Fig. 8. Accuracy Test

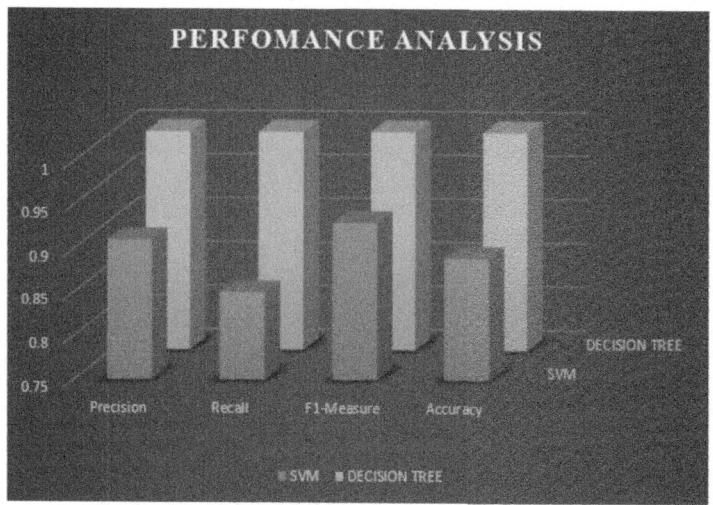

Fig. 9. Performance Analysis

From the graph (Fig. 9) it is depicted that Decision Tree is more reliable for the crop dataset when compared to SVM, the Accuracy of Decision Tree is said to be 99% which exceeds the existing system. In all performance criteria, decision trees yield superior prediction than SVM. Hence optimal system is built.

From the above Fig. 10, the given information compares the performance metrics of a proposed system to an existing system. The metrics used to evaluate the performance of the systems are precision, recall, F1 score, and accuracy. The proposed system has a precision of 0.91, which is higher than the existing system's precision of 0.83. This indicates that the proposed system is better at identifying positive instances.

From the Fig. 11, we contemplate that the performance metrics of a proposed system to an existing system. The metrics used to evaluate the performance of the systems are

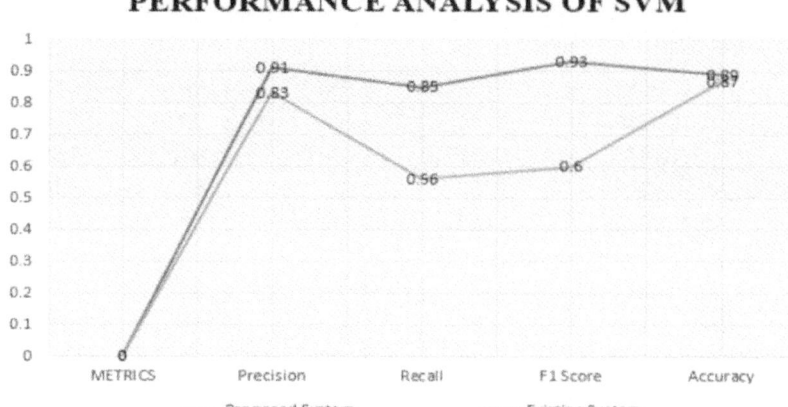

Fig. 10. Comparison graph of SVM over existing system

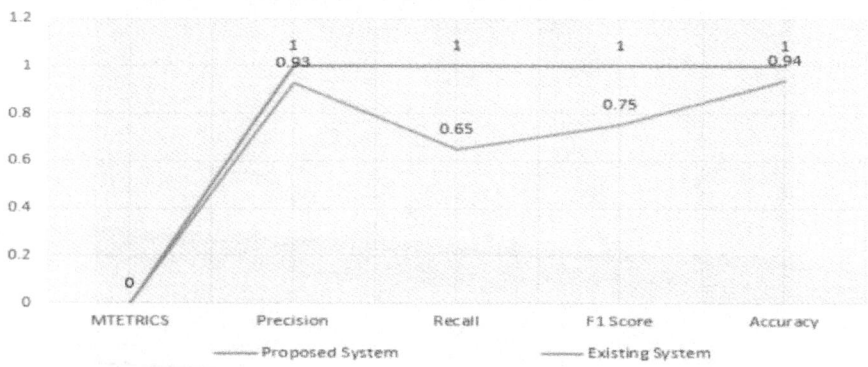

Fig. 11. Comparison graph for Decision Tree Algorithm

precision, recall, F1 score, and accuracy. The overall result depicts that the proposed system works better than the existing system. The proposed system has a precision of 1, which is higher than the existing system's precision of 0.93. Hence optimal system is built.

8 Conclusion and Future Works

Classification of soil is an expanding field of study in the modern period. Several research have offered various strategies for addressing the challenges, including rule-based, statistical, and conventional learning strategies. To define the soil type, however, requires significant time and effort. The system used the Supervised Learning Algorithm to construct a soil-type classification model with increased accuracy and little loss. With the suggested model, we can readily categorize the soil type, and determining classification results requires little time and effort. By using IoT, the system may be implemented in real time and intended outcomes can be achieved.

The proposed system can be enhanced by adding features to the analyzer, such as a webcam to capture the soil picture, which aids in determining the kind of soil. Furthermore, the prototype may be transformed into a mobile application, making it more helpful. Raspberry Pi supports the integration of high-quality sensors, which improves the effectiveness of the equipment. The proposal to improve soil fertility for farmers can also be improved by employing an appropriate IoT device.

References

1. Hamuda, E., Mc Ginley, B., Glavin, M., Jones, E.: Automatic crop detection under field conditions. National University of Ireland (2017)
2. Priyadharshini, A., Chakraborty, S., Kumar, A., Pooniwala, O.R.: Intelligent crop recommendation system using machine learning. In: 5th International Conference on Computing Methodologies and Communication (2021)
3. Patil, N., Kelkar, S., Ranawat, M., Vijayalakshmi, M.: Krushi sahyog: plant disease identification and crop recommendation using artificial intelligence. In: 2nd International Conference for Emerging Technology (2021)
4. Jain, S., Ramesh, D.: Machine Learning convergence for weather-based crop selection. In: IEEE International Students' Conference on Electrical, Electronics and Computer Science (2020)
5. Thilakarathne, N.N., Bakar, M.S.A., Abas, P.E., Yassin, H.: A cloud enabled crop recommendation platform for machine learning-driven precision farming. Faculty of Integrated Technologies, Universiti Brunei Darussalam (2022)
6. Wani, A., Yadav, V., Todankar, R., Wade, P., Malik, S.: Crop prediction using IoT & machine learning algorithm. Int. Res. J. Eng. Technol. (2022)
7. Singh, G., Sharma, D., Goap, A., Sehgal, S., Shukla, A.K.: Machine learning based soil moisture prediction for internet of things based smart irrigation system. In: 5th International Conference on Signal Processing, Computing and Control (2019)
8. Wu, T., Luo, J., Dong, W., Sun, Y., Xia, L., Zhang, X.: Geo-object-based soil organic matter mapping using machine learning algorithms with multi-source geo-spatial data. IEEE J. Sel. Top. Appl. Earth Obs. Remote Sens. (2019)
9. Bakthavatchalam, K., et al.: IoT Framework for Measurement and Precision Agriculture: Predicting the Crop Using Machine Learning Algorithms (2022)
10. Ren, Y., Hen, R.-Y., He, Q., Wang, J.: Soil temperature and humidity monitoring system design for farm land based on ZigBee communication technology. DEStech Trans. Environ. Energy Earth Sci. (2019)
11. Reshma, R., Sathiyavathi, V., Sindhu, T., Selvakumar, K., Sai Ramesh, L.: IoT based classification techniques for soil content analysis and crop yield prediction. In: Fourth International Conference on I-SMAC (2020)
12. Anguraj, K., Thiyaneswaran, B., Megashree, G., Shri, J.P., Navya, S., Jayanthi, J.: Crop recommendation on analyzing soil using machine learning. Turk. J. Comput. Math. Educ. **12**(6), 1784–1791 (2021)
13. Motwani, A., Patil, P., Nagaria, V., Verma, S., Ghane, S.: Soil analysis and crop recommendation using machine learning. In: International Conference for Advancement in Technology (2022)

14. Ayyasamy, S., Eswaran, S., Manikandan, B., Solomon, S.M., Kumar, S.N.: IoT based agri soil maintenance through micro-nutrients and protection of crops from excess water. In: Fourth International Conference on Computing Methodologies and Communication (2020)

15. Sharma, P., Padole, D.V.: Design and implementation soil analyser using IoT. In: International Conference on Innovations in Information, Embedded and Communication Systems (2017)

16. Brouwer, C., Heibloem, M.: Irrigation Water Management: Irrigation Water Needs, manual 6 Reading, ITALY: Food and Agriculture Organization of the United Nations (2011)

17. Ghent, J., McDonald, J.: Facial expression classification using a one against-all support vector machine. In: Proceedings of the Irish Machine Vision and Image Processing Conference (2005)

18. Thilakarathne, N.N., et al.: A cloud enabled crop recommendation platform for machine learning-driven precision farming. Sensors 22(16), 6299 (2022)

19. Wani, A., et al.: Crop Prediction using IoT & Machine Learning Algorithm (2022)

20. Jaspin, K., et al.: Real-time surveillance for identification of fruits ripening stages and vegetables maturation stages with infection detection. In: 2021 6th International Conference on Signal Processing, Computing and Control (ISPCC). IEEE (2021)

Analysis of Social Engineering Security Challenges and Its Emerging Technologies

P. Vidyasri[✉] and S. Suresh

Department of Computer Science and Engineering, SRM Institute of Science and Technology,
Ramapuram Campus, Chennai, India
vidyasrisankar@gmail.com, sureshs12@srmist.edu.in

Abstract. Information Technology is advancing at a record rate in the twenty-first century. It has grown to be an integral component of our daily lives. As the globe develops and more people use ubiquitous computing, cyber-security risks and privacy issues also grow proportionately. Cyber security is crucial to the information technology industry. Cyber-crimes, which are on the proliferate daily, are the first thing that transpire to mind whenever we think about cyber security. In spite of several safety measures, many people are still utterly concerned about cyber security. According to studies, social engineering attacks, which account for a sizable portion of all cyberattacks, are growing more common. By manipulating consumers' minds, social engineers might trick them into revealing sensitive information or compromising their security. One of the best ways to defend against social engineering attacks is to be able to identify them. Phishing is a common method of social engineering attack that is used to start more than 90% of successful hacks and data breaches. Cyber-attack detection and prevention are challenging tasks. This paper provides an extensively analysis of social engineering attacks, including their categories and a pop-up phishing classification system that we developed uses one of the best-known classifiers Support Vector Machine of Supervised Learning Paradigm in Machine Learning and Wrapper-based feature selections that incorporates the metadata of URLs are used extracts elements intended to defeat popular phishing detection techniques in order to provide a security model and to make predictions. Furthermore, this study offers an overview of the problems with phishing attacks now as well as recommendations for future research in this area.

Keywords: Cyber-crime · Cyber Security · Social Engineering · Phishing · SVM · Wrapper-based features

1 Introduction

Massive web users are getting impacted by the cyber threats in recent days. Individuals frequently invest more time on web for facilities like bank transactions, communication, and business, scholastic and business development. Sites like Facebook, tweeter, you tube are habitually utilized by the clients for some reasons. The much of the time got to sites have the record of impressions of the clients. The threatening alerts, diverts the client consideration towards malevolent and unapproved sites without any authentication. [1] The programmers and spammers today, yield an alluring snare for clients to go into the malevolent sites even through a solitary snap. Frequently, a client's visit to the pernicious sites, hence there is higher possibilities of weak applications go into the client framework premises. Phishing sites make various snares for clients, regularly getting to the informal organization sites [2].

1.1 Types of Phishing

The essential component, where the phishing begins from is through constant messages, correspondence through messages, web-based entertainment and other electronic means. A phisher might be depicted as a public asset, for example, long range interpersonal communication sites, diversion giving connections, sites and so on these sources snatch the client data, for example, work, name, email id as well as client interest and exercises log. [3] This data are additionally handled by the programmer in the backend to make counterfeit correspondence through messages, contact number and so forth. The specialty of phishing sites is it seems to be like the first data that an organization or believed association could give. The effect of phishing began from making trust between the client and the email source. The vast majority of the business correspondence messages and occupation affirmation messages in these days are sent with phishing tests [4].

Email phishing is made through trustable sites or association joins created by the programmer or spammer. The objective of the system is to get the client entrust and make a correspondence with them. Through believed data and at least one correspondence subsequent meet-ups the programmer snatches the client data, at times qualification information without affirmed information on the client. The believed correspondences empowers the client to give every one of the information in spite of getting specific advantage. These sort of email-based phishing assaults are all the more habitually occurring in current days. Then again email phishing assaults are sent through joins. By tapping the specific connection partook via the post office, the client become the survivor of losing the control of current organization and caught by the programmer. Inside specific edge of correspondence window, the client certifications are removed quickly by the hacker [5].

Stick Phishing assault incorporates, are basic one, where the client lose cash for the phony advantage presented by the spammer. By gathering the contact number of clients, partner's relatives and different companions, the spammer sends a phony connection on benefits. Whaling assaults were like phishing assaults in which the procedures frequently happen in senior representatives, when normally associated with public area. The aggressors create the data with an elevated degree of registering Technology. These aggressors don't have any significant bearing malevolent deceives and counterfeit connections. Rather they influence customized data in view of the exploration and casualty. [2] Whaling assaults or usually founded on expense forms, revelation of delicate information, research uncovering and so forth.

One more sort of phishing attack is called fake SMS messages that are associated with telephone discussions. In this attack scenario, the programmer misled the Master card subtleties from the organization or bank and deceived the client account. The total subtleties on exchange and certification data and with investors are hacked by the tricksters. Now and again these sorts of phishing assaults can be involved by settling on the decision and by utilizing the telephone keypad

The methodology is to phishing messages is erratic. Mostly these messages are sent indiscriminately to enormous quantities of beneficiaries and depend on the sheer weight of numbers for progress. The more messages are sending by the phishing assailant, the more probable they will find a casualty who will open and tap on their conveyed pernicious connection or connection. There are various sorts of phishing assaults accessible each sort, the aggressor follows the secret to assemble delicate record or other login data from a web-based source. Since the upside of phishing is intended to be that a great many people do online business. Skewer phishing is an assault focusing towards explicit associations, people or business to attempt to get their login certifications, this kind of cybercriminals prior to beginning the assault first assembles data may likewise goal to introduce malware on the designated beneficiaries.

Clone phishing is a kind of phishing assault where the email address is utilized to send misrepresentation to seem to come from a confided in shipper. Then, at that point, the email connection or connection is supplanted with the clone of the first site that the shipper is mimicking. When the cloned site is made, it will permit the aggressor to enter their login certifications. Chief misrepresentation is a sort of trick. It is a profoundly focused on from of lance phishing in which a CEO or another significant level leader is normally designated to fool representatives or others into giving them secret data or cash. The casualties are reached through email, telephone, or online entertainment, and are told to utilize counterfeit sites or different techniques to cause their trick to seem authentic.

BEC is a particular kind of digital assault where the aggressor utilizes email to fool representatives into imitating an organization chief or undeniable level worker with the expectation of cheating or extricating delicate information, then, at that point, moving cash or delicate organization data to them.

Spear phishing is a kind of phishing in which bogus emails are mailed to specific businesses with the aim of obtaining credential data [6]. A successful phishing campaign necessitates the use of a few various resources as well as some setup. Inducement, Impersonation, and access- control bypass techniques are among its approaches [9].

1.2 Identification of Attacks

In order to prevent the phishing assaults in email services, system need to be more perspective of selection techniques in which they use to grab user's authorized information. Some of the suggested techniques are listed below.

Collection of Personal Credentials: This technique utilizes personal information like credit card number or account passwords. They may provide some secure link, but it redirects to fake website and asking some more account information for verification. This technique will establish hazard through some phishing emails.

Priority Creation: These phishing emails will emerge us to take frequent step to avoid negativity and to claim different compromise regarding accounts. Spoof email addresses:

It imitates some legitimate companies like bank and other financial sectors. These phishing emails can use same logos and branding of well-founded organization. Links or attachments: These phishing emails will be similar to some legitimate sources but using different URL. It can steal your information by averting to infected websites.

- The proposed approach considers Meta data from website links. The dataset considered here is collected from Kaggle Phishing websites dataset.
- In spite of pre-process the link data, perform feature extraction through wrapper method of feature extraction. Based on the updated features from the WFE method, the classification of legitimate links and normal links are classified using support vector machine algorithm.
- Higher the correlation of features between the test website link, with training meta data, the attack scenario is being detected in the early stages.
- The proposed approach considers the most sensitive phishing attack called pop-up message-based phishing probes. The interpretation of the system is measured using accuracy, recall, precision, F1Score etc.

The rest of the paper is organized as detailed study of existing works in Section 2. The system tool selection is explained in Section 3. Further the proposed system design is discussed in Section 4. The obtained outputs are discussed in Section 5. Followed by the conclusion and future enhancement is provided.

2 Literature Survey

[7] M. N. Alam et al., (2020) In globalised circumstances, there is a development in digital transformation due to magnification of cyber security attacks. World getting more and more digitized hence it acts as a gateway to all other Cyber attackers. At first, these threats are familiar in phishing which can seize all the secret information of user. These can gain more access to corrupt all digital assets and networks. Along with that, phishing can also be easily influenced by hackers. Random forest (RF) and decision tree (DT) are combined form of machine learning (ML), algorithms. The main purpose of designing this model is to detect the phishing attacks. Kaggle have been considered as an acceptable dataset for ML processing. Principal component analysis (PCA) is one of the feature selection algorithms used in the proposed system. Its function is mainly to analyse the attributes of dataset. At last, we concluded that this random forest algorithm can achieve an accuracy of about 97%.

[8] Salahdine F et al., (2019) The new proposal involves machine learning for the detection of phishing attack. Over 4000 phishing email resources are collected and analysed from the University of North Dakota. Huge datasets are built to model these attacks and also it selects 10 features related to these attacks were established for modelling this network. When compared to previous studies, this particular machine learning algorithms undergo training, validating and testing process using these datasets. Four factors, including the likelihood of detection, the likelihood of miss-detection, the likelihood of a false alarm, and accuracy, can be predicted using this model to demonstrate the effectiveness of the detection methodology.

[10] C. Pham et al., [2018] Based on a planned neuro-fuzzy framework, the author provided a system with Web traffic features and uniform resource locator capabilities to identify phishing websites. We create an anti-phishing model based on the novel method of fog computing promoted by Cisco in order to transparently monitor and defend fog users from phishing assaults. The trial results of our proposed solution, which were based on a significant dataset collected from real phishing cases, have shown that our system can effectively prevent phishing attacks and enhance network security.

[11] L. R. Kalabarige et al., (2022) According to the experimental findings, the suggested model performed admirably when tested against various datasets, with an accuracy range of 96.79% to 98.90%. With data from UCI(D1), Mendeley 2018(D2), and Mendeley 2020, the suggested model is assessed (D3, D4). With the D1 dataset, the suggested model has a detection rate of 97.66% and a D2 dataset accuracy of 98.8%.

[12] C. Esposito et al., (2022) The author presented a system in which the trustworthiness of social media users is discussed. The system considers various samples of social media data, and classifies it into legitimate user and malicious user etc. In the current massive internet user's community, the entry of order malicious user is obvious. The presented approach considers Fuzzy logic system in order to classify the presence of authenticated user and non-authenticated user.

[13] A. El Aassal et al., (2020) The author presented a systematic benchmark evaluation study in which the imbalanced dataset and its robustness are validated. The system considers extensive feature extraction from the given dataset and further classifies the legitimate user and unauthorized user to identify the attackers, that create fake authentication.

[14] S. Eftimie et al., (2022) The study discusses the psychological aspects of phishing attacks that focus on getting the user personal data and leverage the multiple vulnerable activities. The author discussed binary logistic regression based cyber security problems. The phishing attacks are attracted by the users through psychological aspect since the email attacks are attracted by the user with different perspectives.

[15] W. Ali et al., (2020) detection of phishing websites, in order to provide early warning is important to secure the individual data from the massive cloud. Particle swarm optimization technique is implemented by considering various features of the website links. The proposed result on legitimate user and normal user is identified by the positive and negative rates of the classification model.

[16] Y. A. Alsariera et al., (2020) the author presented a comparative analysis of four different learning algorithms namely, AdaBoost algorithm, Bagging tree model, rotation forest, Logi-Boost model etc. Through detailed attributes derived from the websites' meta data, AI-based meta learners outperform human analysts at identifying phishing websites.

Considering various existing frameworks on machine learning for phishing offensive detection, the serious impacts in the social media networks and its drawbacks are required to get detected in the early stages. Prevention of such malicious attacks also important to secure the user data.

3 System Design

The increasing numbers of users in internet connected social community create various threatening problems in recent days. Not all the websites are developed with strong authentication protocols. The impact of cyber-attacks through these loop holes are obvious in current internet community. The proposed approach considers the sensitive issue on phishing attacks that grab the attention of users easily. Pop-up notification-based phishing probes are occurring frequently in many familiar websites. People spend lots of time in internet browsing, education, entertainment and business purposes. The Pop-up message enabled notification connect the user credential to the common window for long duration. In the backend, probing is induced by the malicious content coming over the meta data of the web site link. Using Python based machine learning algorithm and hybrid approach on wrapped feature extraction, the robust model is created. The protocol model is tested with Google Collab, to show the process of working model significantly.

3.1 Data Collection

The Phishing dataset is collected from publicly available website Kaggle.com, in which the recorded data are organized in a .CSV file. The dataset contains numerous website URLs around 11000+ website link with labelled data. The data is precisely organized as +1 as phishing website, −1 as not a phishing website. The meta data in the website link is having a pattern of information specifies the presence of phishing attacks.

4 Methodology

4.1 System Architecture

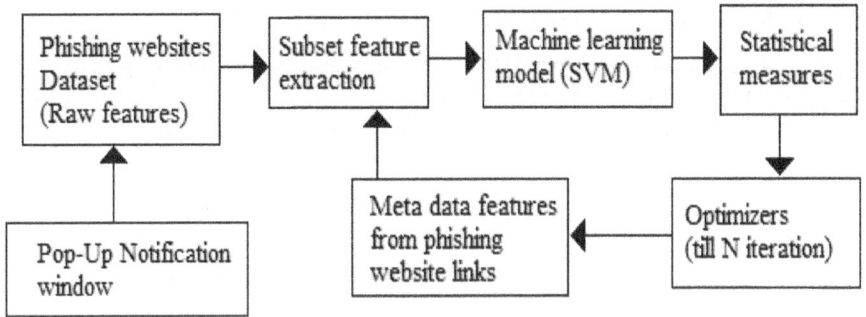

Fig. 1. Proposed Wrapper based feature extraction

Figure 1 Shows the Proposed wrapper-based feature extraction method incorporated with SVM model. The proposed approach considers the Pop-up notification message as a sensitive case on probing the user data in socially connected networks.

4.2 Wrapper Based Feature Extraction (WFE)

Wrapper based feature extraction is the unique method on fitting the data patterns with specific machine learning algorithm. Here support vector machine is considered for iterative fitting process. The wrapper-based approach act as an optimized model for deriving the existing dataset into featured dataset through iterative fitting and updating process. The process act as a benchmark creation model on various existing datasets. The basic approach of the WFE technique is based on greedy search algorithm. The model considers all possible feature combinations against the features for perfect fitting. WFE method also utilized for forward selection process, backward elimination process and Bi-directional elimination process etc. In python IDE, through Sklearnkit library model, the WFE is developed.

4.3 Support Vector Machine (SVM) Classifier

SVM is a class of compact analysis model in which limited memory space is occupied. Based on relatively low dependency data, SVM incorporates the correlation score between the training data and testing data. The initial process performs the learning process, once the fitting is completed; the prediction process is relatively fast and reliable. The large samples of data require analysis time significantly higher. The SVM with tuneable Kernel adopts the challenges faced by the large dataset. The SVM model consumes more computational cost and strongly relies on the kernel constants. Further WFE with SVM here hybrid as W-SVM provides probabilistic interpretation. Keeping these constraints as important challenges with existing SVM model, the Novel W-SVM is created here. The proposed W-SVM also achieves reduced computation time and ensures accurate data, tested with Phishing data detection.

4.4 Implementation Summary

The input dataset is pre-processed in spite of removal of unwanted junk values apart from meta-data. The website links are often processed with WFE model with repeated iterations to perform the fitting process with SVM. Whenever a pop-up notification occurs with the testing window, the URL is extracted for analysis. The link attacked with the Pop-up notification is extracted and pre-processed. The meta data present in the new URL link is analysed with the existing training data. For creating the Novel W-SVM model, the dataset is split up into 15% as testing data, 70% as training data and 15% as validation data. The model is initially tested and trained with the part of the training data. During the testing phase, the test data coming up from the Pop-Up notification is considered. In spite of maximum correlation persist with the training data, incorporating with relevant labels. The presence of phishing attack is classified. The presence of phishing or malicious context is labelled as class 1, where the normal URL is labelled as class −1. The performance of the system is measured using accuracy, precision, recall and F1Score.

4.5 Performance Measure

The proposed W-SVM technique analyse the dataset opted for training. The performance of the model created is analysed using parameters such as accuracy, precision, recall and F1score. The parameters are calculated by the expressions below.

Accuracy = TN + TP/TN + FP + FN + TP
Precision = TP/FP + TP
Recall = TP/FN + TP
F1Score = 2*[Recall* Precision/(Recall + Precision)]

5 Results and Discussions

5.1 Sample Input Data

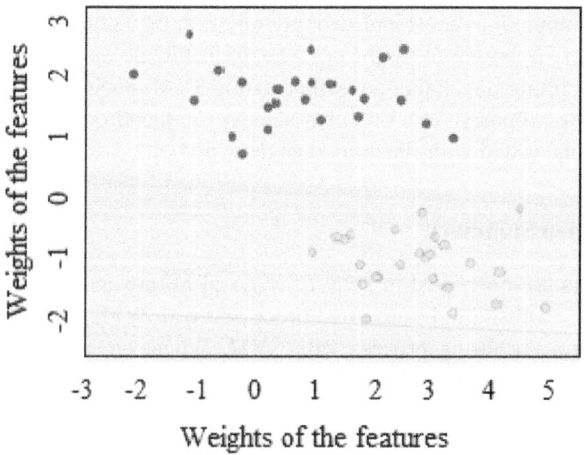

Fig. 2. Input raw dataset feature weights

Figure 2 Shows the linear analysis of features extracted from WFE in the beginning stage. Creation of model initiated with the separated data before fitting. After the iterative learning process, probabilistic analysis these data are further tuned.

Figure 3 Shows the W-SVM classification result after the N update into the kernels. The SVM process separate the data into left panel as class 1 and right panel as class −1. The fetched data undergoes higher dimensional probabilistic update until the data fit for the unbalanced classifier. The data is separated by the kernels using non-linear principle.

Orientation of feature weights at initial time *iter* = *n*

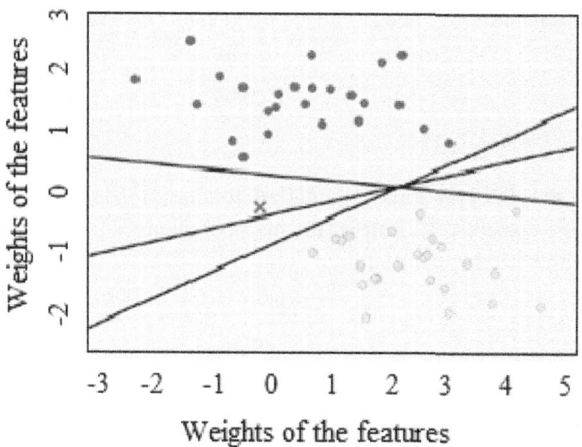

Fig. 3. W-SVM classification results

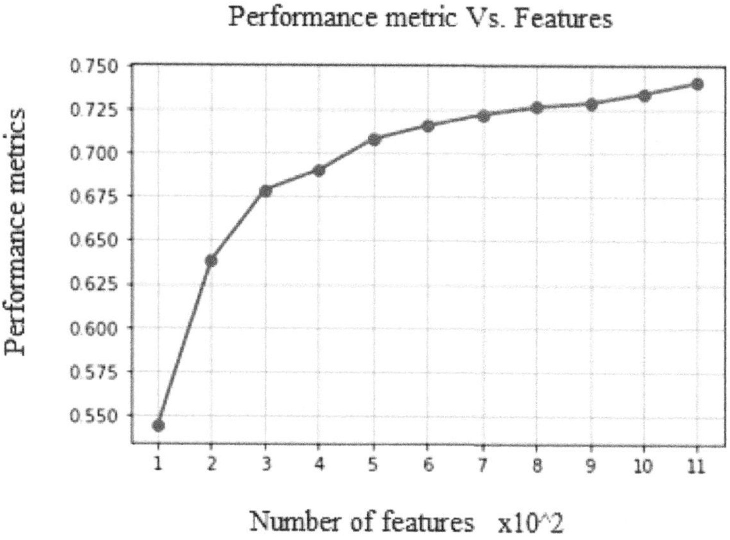

Performance metric Vs. Features

Fig. 4. WPE feature update with performance metric score

Figure 4 shows the performance metrics score of proposed W-SVM model with respect to number of features and performance metric score or correlation match score.

Table 1 Quantitative measure of the proposed W-SVM model with respect to accuracy, precision, recall, and F1Score. The classification of Phishing Meta data presence is determined by Class 1 and normal URL meta data is determined by Class −1. The proposed W-SVM model achieved 96% accuracy, Precision = 0.95 for Class −1, and

Table 1. Quantitative measures

Class	Precision	Recall	F1Score	Accuracy
−1	0.95	0.95	0.95	0.96
1	0.96	0.96	0.96	0.96

0.96 for Class 1. Recall is 0.95 for class −1 and 0.96 for class 1. F1score is 0.95 for class −1 and 0.96 for class 1 respectively. Further the major challenges faced by the proposed approach is by dealing with the un structured data.

6 Conclusion

Novel execution of phishing site discovery framework is created here. The current framework considers an adequate informational index from a particular site and considers the informational index URL for investigation. Here the element extraction includes, URL based wrapper-based feature extraction process, Meta data-based feature extraction is done. The component separated information or split up into preparing information and testing information to make a model utilizing Wrapper feature based Support vector machine (W-SVM). The novel approach classifies the meta-URL coming from the pop-up notification into class 1 as phishing present and class −1 as phishing not present. The system achieved the accuracy of 96%, precision 0.95, recall 0.95, F1score 0.95 significantly. Further the current framework is further developed by growing profound learning calculations and move learning ways to deal with work on the accuracy, detailed feature extraction using deep learning is recommended.

Conflict of Interest. On behalf of all authors, the corresponding author states that there is no conflict of interest.

References

1. Sun, B., et al.: Leveraging machine learning techniques to identify deceptive decoy documents associated with targeted email attacks. IEEE Access **9**, 87962–87971 (2021). https://doi.org/10.1109/ACCESS.2021.3082000
2. Le Blond, S., Gilbert, C., Upadhyay, U., Rodriguez, M.G., Choffnes, D.: A broad view of the ecosystem of socially engineered exploit documents. In: Proceedings of Network and Distributed System Security Symposium, pp. 1–15 (2017)
3. Guo, Z., Cho, J.-H., Chen, I.-R., Sengupta, S., Hong, M., Mitra, T.: Online social deception and its countermeasures: a survey. IEEE Access **9**, 1770–1806 (2021). https://doi.org/10.1109/ACCESS.2020.3047337
4. Tankard, C.: Advanced persistent threats and how to monitor and deter them. Netw. Secur. **2011**(8), 16–19 (2011)
5. Nathezhtha, T., Sangeetha, D., Vaidehi, V.: WC-PAD: web crawling based phishing attack detection. In: 2019 International Carnahan Conference on Security Technology (ICCST), pp. 1–6 (2019). https://doi.org/10.1109/CCST.2019.8888416

6. Baig, M.S., Ahmed, F., Memon, A.M.: Spear-phishing campaigns: link vulnerability leads to phishing attacks, spear-phishing electronic/UAV communication-scam targeted. In: 2021 4th International Conference on Computing & Information Sciences (ICCIS), pp. 1–6 (2021). https://doi.org/10.1109/ICCIS54243.2021.9676394

7. Alam, M.N., Sarma, D., Lima, F.F., Saha, I., Hossain, S.: Phishing attacks detection using machine learning approach. In: 2020 Third International Conference on Smart Systems and Inventive Technology (ICSSIT), pp. 1173–1179 (2020). https://doi.org/10.1109/ICSSIT48917.2020.9214225

8. Salahdine, F., Kaabouch, N.: Social engineering attacks: a survey. Future Internet. 11(4), 89 (2019). https://doi.org/10.3390/fi11040089

9. Chen, S., Fan, L., Chen, C., Xue, M., Liu, Y., Xu, L.: GUI-squatting attack: automated generation of android phishing apps. IEEE Trans. Dependable Secure Comput. 18(6), 2551–2568 (2021). https://doi.org/10.1109/TDSC.2019.2956035

10. Pham, C., Nguyen, L.A.T., Tran, N.H., Huh, E.-N., Hong, C.S.: Phishing-aware: a neuro-fuzzy approach for anti-phishing on fog networks. IEEE Trans. Netw. Serv. Manage.Netw. Serv. Manage. 15(3), 1076–1089 (2018). https://doi.org/10.1109/TNSM.2018.2831197

11. Kalabarige, L.R., Rao, R.S., Abraham, A., Gabralla, L.A.: Multilayer stacked ensemble learning model to detect phishing websites. IEEE Access 10, 79543–79552 (2022). https://doi.org/10.1109/ACCESS.2022.3194672

12. Esposito, C., Moscato, V., Sperlí, G.: Trustworthiness assessment of users in social reviewing systems. IEEE Trans. Syst. Man Cybern. Syst. 52(1), 151–165 (2022). https://doi.org/10.1109/TSMC.2020.3049082

13. El Aassal, A., Baki, S., Das, A., Verma, R.M.: An in-depth benchmarking and evaluation of phishing detection research for security needs. IEEE Access 8, 22170–22192 (2020). https://doi.org/10.1109/ACCESS.2020.2969780

14. Eftimie, S., Moinescu, R., Răcuciu, C.: Spear-phishing susceptibility stemming from personality traits. IEEE Access 10, 73548–73561 (2022). https://doi.org/10.1109/ACCESS.2022.3190009

15. Ali, W., Malebary, S.: Particle swarm optimization-based feature weighting for improving intelligent phishing website detection. IEEE Access 8, 116766–116780 (2020). https://doi.org/10.1109/ACCESS.2020.3003569

16. Alsariera, Y.A., Adeyemo, V.E., Balogun, A.O., Alazzawi, A.K.: AI meta-learners and extra-trees algorithm for the detection of phishing websites. IEEE Access 8, 142532–142542 (2020). https://doi.org/10.1109/ACCESS.2020.3013699

Continuous Sign Language Recognition Using Holistic Key Points

Sudiksha Ghosh[(✉)], Aishwarya Mocherla[(✉)], and G. Sivashankar

Department of Computational Intelligence, College of Engineering and Technology, Kattankulathur, Chennai, India
sg6323@srmist.edu.in, aishwaryamocherla@outlook.com

Abstract. The International Federation of the Deaf estimates that there are more than 70 million deaf persons worldwide. Almost 80% of them reside in underdeveloped countries [1] While there are many platforms for translating our spoken languages, there aren't many for understanding sign language. This is an important issue because sign language is used to communicate with deaf individuals. To address this issue, we decided to develop a Continuous Sign Language Recognition Using Holistic Key Points. The system will typically consist of a camera or other input device for capturing sign language gestures, a feature extraction and classification module for identifying and interpreting the gestures, and a real-time display or output mechanism for presenting the interpreted sign language to the user. This system can be used to facilitate communication between individuals who use sign language and those who do not. A real-time sign language translation system can make it easier for deaf individuals to access information and communicate with others. A real-time sign language estimation system can be used in educational settings to help students who use sign language to participate more actively in class and understand lectures and other materials. By making it easier for individuals who use sign language to communicate with others, this system can help to promote social inclusion and reduce barriers to communication.

Keywords: Sign language detection · computer vision · artificial intelligence

1 Introduction

Communication is a very important aspect of a person's life. However, for the hearing-impaired, this happens to be the most challenging part. Sign language is not commonly taught as any other language in educational institutes, which leads to segregation among people. On a daily basis, most deaf people face issues with communication, ranging from difficulty in understanding things to being deprived of good quality education and other opportunities. This creates barriers in society, leading to the stunted professional and social development of these people.

In educational settings, deaf students face additional challenges. They may struggle to understand lectures and other materials, leading to a lower level of engagement in class. Furthermore, the lack of trained interpreters can limit their access to education and limit their opportunities for personal and academic development. Additionally in workspaces, well-deserving individuals tend to miss opportunities because of not being able to communicate effectively.

Unfortunately, it also leads to an inability to participate in meetings, group discussions and decision-making.

To address this pressing issue, we propose the development of a Continuous Sign Language Recognition Using Holistic Key Points. This system aims to improve communication between deaf individuals and the hearing population by providing a real-time translation of sign language into text. By making sign language more accessible and widely understood, we hope to promote social inclusion and reduce barriers to communication for deaf individuals.

A camera will be used as an input device to capture sign language gestures, a feature extraction and classification module will be used to identify and interpret the gestures, and a real-time display or output mechanism will be used to show the interpreted sign language to the user. In order to effectively estimate sign language in real-time, we will first collect data from videos, captured previously, depicting sign language gestures. This data will be stored for processing, to beused in our model. Next, we will train a deep neural network, utilising Long Short-Term Memory (LSTM) layers, to accurately predict the temporal components of sign language gestures. Finally, we will integrate all components using OpenCV to create a real-time sign language estimation system. This system will utilise the data and model we have developed to provide real-time interpretation and display of sign language gestures, enabling effective communication between individuals who use sign language and those who do not. Overall, our approach will allow us to create a robust, real-time sign language estimation system that can be used to improve communication and reduce barriers for individuals who use sign language.

2   Related Work

An active area of research in the field of computer vision has been Sign language recognition. There have been numerous efforts to develop systems for recognizing and translating sign language into spoken language. One early approach was the use of hidden Markov models (HMMs) for sign language recognition [2]. These systems modelled sign language gestures as sequences of positions and orientations of the hand and fingers and used HMMs to recognize and classify the gestures. Another approach for sign language recognition is using 3D depth cameras to capture sign language gestures. In a study by Mehrotra et al. (2015), a system was developed using a Kinect sensor to recognize Indian Sign Language (ISL) gestures. [3]. The system achieved an accuracy of 86.16% on the ISL dataset, demonstrating the effectiveness of using depth information for sign language recognition.

Another method detected ASL actions using convolutional neural networks (CNNs) [4]. A dataset of 27,000 images helped the researchers accomplish an astounding accuracy of 99.7%. However, this technique can only recognise individual static images rather than a continuous video may be a drawback. We also reviewed studies [5] that employed recurrent neural networks (RNNs) to recognise gestures in British Sign Language (BSL).

Using a dataset of 3000 hand motions, the researchers' accuracy was 98.3%. This method can recognise continuous sign language, but in order to achieve high accuracy, it needs a big dataset of hand gestures. Lastly, Deep Learning was employed in a study [6] to identify Indian Sign Language (ISL) motions. Using a collection of 10,000 images, the researchers' accuracy was 98.9%. While this method can recognise individual still images rather than a continuous video, it still has the same limitation as the earlier study.

Overall, the aforementioned works show the possibility of using computer vision and machine learning methods to identify sign language gestures. Our research seeks to build on these methods by creating a real-time sign language recognition system capable of recognising a broader variety of signs and gestures.

3 Proposed Architecture

The flowchart in Fig. 1 depicts our proposed system architecture. There are three main steps: data preprocessing, feature extraction, and model training. First, the sign language videos are preprocessed to extract key points using the mediapipe holistic approach. These key points represent the positions of the hands, arms, and body during sign language gestures. Next, these key points are used as features to train a Long Short-Term Memory (LSTM) model, which is a type of recurrent neural network (RNN) known for its ability to capture temporal dependencies. The LSTM model is trained on a labelled dataset of sign language videos with corresponding sign language labels. During the training process, the LSTM model learns to recognize the sequential patterns of sign language gestures from the extracted key points. Once the model is trained, it can be used to make real-time predictions on new sign language videos by inputting the extracted key points from these videos.

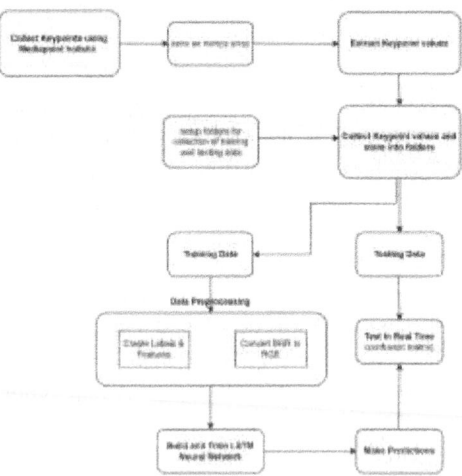

Fig. 1. The Proposed System Architecture

4 Methodology

Our research methodology entails implementing a deep learning model capable of accurately recognising sign language gestures in real time.

4.1 Establish Landmarks from Keypoint Values Using MediaPipe Holistic

The first step involves keypoint extraction using MediaPipe Holistic, a technique that identifies significant points in the data that are useful to our system. This step is essential as it allows us to extract key features from the input video data, which are then used by the machine learning model to make accurate predictions.

To begin, we ensure that we can access the webcam using OpenCV. We open the webcam, read the feed, and show the feed. This ensures that we can access the real-time video feed and use it as input for our model. Next, we take images from our feed and convert BGR to RGB for colour conversion. We then use the converted images to make detections using MP Holistic, which accesses the Mediapipe model. This allows the model to view various landmarks, such as those on the face and hand. However, it is important to note that the hand needs to be in the frame for this to work accurately.

Once we have made the detections, we draw these landmarks and connect them, such as the right shoulder to the right elbow. We use Matplotlib to visualise these landmarks in the real-time video feed, allowing us to see the detected features and confirm that the system is working as expected. Overall, this step is crucial for accurately extracting key features from the input video data and preparing it for use by the machine learning model.

4.2 Collect Keypoint Values for Training and Testing

In the next step, we focus on extracting the keypoint values associated with the landmarks obtained in the previous step. Here, we concatenate the values of the landmarks obtained in the previous step into numpy arrays. These arrays consisted of the x-coordinate, y-coordinate, and visibility of each landmark. To ensure that our model was not subject to errors during the processing of keypoints, we flattened and formatted the arrays to ensure ease of comprehension. We also anticipated potential errors that may arise when parts of the body were not detected in the frame of the camera. To address this issue, we replaced the array for those landmarks with an empty array. For instance, if the right hand was not detected in the frame, we replaced the corresponding array with an empty array to prevent any errors from occurring in the program. We applied this process for all sets of landmarks including right and left hand, pose, and face landmarks. Finally, we concatenated all the extracted keypoints from hands, pose, and face to form the final dataset that was used for training and testing our model (Fig. 2).

4.3 Build and Train LSTM Neural Network

We decided to capture the data by constantly looping the video from the webcam feed. To capture the data, we used a keypoint pattern recog as it is more resilient and better at

Fig. 2. Collecting Videos for Dataset

detecting actions with different backgrounds. Once the data was collected, we preprocessed it and created labels and features to be fed into the model for training. This step involved normalising the data and converting the keypoint values into sequences.

For model building and training, we used a Long Short-Term Memory (LSTM) neural network, which is well-suited for processing sequential data like our keypoint values. LSTMs have a unique architecture that allows them to capture dependencies and patterns across time, making them an effective choice for our project. We trained the model using the preprocessed data and labels, adjusting parameters as needed to improve performance (Figs. 3 and 4).

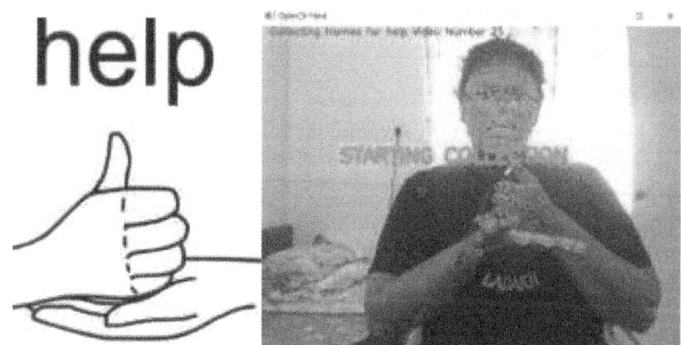

Fig. 3. Collecting sample video for 'help' in ASL

4.4   Make Predictions and Save Weights

After training the LSTM neural network, the next step was to use it to make predictions on new test data. To achieve this, the test data was passed through the trained mode. The predicted results were then stored to be used later. To verify the accuracy of the model's predictions, the predicted results were compared with the true test data. The

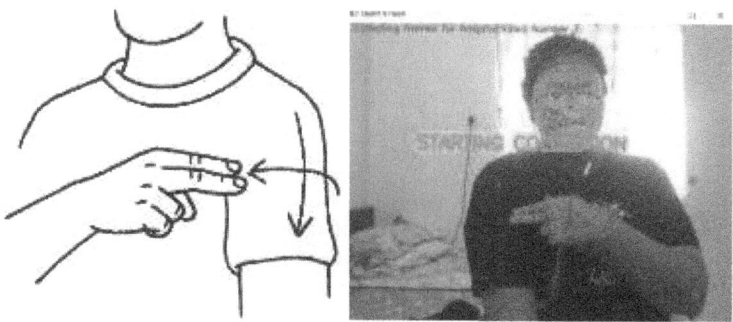

Fig. 4. Collecting sample video for 'hospital' in ASL

model's weights were then saved to ensure that the trained model could be used for future predictions without needing to retrain it.

4.5 Evaluation Using Confusion Matrix

The evaluation process was conducted by making additional predictions on the test set using the trained model and then comparing these predictions to the true labels using a multi-label confusion matrix. The confusion matrix allows for the evaluation of true positives, true negatives, false positives, and false negatives for each label, thus providing insight into the performance of the model.

The model's accuracy was then determined using scikit-learn's accuracy_score function. A high accuracy score indicates that the model has performed well on the test data. The evaluation was then repeated on the training data to gain a broader perspective on the performance of the model. The model's performance was found to be reasonably well. Overall, the evaluation process provided valuable insight into the performance of the model and helped to validate its accuracy.

5   Results and Discussion

Our system achieved a remarkable accuracy. We evaluated our system using a confusion matrix. This matrix displays the total number of true and false positives and negatives and their predictions for each class. In our evaluation, we used a testing dataset of 49860 sign language videos for each sign, each 30 seconds long.

The LSTM model trained on the extracted key points achieved excellent accuracy in recognizing sign language gestures in real time. The holistic keypoint approach captures the key features of sign language gestures in a comprehensive

and accurate manner. Unlike other methods that rely on image or pixel-based processing, the holistic keypoint approach focuses on tracking and analysing the motion and orientation of key body parts, such as the hands, arms, and torso (Fig. 5).

Another advantage of the holistic keypoint approach is its ability to recognize signs in real-time. This is because the keypoint tracking is based on motion, making it more suitable for dynamic sign language gestures. Moreover, the holistic keypoint approach

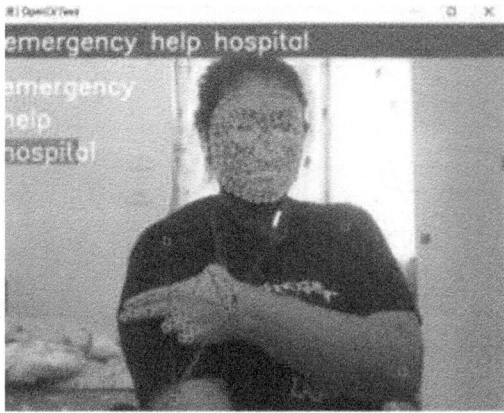

Fig. 5. Real time prediction

is scalable, meaning it can be used to recognize sign language gestures in longer conversations, as it is not limited to specific image frames. This makes it suitable for real-world applications, such as assistive communication devices, sign language translators, and educational tools.

The Continuous Sign Language Recognition system using holistic key points and the LSTM model has shown remarkable accuracy and robustness in recognizing sign language gestures in real time. This research has the potential to improve the accessibility and inclusivity of sign language communication, benefiting millions of people around the world who rely on sign language as their primary mode of communication.

6   Conclusion

In conclusion, this research project has successfully demonstrated the effectiveness of using the holistic keypoint approach to build a robust Continuous Sign Language Recognition system. By leveraging the LSTM model and keypoint tracking, the system achieved exceptional accuracy and robustness in recognizing a wide range of sign language gestures. The holistic keypoint approach has proven to be an idea, as it accurately captures the key features of sign language gestures, is scalable for longer conversations and enables real-time recognition. These advantages make it a promising solution for practical applications, such as assistive communication devices and educational tools.

The outcomes of this research can aid in the development of a system which has the potential to improve communication accessibility and inclusivity for the deaf community. Future research may examine the system's performance in various sign language dialects and broaden the dataset used for training and evaluating.

References

1. International Day of Sign Languages. International Day of Sign Languages (n.d.). https://www.un.org/en/observances/sign-languages-day. Accessed 23 Mar 2023

2. Starner, T., Pentland, A.: Real-time American sign language recognition from video using hidden Markov models. In: Proceedings of International Symposium on Computer Vision - ISCV, Coral Gables, FL, USA, pp. 265–270 (1995). https://doi.org/10.1109/ISCV.1995.477012
3. Mehrotra, K., Godbole, A., Belhe, S.:Indian sign language recognition using kinect sensor. In: Kamel, M., Campilho, A. (eds.) ICIAR 2015. LNCS, vol. 9164, pp. 528–535. Springer, Cham (2015). https://doi.org/10.1007/978-3-319-20801-5_59
4. Pu, J., Chen, X.: Real-time american sign language recognition using convolutional neural networks. IEEE Trans. Hum.-Mach. Syst. **48**(2), 116–122 (2018)
5. Li, X., Li, H., Zhang, L.: Sign language recognition using recurrent neural networks. IEEE Access **5**, 14804–14814 (2017)
6. Purohit, P., Patel, K.: Indian sign language recognition using deep learning. Int. J. Adv. Res. Comput. Sci. Softw. Eng. **8**(6), 556–562 (2018)
7. Taskiran, M., Killioglu, M., Kahraman, N.: A real-time system for recognition of american sign language by using deep learning. In: 2018 41st International Conference on Telecommunications and Signal Processing (TSP), Athens, Greece, pp. 1–5 (2018). https://doi.org/10.1109/TSP.2018.8441304
8. Lee, C.K., Ng, K.K., Chen, C.H., Lau, H.C., Chung, S.Y., Tsoi, T.: American sign language recognition and training method with recurrent neural network. Expert Syst. Appl. **167**, 114403 (2021)
9. Bantupalli, K., Xie, Y.: American sign language recognition using deep learning and computer vision. In: 2018 IEEE International Conference on Big Data (Big Data), pp. 4896–4899. IEEE (2018)
10. Rao, G.A., Syamala, K., Kishore, P.V.V., Sastry, A.S.C.S.: Deep convolutional neural networks for sign language recognition. In: 2018 Conference on Signal Processing and Communication Engineering Systems (SPACES), pp. 194–197. IEEE (2018)
11. Kumar, E.K., Kishore, P.V.V., Kumar, D.A., Kumar, M.T.K.: Early estimation model for 3D-discrete Indian sign language recognition using graph matching. J. King Saud Univ.-Comput. Inf. Sci. **33**(7), 852–864 (2021)
12. Rastgoo, R., Kiani, K., Escalera, S.: Hand sign language recognition using multi-view hand skeleton. Expert Syst. Appl. **150**, 113336 (2020)
13. Kumar, P., Roy, P.P., Dogra, D.P.: Independent Bayesian classifier combination based sign language recognition using facial expression. Inf. Sci. **428**, 30–48 (2018)
14. Amrutha, K., Prabu, P.: ML based sign language recognition system. In: 2021 International Conference on Innovative Trends in Information Technology (ICITIIT), pp. 1–6. IEEE (2021)
15. Kishore, P.V.V., Kumar, D.A., Sastry, A.C.S., Kumar, E.K.: Motionlets matching with adaptive kernels for 3-D Indian sign language recognition. IEEE Sens. J. **18**(8), 3327–3337 (2018)
16. Ko, S.K., Kim, C.J., Jung, H., Cho, C.: Neural sign language translation based on human keypoint estimation. Appl. Sci. **9**(13), 2683 (2019)
17. Pathak, A., Kumar|priyam|priyanshu, A., Chugh, G.: Real time sign language detection. Int. J. Mod. Trends Sci. Technol. **8**, 32–37 (2022). https://doi.org/10.46501/IJMTST0801006
18. Kurhekar, P., Phadtare, J., Sinha, S., Shirsat, K.P.: Real time sign language estimation system. In: 2019 3rd International Conference on Trends in Electronics and Informatics (ICOEI), pp. 654–658. IEEE (2019)
19. Garcia, B., Viesca, S.A.: Real-time American sign language recognition with convolutional neural networks. Convolutional Neural Netw. Vis. Recogn. **2**(225–232), 8 (2016)

20. Shenoy, K., Dastane, T., Rao, V., Vyavaharkar, D.: Real-time Indian sign language (ISL) recognition. In: 2018 9th International Conference on Computing, Communication and Networking Technologies (ICCCNT), pp. 1–9. IEEE (2018)
21. Li, D., Rodriguez, C., Yu, X., Li, H.: Word-level deep sign language recognition from video: a new large-scale dataset and methods comparison. In: Proceedings of the IEEE/CVF Winter Conference on Applications of Computer Vision, pp. 1459–1469 (2020)

A New Hybrid Ensemble Learning-Based Malware Detection Technique

Sanskriti Bansal, D. Ruby$^{(\boxtimes)}$, and Rajat Bargoti$^{(\boxtimes)}$

VIT, Vellore, India
`ruby.d@vit.ac.in`, `rajatbargoti2020@gmail.com`

Abstract. One of the most challenging difficulties that businesses and institutions face is information security. In recent years, the frequency and magnitude of cybercrime have increased, with innovative ways to access, manipulate, and delete data or render information systems disabled, appearing on a daily basis. Malware is one type of intrusion into information systems that process confidential data. For a long time, malware has posed a serious threat to computer system security. A hacker injects malware into a computer system, acquiring complete or partial accessibility to the system's critical data. Conventional detection methodologies based on dynamic as well as static analysis have limited effectiveness due to the rapid advancement of anti-detection technologies. AI-based malware detection has grown in popularity in recent years because of its superior prediction ability. As a result, this research presents a hybrid ensemble classification-based approach for information security and malware detection. A stacked ensemble of 5 homogenous machine learning algorithms performs the first stage classification, where each model is itself ensembled 5 times individually, while in the the final stage, classification is carried out by assembling these 5 ensembled models together, resulting in a meta-learner. Individually, the following machine learning algorithms are utilized for baseline comparison: K-Nearest Neighbors, Support Vector Machine (SVM), Logistic-Regression, Naive Bayes and Decision Tree. Further, Malware Classification using PE headers (ClaMP) dataset was tested, and the findings are reported. It has been found that the proposed hybrid ensemble model achieves the optimum performance.

Keywords: machine learning · malware detection · cyber security · hybrid ensemble learning

1 Introduction

Many Softwares provide us with the resources we need but some of them might contain potential danger in form of malicious malwares. In 2020, the number of new malware samples detected by cybersecurity companies reached an all-time high of 127.5 million, as reported by AV-TEST, A leading independent IT security institute, Cybersecurity Ventures estimates that the cost of cybercrime worldwide would reach $10.5 trillion by 2025, with malware being one of the leading causes of these costs. When it comes to detection of these malwares and other hazardous unwanted programs, the sample is just

© The Author(s), under exclusive license to Springer Nature Switzerland AG 2024
P. Dassan et al. (Eds.): ICICSCNT 2023, CCIS 1970, pp. 235–249, 2024.
https://doi.org/10.1007/978-3-031-75957-4_20

too big to process and constant innovations in current technology is required to move a step forward in search of a safer internet environment.

The most common method for safeguarding themselves from malwares for a normal user is an antivirus software, now the services and level of security provided by them differs on what technology they work on. Traditional antivirus techniques frequently employ heuristics and signature-based detection to find and remove malware. Nowadays, certain anti-virus businesses may identify previously acknowledged malware that certain anti-malware software has found. Nevertheless, it is unable to identify fresh malware that lacks a signature as well as polymorphic malware that can change its signatures. Combining heuristic analysis with machine learning strategies to increase detection effectiveness is one way to address this problem. Experience has shown that the traditional method of malware detection based on signature analysis fails to identify unidentified computer infections. Users must regularly and continuously update anti-virus databases to maintain the appropriate degree of protection. However, the time it takes antivirus organizations to react to the appearance of next generation malware be it detection or signature building or a mix of both can vary widely, taking anything from a short time frame of a few hours to a lengthy time frame of a few days. This time frame is enough for malicious new software to do irreparable harm.

Including all the benefits we gain by using the signature approach, the heuristic analysis is utilized to overcome this issue. Simultaneously, the file can then be classified as "potentially dangerous" depending on its behavior; this approach is a dynamic method or an examination of structure; this approach is a static approach. Static analysis is often divided into two stages: training and results application (detection of virus programs). A sample of infected files containing viruses and another one termed "clean" which has legal files is produced during the training step. Certain indications in the file structure distinguish each one as illegitimate or legitimate. Then, for each file, a list of attributes is constructed. Following that, the most important features (features that yield information) are chosen, while unnecessary, irregular and non-required features are removed. Feature attributes are retrieved from the scanned file during the detection stage. Heuristic algorithms designed particularly to detect unknown malware have a significant error rate. However, based on behavior on the model, finally cloud based malware detection technologies are tested to be effective.

Current research in security focuses on developing shielding methods and powerful algorithms capable of detecting and neutralizing new malware, so not only increasing computer security but saving the customer from the need for frequent security software updates. With the advancement of malware writing and manufacturing processes, the number of gray lists is always expanding. As a result, intelligent approaches for detecting malwares automatically are currently needed. Naturally AI finds its use here as well, Machine learning-based methods for malware detection have shown great promise in improving the accuracy and effectiveness of malware detection systems. Compared to conventional signature-based and heuristics-based methods for malware detection, machine learning-based techniques provide a number of benefits. They are more scalable and flexible to shifting threat environments, can identify novel and previously unidentified kinds of malware, and are less likely to produce false positives.

The research here expands on the use of Hybrid Ensemble Learning. Hybrid ensemble learning is a machine learning methodology that combines the advantages of several ensemble learning approaches, in order to increase the precision and durability of a prediction model. The predictions of many models are combined through ensemble learning to provide a final forecast that is more reliable and accurate than any single model. This concept is expanded upon by hybrid ensemble learning, which combines various ensemble learning techniques to produce a more potent prediction model.

2 Related Work and Research

Md Shoaib Akhtar with Tao Feng in their research titled, "Malware Analysis and Detection Using M.L. Algos", (2022), used a dataset given by the Canadian Institute for Cybersecurity, which comprised log data from 51 distinct varieties of malware. The data was pre-processed by utilizing PEiD software to unpack the executables in a secure environment. To avoid overfitting, feature extraction was used to construct a smaller set of features from the larger set. Feature selection was performed in order to choose the most valuable characteristics in order to improve accuracy, simplify the model, and reduce overfitting. The feature rank technique was extensively utilized to identify the appropriate features for developing malware detection models. They discovered that DT had a greater accuracy than other learning algorithms (K-Nearest-Neighbor, Naive Bayes, Convolutional Neural Network, Random Forest, and S.V.M.: Support Vector Machine) based on the confusion matrix.

P. S. T. Shashank, S. M. Elahi, S. S. Sumanth, B. Dhanraj and G. SuryaNarayana published their work titled, "File based malware detection using ensemble method", (2021) The classification of a specific piece of malware into a class chosen from nine classes—Kelihos version 1, Gatak, Simda, and Obfuscator—was proposed. Among the names are Kelihos version 3, Lollipop, TracurACY, Vundo, and Ramnit. For categorisation, they used the XGBoost model. They discovered that the ensemble technique produces outcomes that are more precise and effective than those from a single independent mode.

Farhat Lamia Barsha, Mohammad Jobair Hossain Faruk, Hossain Shahriar, Maria Valero, and Shahriar Sobhan, Md Abdullah Khan, Michael Whitman, Alfredo Cuzzocreak, Dan Lo, Akond Rahman, and Fan Wu In their research titled "Malware Detection and Prevention Using Artificial Intelligence Techniques" (2021), examined future malware detection methods, preventative measures, and the possibility of artificial intelligence.. They highlighted the shortcomings of current techniques and offered a thorough analysis of malware detection and prevention techniques based on Artificial Intelligence. They stated about AI being used as a promising field for the creation of anti-malware systems for the purpose of identifying and preventing malware assaults or security issues associated with software applications that might lead to a technological utopia.

Robertas Damaševi˘cius, Jevgenijus Toldinas, Algimantas Ven˘ckauskas and Šarunas Grigaliunas in their study titled, "Ensemble-Based Classification Using Neural Networks and Machine Learning Models for Windows PE Malware Detection", (2021) proposed an ensemble categorization-based malware detection method. Stage one classification is carried out by a stacked ensemble of dense (fully connected) and convolutional neural networks (CNN), while stage two classification is carried out by a meta-learner. The classification strategy that included an ensemble of dense-ANN and 1 D CNN models as well as the additional Trees method as a meta-learner outperformed other machine learning classification methods.

Jagsir Singh and Jaswinder Singh in their paper titled "Detection of malicious software by analyzing the behavioral artifacts using machine learning algorithms", (2020), proposed a malware classifier using dynamic features. The text mining technique was used to process the PSI feature. To account for randomness in malware detection, Shannon entropy was calculated over dynamic properties of PSI and API call. The experimental results reveal that utilizing ensemble machine learning techniques, malware detection is 99.54% accurate.

Mahdi Rabbani, Reza Khoshkangini, Yong li Wang and Hamed Jelodar in their work titled, "A hybrid machine learning approach for malicious behavior detection and recognition in cloud computing", (2020), modeled user behavior and detects harmful conduct by utilizing a particle swarm optimization-based probabilistic neural network (PSO-PNN). The suggested system is divided into two modules namely data preprocessing and recognition. The data pre-processing module extracts useful information for learning and modeling. The training and prediction phases are included in the recognition module. The proposed method is validated using the UNSW-NB15 dataset, which includes real-time network traffic from thousands of users. The evaluation results show that the suggested method has potential for use in security observing and spotting risky actions.

Yong Qi, Jinpei Yan and Qifan Rao in their research titled, "Detecting Malware with an Ensemble Method Based on Deep Neural Network", (2018), used an LSTM (Long Short-Term Memory) network for learning features from the opcode sequence. The next stage was malware classification which was performed by stacking the two learnt features and then classifying malware using an ensemble method. MalNet, the suggested malware detection system, had a high validation accuracy of 99.88%. MalNet beat most related efforts with a detection accuracy of 99.36% in a malware family classification experiment done on 9 malware families. In terms of detection efficiency, the paper claimed that MalNet surpassed two groundbreaking results on the Microsoft malware dataset.

3   Methodology and Models Used

It has been demonstrated that hybrid ensemble learning has considerable potential for enhancing the reliability and accuracy of machine learning-based detection systems. Hybrid models have the potential to identify novel and previously unidentified forms of malware while lowering false positives and enhancing the scalability of the detection system by integrating several ensemble learning techniques.

Machine learning approaches can detect malware in a dual step procedure: extracting some features from the input set and then selecting chosen ones that represent the data more accurately. The proposed model consists of a heterogeneous mix of weak learners which are ultimately used to create hybrid ensemble models. First step, data pre-processing and feature-selection; second step, model-engineering, this step entails the subsequent steps: data collection followed by processing; third step, reducing dimensionality; fourth step, individual exploration of models; fifth step, development of the hybrid ensemble technique; sixth step, testing the final model and performance evaluation. The system methodology's steps are broken down into more detail in the following subsections, but the flow is depicted in the picture below (Fig. 1).

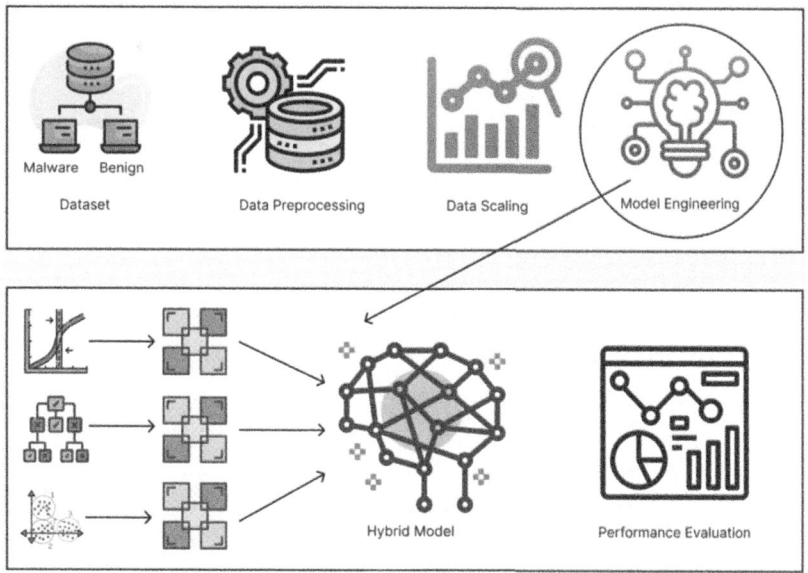

Fig. 1. Outline of proposed model methodology

3.1 Collection and Processing of Data

The dataset utilized in the proposed model incorporates malicious as well as benign samples from the (ClaMP) dataset i.e. Classification of Malware with PE headers, that was downloaded from Kaggle. There are 2722 malware instances whereas 2488 benign incidents in the ClaMP Integrated dataset. DOS image header features, File header features, some other characteristics, and some derived features are among the 69 features in the collection.

Nevertheless, we only used 68 features, which are all numerical, in the study as one of the features, "packer type", is of string format and hence, was not used. The usual scaling approach was used to scale the numerical features. The ensemble classification model was built using these features, and a class label (0 as benign and 1 as malignant) was also defined.

3.2 Reducing Dimensionality

To reduce attribute dimensionality, the StandardScaler() function in Python was utilized. It standardized characteristics by eliminating mean and then sizing to unit standard deviation. The standard value of a sample 'h' is computed as follows:

$$s = (h - m)/sd \tag{1}$$

where *'m'* denotes mean of the training samples and it is zero if *with_mean=False*, whereas *'sd'* denotes standard deviation of the training samples and it is one if *with_std=False*.

3.3 Individual Model Exploration

Logistic Regression
A typical statistical model for binary classification problems that forecasts the likelihood of a binary result is called logistic regression. The logistic regression model is built on top of the mathematical logistic function, or sigmoid function. All real-valued inputs are changed to a value between 0 and 1 using following formula:

$$f(a) = 1/(1 + e^{-a}) \tag{2}$$

where *a* is a linear combination of model parameters and is represented as:

$$a = b0 + b1 * x1 + b2 * x2 + \ldots + bn * xn \tag{3}$$

Here, the model parameters (or coefficients) are *b0, b1, b2, ..., bn,* while the input characteristics are *x1, x2, ..., xn*. Finding the values of *b0, b1, b2, ..., bn* that best matches the training set of data is the aim of logistic regression.

By determining the values of *b0, b1, b2, .., bn* that maximizes the likelihood of the training data, maximum likelihood estimation technique can be utilized to train logistic regression model. Gradient-descent or other optimization methods are often used for this.

After the model has been trained and running it can predict the probability value of the binary outcome, the model predicts a positive outcome if the probability is higher than a certain threshold, which is typically *0.5*; otherwise, it predicts a negative outcome.

Decision Tree
A simple decision tree approach has been used that involves finding the information gain or the gini impurity to determine the best feature to split the data and construct the tree. Assume that one wishes to predict a binary target variable *Y* (like "Yes" or "No") from a dataset of *N* data points. *K* input characteristics, identified as *x1, x2, ..., xk*, are present in the dataset. Following the data's separation on a specific characteristic, the information gain calculates the target variable *Y's* decreased entropy (or uncertainty). The definition of entropy is:

$$H(x) = -p(x = 1) * log\,2(p(x = 1)) - p(x = 0) * log\,2(p(x = 0)) \tag{4}$$

where $p(x = 1)$ and $p(x = 0)$ are the probabilities of *x* being 1 and 0, respectively.
Now the information gained for the a feature *Yk* would be given by –

$$IG(Yk) = H(X) - sum(wi * (X|Yk = yi)) \tag{5}$$

where wi is the percentage of data points in the subset, Yk has the value yi, and $H(X|Yk = yi)$ is the entropy of X for that subset of data. The optimal characteristic to separate the data is the one with the largest information gain.

After the best feature has been chosen, the process is repeated recursively until a stopping condition is met, at which time the data is split into subsets according to the feature's possible values. The prediction is made based on the majority class of the data points that reached that leaf node when making a forecast for a new data point. The decision tree follows the path from the root node to a leaf node dependent on the values of the input attributes.

Support Vector Machine

To decide which hyperplane best classifies the data, Support Vector Machines (SVMs) employ optimization to solve a convex quadratic programming problem. Given a set of training samples like (c1, d1), (c2, d2), …, (cn, dn), where ci represents ith feature vector and di is its associated label, SVMs are especially made to determine a hyperplane which separates the data into two classes with the greatest margin of error.

The equation: $w^T a + m = 0$ defines the hyperplane.

where a is the input vector, m is the bias term, and w represents weight vector perpendicular to the hyperplane. Whatever side of the hyperplane the input vector x is on is determined by the sign of $w^T a + m$. The margin is the separation among the nearest data point from either class and the hyperplane.

SVMs tackle a convex quadratic programming problem of the following type to choose the best hyperplane:

$$minimize(1/2)\|w\|^2 + C \; sum_i = 1^n \, a \, i \tag{6}$$

$$subject \; to \; bi \, (w^T ai + m) \geq 1 - ai, i = 1, 2, \ldots, n \tag{7}$$

where ai are slack variables that permit some data misclassification, bi is the label of the ith training sample, lastly C, a regularization parameter, is used to regulate the trade-offs between raising the margin and lowering error in classification.

Under the restriction that every training example be properly categorized, the objective function $(1/2)\|w\|^2$ corresponds to maximizing the margin, or alternatively, $bi \, (w^T ai + m)$ 1 for every i. In order to accommodate some margin breaches, which are fined by the regularization parameter C, the slack variables ai are added. At last, support vectors are combined linearly to provide the solution w to the optimization issue using Lagrange multipliers.

K-Nearest Neighbor (KNN)

For both classification and regression issues, the K closest neighbor supervised machine learning model is used. It operates under the tenet that the K nearest training set data points may be used to forecast the label or value of an input point.

Mathematically, KNN works as follows:

1) Given a training set of labeled data using the following notation: $(a1, b1), (a2, b2), \ldots, (an, bn),$ where ai is the ith feature vector and bi is the label or value that corresponds to it.

2) The K closest neighbors in the training set are identified for a new input point x using a distance metric like the Manhattan distance or the Euclidean distance.
3) The projected label or value for the now input point x is then determined by averaging the K-nearest neighbors' replies to classification tasks or the K-nearest neighbors' responses to regression tasks, depending on the task.

Classification Model:
Based on a distance metric like the Manhattan distance or the Euclidean distance, the K-nearest neighbors in training dataset are discovered for a new input point a. Let N represent the collection of the K closest neighbors' indices.

Naive-Bayes Model

Naive-Bayes is a popular probabilistic based machine learning approach which is utilized to solve classification problems. It is found in the Bayes theorem and the presumption that, given the class label, the characteristics are conditionally independent.

Naive Bayes' mathematical method operates as follows:

1) Given a training set of labeled data, where $(a1, b1), (a2, b2),..., (an, bn)$ represent the ith feature vector and associated class label, respectively.
2) And given the characteristics, the posterior probability of each class label is calculated for a fresh input point x as follows:

$$P(bi|a) = P(a|bi) * (P(bi)/P(a)) \tag{8}$$

where $P(a \mid bi)$ is the likelihood of the characteristics of a given class bi, $P(bi)$ is the prior probability of class bi, and $P(a)$ is the evidence probability. $P(bi \mid a)$ represents posterior probability of class yi given the features a.
3) The new input point x is then given the projected class label by:

$$y = argmax_i P(bi|a) \tag{9}$$

where $argmax\_i$ refers to the class label for which the posterior probability is maximized.

As the characteristics are presumed to be conditionally independent, given the class label by naive bayes method, the likelihood may be factored as follows:

$$P(a|bi) = \Pi_j P(aj|bi) \tag{10}$$

the probability distribution of the jth feature given class yi is $P(aj \mid bi)$.

The frequency of each and every class in the training dataset can be utilized for estimation of the prior probabilities $P(bi)$, whilst maximum likelihood estimation or Bayesian estimation can be used to estimate the likelihoods $P(aj|bi)$.

3.4 Developing Hybrid Ensemble Model

The goal of hybrid ensemble learning is to take use of the strengths of many model types while overcoming each one's specific flaws. A decision tree, for instance, could perform well on some types of data but badly on others, whereas a neural network might perform well on many types of data but necessitate more training data.

A hybrid ensemble learning model can offer higher overall performance and more flexibility than a single model by merging numerous models. Combining models may be done in a number of ways, such as bagging, boosting, and stacking.

The training datasets are reorganized, and each rearranged set is then added a base classifier to create an ensemble of basic classifiers. After that, methods like resampling and reweighting are used to arrange training datasets in various ways. Following that, an advanced ensemble classifier is created by utilizing the stacked ensemble technology where effects produced by all those base classifiers are merged, and the created model learns to add-up or integrate predictions from various base models more successfully.

A dataset is used to train a number of models. The output from each model is then processed to produce a new dataset. Each sample of the current dataset is then linked to the most genuine value that it is meant to imitate. Finally, the dataset and meta-learning technique are used in conjunction to get the desired final output.

Base models are sometimes termed as level-0 models, while a level-1 model that incorporates base model projections acts as a meta-learner (or generalizer). The foundation models consist of forecasts and models that suit the training data. A meta-learner is a kind of classification model that has been set up to combine the predictions of the underlying model (level 1 model). Basic models of the decisions taken provide information to the meta-learner. The training dataset's input and output value pairings are utilized to fit made the meta-learner, along with proposed outputs given by these projections, and a new batch of data which was previously unutilized is used to train the basic models.

Ensemble learning algorithm can be broadly divided into three stages or phases –

A - building ensemble:
   a. Select base learners;
   b. Select a meta-learning algorithm.
B - Training:
   a. training dataset represented as H1, H2,..., Hm, where Hm is the number of samples, should be used to train each of the N base learners;
   b. Cross-validate each of all basic learners k times and write the cross validated predictions as "P1, P2, ..., PN";
   c. By merging the cross validated result from the basic learners, create a new feature matrix. Using the features and predictions from the base-level classifiers (H1, H2,..., Hm, P1),..., (H1, H2,..., Hm, P2),..., train the meta-learner on this newly made data (H1, H2,..., Hm, PN). Use the meta-learner with fundamental learning models to get predictions about unknown data that are more accurate.
c - Test new data:
   (a) Keep track of the basic learners' output selections;
   (b) Deliver ensemble decisions based on base-level decisions to the meta-learner.

In the proposed hybrid ensemble technique, homogeneous ensembles of Logistic-Regression, KNN, Decision-Trees, SVC, Naive-Bayes, taken individually, have been considered as the weak learners which are further stacked to create the hybrid ensemble model.

3.5 Performance Evaluation

The suggested ensemble learning model's performance was assessed utilizing the Leave-One-Out-Cross-Validation or (LOOCV) method, utilizing a 10-fold cross validation in order to ascertain its classification potential.

Before calculating the (TP) true positive, (FP) false positive, (TN) true negative, (FN) false negative values, the true labels and anticipated labels were compared. The values for recall, accuracy, precision, error-rate and F-score were determined as follows:

-

$$Accuracy := (TP + TN)/(TP + FP + FN + TN) \tag{11}$$

-

$$False\ positive\ rate\ FPR := \sum_{i=1}^{m} [a(xi) = +1][yi = -1]/\sum_{i=1}^{m} [yi = +1] \tag{12}$$

-

$$True\ positive\ rate\ (TPR): TPR = \sum_{i=1}^{m} [a(xi) = +1][yi = +1]/\sum_{i=1}^{m} mi = 1[yi = +1] \tag{13}$$

is a measure of recall

-

$$False\ negative\ rate\ (FNR): FNR = \sum_{i=1}^{m} [a(xi) = -1][yi = +1]/\sum_{i=1}^{m} [yi = +1] \tag{14}$$

-

$$True\ Negative\ Rate\ (TNR): TNR = TN/(TN + FP) \tag{15}$$

is a measure of specificity

-

$$Precision: TPR/(TPR + FPR) \tag{16}$$

-

$$F1\ Score: 2((\ Precision \times Recall\ )/(\ Precision + Recall\ )) \tag{17}$$

The proportion of accurate agreement in the test-dataset is represented bDy p0, whereas the proportion of agreement anticipated by random selection is represented by pe.

4   Implementation and Results

4.1   Dataset and Environmental Information

Dataset utilized - malicious and benign samples from ClaMP, Downloaded from Kaggle. Distribution - 2722 malware instances and 2488 benign incidents.

Features - DOS image header features, File header features, some other characteristics, and some derived features are among the 69 features in the collection.

Omitted Features - "packer type" is of string format. Class Label - 0 as benign and 1 as malignant.

Using Python's Scikit-learn packages, the machine learning algorithms were created based on the dataset's features. The experiments were performed on a PC running Windows 11 64-bit OS.

4.2   Implementation

The confusion matrices of each individual model along with the hybrid ensemble model are presented in the below Fig. 2:

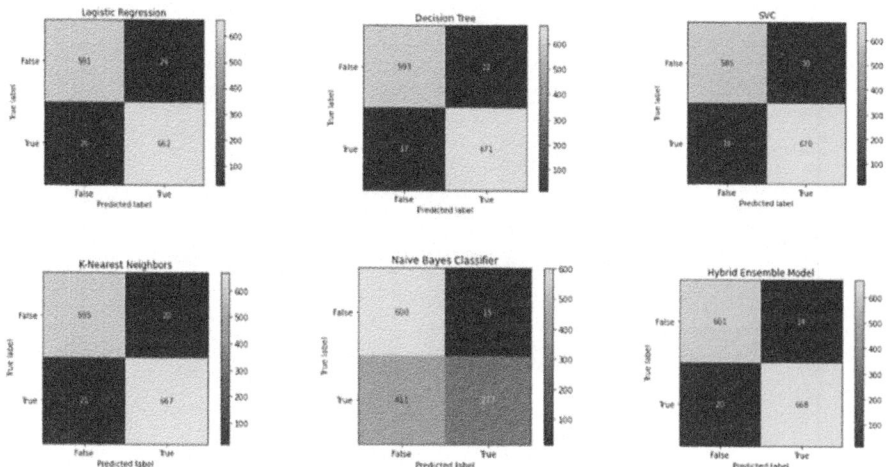

Fig. 2. Confusion Matrices of ML Models

The below table gives the measure of accuracy, precision, recall, specificity, FPR, FNR and F1 Score (Table 1).

It can be inferred from the table that the hybrid ensemble model gives the maximum accuracy of **97.5%** followed by the decision tree at 97.1%. It achieves maximum precision of about **98%** denoting that the proposed model is good in predicting correct results. The false positive and false negative rates have been reduced significantly as compared with other models, achieving 0.026 and 0.025 respectively. Performance was assessed in this study using 10-fold cross-validation. It is performed to avoid overfitting and achieve more validation on the performances of each model.

Table 1. Performance comparison of proposed model with other models

Model	Accuracy	Precision	Recall	Specificity	FPR	FNR	F1 score
Logistic Regression	0.957	0.965	0.961	0.961	0.038	0.037	0.963
Decision Tree	0.971	0.968	0.964	0.964	0.029	0.029	0.972
SVC	0.963	0.957	0.962	0.951	0.037	0.036	0.965
KNN	0.966	0.971	0.968	0.967	0.031	0.033	0.970
Naive Bayes	0.680	0.948	0.688	0.975	0.311	0.312	0.565
Hybrid Ensemble	**0.975**	**0.979**	**0.974**	**0.977**	**0.026**	**0.025**	**0.975**

The below figure (Fig.3) depicts the precision of the model as achieved on the CLaMP dataset.

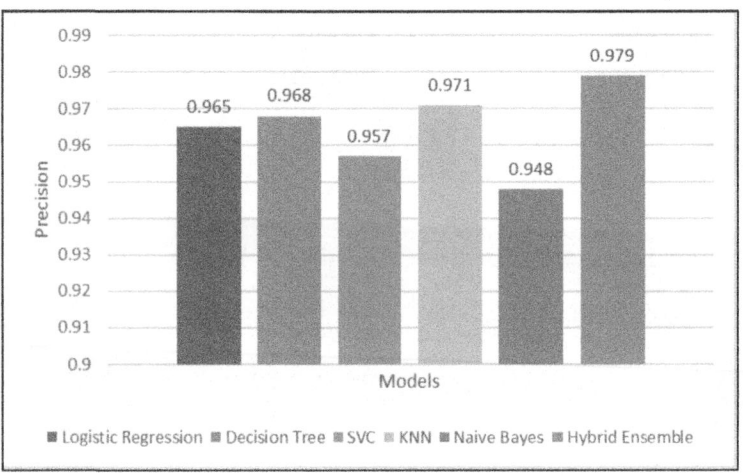

Fig. 3. Performance evaluation based on Precision

The accuracy measures and comparisons are given in the below figure (Fig. 4). The maximum accuracy is achieved by the proposed hybrid model as 97.5%.

Instead of just evaluating the model's accuracy, the researchers chose the best model using the F1-score (as depicted in Fig. 5). The proposed model achieves F1 score as 0.975 which is the best among all other models.

A point to note is that the datasets containing significant class imbalance cannot be solely interpreted by accuracy factor as in that case the model would perform well in classification but would predict incorrectly for the minority and main classes. The

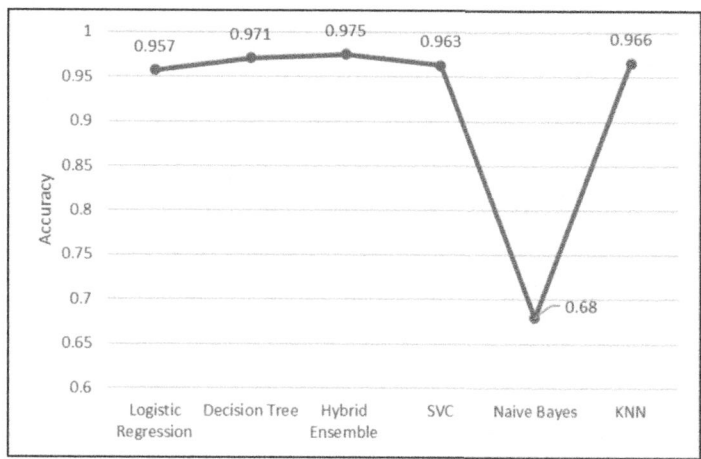

Fig. 4. Performance evaluation based on Accuracy

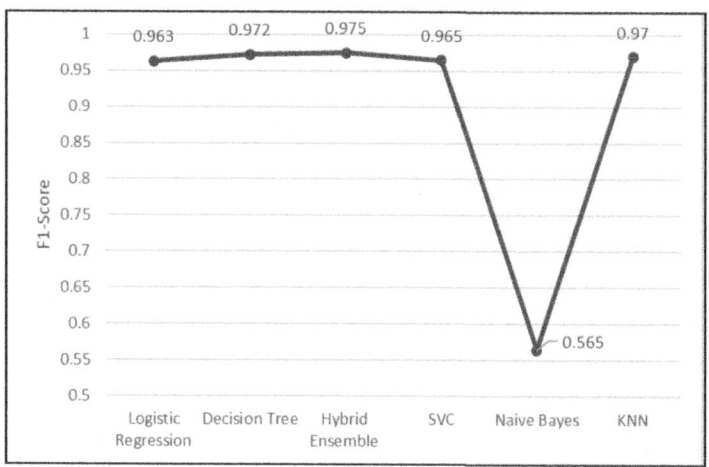

Fig. 5. Performance evaluation based on F1-Score

F1-score stops this kind of behavior by calculating each mark's metrics and unweighted average. To conclude, the proposed hybrid ensemble model dominates in every aspect.

5 Conclusion

Hybrid ensemble learning has shown promise in enhancing the reliability and accuracy of detection systems when used for malware detection. Hybrid ensemble models may efficiently identify a variety of malware kinds and reduce false positives by fusing the capabilities of several learning methods. This method has also demonstrated promise in tackling unbalanced datasets and idea drift, two issues that frequently arise in malware identification.

The output and findings of this research show that the hybrid ensemble technique that has been suggested, outperforms both individual classifiers and conventional ensemble techniques like bagging and boosting. The model considerably enhances the malware detection system's performance on the testing dataset, achieving low false-positive rates and a higher detection rate relatively.

In this research, it has been shown that hybrid ensemble learning with weak learners can improve malware detection systems' performance. The findings demonstrate that an ensemble framework consisting of logistic-regression, KNN, SVN, decision-tree and naive-bayes may achieve high detection rates while preserving low false positive rates. The suggested method can also handle changes in data distribution and new malware threats.

Further research may fully explore the possibilities of the hybrid ensemble learning technique with weak learners for malware identification. By enhancing the performance of the ensemble model, the safety and security of computer systems, as well as their defense capabilities against malware attacks, can be improved.

References

1. Li, S., Zhou, Q., Zhou, R., Lv, Q.: Intelligent malware detection based on graph convolutional network (2021)
2. Faruk, M.J.H., et al.: Malware detection and prevention using artificial intelligence techniques. In: IEEE International Conference on Big Data (2021)
3. Singh, J., Singh, J.: Detection of malicious software by analyzing the behavioral artifacts using machine learning algorithms (2020)
4. Alkahtani, H., Aldhyani, T.H.H.: Artificial intelligence algorithms for malware detection in android-operated mobile devices. Sensors (2022)
5. Aslan, Ö., Yilmaz, A.A.: A new malware classification framework based on deep learning algorithms. IEEE Access (2021)
6. Baptista, I., Shiaeles, S., Kolokotronis, N.: A novel malware detection system based on machine learning and binary visualization. In: IEEE ICC (2019)
7. Sewak, M., Sahay, S.K., Rathore, H.: Comparison of deep learning and the classical machine learning algorithm for the malware detection. IEEE Xplore (2018)
8. Yoo, S., Kim, S., Kim, S., Kang, B.B.: AI-HydRa: advanced hybrid approach using random forest and deep learning for malware classification. Elsevier (2021)
9. Damaševičius, R., Venčkauskas, A., Toldinas, J., Grigaliūnas, Š.: Ensemble-based classification using neural networks and machine learning models for windows PE malware detection. Electronics (2021),
10. Shenfield, A., Day, D., Ayesh, A.: Intelligent intrusion detection systems using artificial neural networks. Science Direct (2018)
11. Wang, W., Zhu, M., Zeng, X., Ye, X., Sheng, Y.: Malware traffic classification using convolutional neural network for representation learning. IEEE Xplore (2017)
12. Tahir, R.: A study on malware and malware detection techniques. Int. J. Educ. Manage. Eng. (2018)
13. Choi, S.: Combined kNN classification and hierarchical similarity hash for fast malware detection. Appl. Sci. (2020)
14. Lee, J., Kim, J., Kim, I., Han, K.: Cyber threat detection based on artificial neural networks using event profiles. IEEE Xplore (2019)

15. Van der Laan, M.J., Polley, E.C., Hubbard, A.E.: Super learner. Stat. Appl. Genet. Mol. Biol. **6** (2007)
16. Marín, G., Caasas, P., Capdehourat, G.: DeepMAL - deep learning models for malware traffic detection and classification. IEEE Access (2020)
17. Martins, N., Cruz, J.M., Cruz, T.J., Abreu, P.H.: Adversarial machine learning applied to intrusion and malware scenarios: a systematic review. IEEE Access (2020)
18. Samantray, O.P., Tripathy, S.N.: An opcode-based malware detection model using supervised learning algorithms. Int. J. Inf. Secur. Priv. (2021)
19. Asam, M., Hussain, S.J.: Detection of exceptional malware variants using deep boosted feature spaces and machine learning. MDPI (2021)
20. Rabbani, M., Wang, Y., Khoshkangini, R., Jelodar, H.: A hybrid machine learning approach for malicious behavior detection and recognition in cloud computing (2020)
21. Sharma, S., Challa, R.K., Sahay, S.K.: BITS Pilani, K K Birla Goa (2019)

Computerized Number Plate Detection and Generalized Character Recognition Using YOLOv8 Model

R. Saranya[✉], S. Sudhakar, T. Priyadharshini, and V. Geethalakshmi

Department of Electronics and Communication Engineering, Vel Tech High Tech Dr. Rangarajan and Dr. Sakunthala Engineering College, Avadi, Chennai, India
saranyaravi892@gmail.com

Abstract. In recent years, Automatic Number Plate Recognition (ANPR) have been proposed and used in the car parking slots, highway monitoring system, traffic regulation etc. Detecting the number plate in the high traffic and also in dusty climate is a challenging work. Country like India with more population, it is not an easy task to identify the number plate with the recognition of characters irrespective of different size plate in the captured image. In this research, implementation of a State-of-the-Art YOLOv8 model for number plate detection and EasyOCR for character recognition. Dataset with number plate images is manually labelled and preprocessed before fed into the deep learning model. For the text localization and individual character recognition, detected Region of interest (ROI) is applied to preprocessing phase which includes Binarization, Denoising the image. Individual characters is extracted and the output is given in text format. The accuracy of this deep learning model is analyzed by Mean Average Precision (mAP) which is 93.7% and recall score is 94.7%.

Keywords: ANPR · YOLOv8 · OCR · Mean Average Precision (mAP)

1 Introduction

Automatic Number Plate Recognition automates a method of identifying the number of the licence plate of the vehicle in real-time without human intervention. There has been an increase in cars, crime, and an unmanageable task intraffic management system that has led to a number of real-life applications of this technology since its advent a few years ago. The ANPR proposed in [1, 8] has also been a leading research topic for deep learning enthusiasts due to its cost-effective solution to many traffic surveillance tasks. A rise in criminal activity in public spaces and more traffic accidents have occurred on the freeways. The need for better traffic management grows as traffic volume increases. Road accidents are frequently caused by high speed vehicles, which are-also hard to find manually by human beings due to their high speed. Thus, devising an ANPR system which automates that method is indispensable in order to punish vehicle owners who are speeding and breaking the traffic rules. With the help of the ANPR system, the figure is scanned, pre-processed, and the characters of the number plate are recognized without intervention from humans. In addition to providing vehicle location information as in [2], it provides reliable and efficient traffic management solutions. By monitoring vehicles on the road using an ANPR system, it would be possible to decrease road accidents and regulate traffic efficiently.

Using this system, fast-moving vehicles are automatically scanned for license numbers, which allows offenders to be caught and penalized. A vehicle surveillance system, an automated toll collection system, and an automated vehicle parking system are also important uses for the ANPR system. In order to ensure that vehicles are parked only in their assigned spots, the autonomous parking system uses the number plate detection for recognizing the incoming vehicles. This is very useful when there is an outbreak of a pandemic and there is no human interaction involved to gather vehicle information. There are five steps to the recognition challenge as in [7], including picture recognition, image preprocessing, license plate localisation, character segmentation, and optical character recognition. It is currently not possible to achieve 100 percent accuracy in ANPR systems because of the number of steps involved and the dependencies [9] between each step. In addition, there are other technical challenges involved, among them are varying illumination [8] or background patterns, picture quality, and the speed at which a machine learning model can be training and testing is higher as in [10]. There are so many ways in which these factors affect numberplate recognition. For the latter, factors such as place, amount, range, font, color, or inclination further the development of an ANPR system that is consistent and reliable. Motorological techniques are useful for gathering accurate information and obtaining accurate results.

2 Literature Review

The problem statement has been the subject of numerous research works for decades. The following are some of the techniques studied in order to learn about earlier ANPR systems are discussed below:

N.Palanivel et al presented Faster R-CNN in [1] to detect the number plate through the camera placed in the traffical areas. The number plate is detected from the captured video using the segmentation of the frame and interpolation of the image. From the obtained image the number is recognized as text form using a technique called optical character recognition. The obtained numbers are provided as input to the base data for retriving the details like name of the vehicle, address, name of the owner, phone number etc.

Atikuzzaman et al proposed a method to detect the license plate of the vehicle using the real-time which is specially made to work based on the videos or images captured through the camera. The project consists of three steps which are plate detection, class letter segmentation and recognition. These steps are done by using HAAR feature based classifier in [2] for detecting license plates. Convolution Neural Network is a technique used for the recognition of letters and numbers. The results are achieved by this method from the collected dataset.

Shreya Raj et al implemented a method to detect the number plates using the Yolov5 algorithm. The object detection is done in [3] using the Yolov5 algorithm and the characters are recognized by technique, called Convolution Neural Network. In [3] obtained output after the detection is processed and separated into single characters by use of contouring and image processing techniques.

Prajawal et al proposed the project which aims to detect the rider without helmet who are violating the traffic rules and the vehicle license plate is also detected and extracted. The major principle used in [4] is detecting the object which is configured at three levels. In first level, objects are detected based Yolov2 and in the second level it is done by Yolov3. After that the license plate number is obtained in the text format by optical character recognition.

Qudes et al presented a idea in [5] to detect and recognize the number plate using Yolov4 algorithm. The process in [5] is splitted into four stages which are image preprocessing, segmentation, noise removal and deep learning. Proceeding image preprocessing and segmentation, license plate is detected through the figure captured using Yolov4 algorithm. The OCR technique is used for character extraction.

Sinha et al provided an examination of several possible solutions to the problem of unpaid vehicle fines is presented in [6] paper. Using specially trained YOLOv3 algorithm and Tesseract OCR in [6] a system that integrates ANPR from beginning to end was built using convolutional neural networks. With this system, objects can be detected, localized, images can be processed, and OCR can be performed. Testing can be conducted in real-world environments using a physical camera integrated with the system. The system in [6] was trained on a large collection of vehicle photos regressively to produce accurate findings.

3   Proposed System

We developed a deep-learning-based model for automated number plate number detection and recognition. Our entire project has two phases: Number plate detection using a deep learning model and Text extraction using Optical Character Recognition (OCR).

3.1   Developing a Deep Learning Model

The first phase of our proposed system is developing a model called a state ofart Yolov8 detection model which is able to detect the number plate from the frame. It involves seven steps namely Data Collection, Data Labeling, Data preprocessing, Training Deep Learning Model, Data Validation, Testing Deep learning model, Number plate detection which is shown below in Fig. 1.

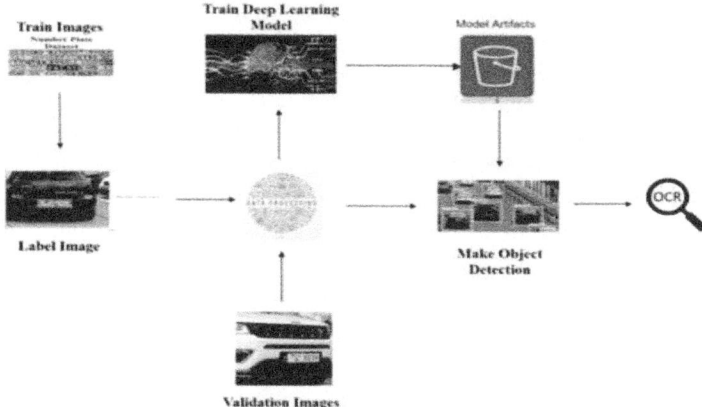

Fig. 1. Deep Learning Model

3.2 Dataset Collection

Dataset preparation is the first step in developing a deep learning model. We collected the data from various sources such as our dataset consist of more than 200 images of vehicle with number plate as shown in Fig 2. The dataset is the important step which is used to train our deep learning model. All the collected images are stored in a dataset folder named images.

Fig. 2. Dataset Collection

3.3 Data Labeling

In order to train our machine learning model we need to do the data labeling which helps to identify the data points in the training dataset images. Once we label the data the training data will be saved in the folder structure consisting of image and .txt file. From the image annotation process we get center position of the bounding box and the width(w) and height(h) of the bounding box is obtained as shown in Fig. 3.

The .txt file structure is in the form of [class, centre_x, centre_y, w, h]

Fig. 3. Data Labeling

3.4 Data Preprocessing

In the previous step after image annotation or labelling process we need to the data preprocessing step which will help to convert the raw data into a understandable format and that will be applied to our machine learning model. In this we first imported our libraries and dataset. After importing all the dependencies, the next step is to extract the dependent and independent variable. Based on filling the mean value of the attribute we splitted our dataset into training set and testing set.

3.5   Number Plate Detection

In this project, to detect the vehicle's number plate, we trained the ultralyticsyolov8 model on our collected dataset. As a first step we started to train our dataset with dimensions of the vehicle number plate. The operation of the yolov8 model is to create the bounding boxes of the number plate on the images. Our yolov8 model is proficient in detection of diverse or manifold of number plate images at once. The training for this yolov8 model has been done with 120 epochs. For inauguration of weights, we used the yolov8 weights that was obtained from ultralytics. The trained model is exported to onnx to test our output with multiple image in different IDEs. The block diagram ofyolov8 model is shown in Fig. 4.

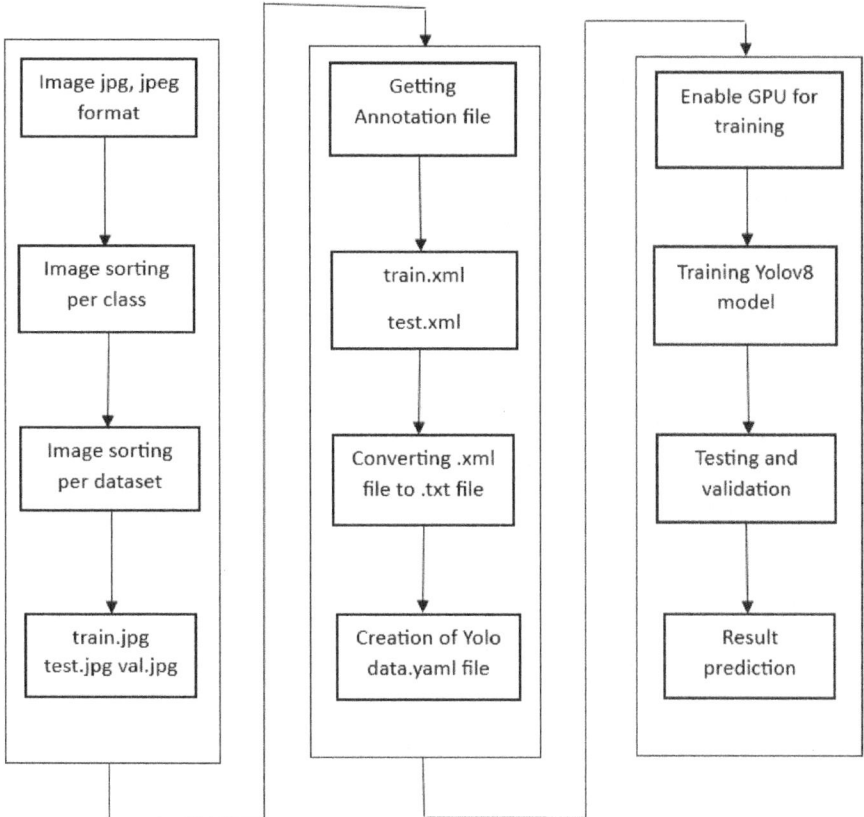

Fig. 4. Proposed system workflow

4   Text Extraction Using Optical Character Recognition (OCR)

Optical Character Recognition is the technique which is used to extract the text in the number plate detected images. In the detected number plate images, to extractthe number plate details, EasyOCR engine is used to extract those text from the images and return it in the text format. The entire process of this is shown Fig. 5.

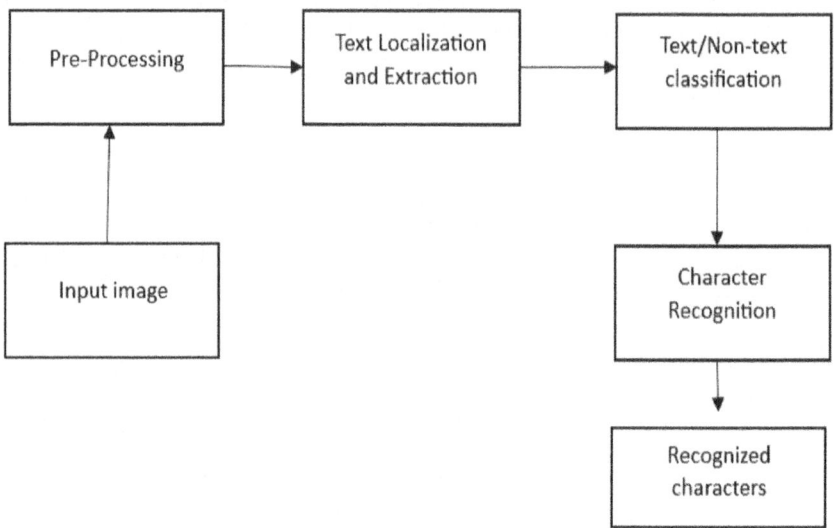

Fig. 5. OCR Architecture

As the first step, the OCR software checkout the image that has been detected and identify the dark parts. As the next step, processing will be done where it will invent the numerical digits and alphabets in the dark regions of the image. To remove the non text artefacts, Image Binarization is done where the image is converted to binary image so that the text localization is effectively done. Character classification is used to identify the individual characters and compared with the database. Now the output image which detected the number plate is applied to OCR engine to obtain the results.

5   Result and Discussion

In our paper, Number plate data sets contains most of the Indian states are used for experiments. All the images which is collected are stored in .jpg and .jpeg format. They are manually labeled with LabelImg, an image annotation tool. After the images are labeled, an XML file corresponding to the same file name is generated. The XML file is converted to txt file which contains the class probalility, centre_x, centre-y, width and height of each image. The data sets are divided into three groups, including trainset, validation set and test set. In our experimental results, it has been observed that YOLO v8 method is achieving faster recognition rate.

Fig. 6. Number Plate Detection Results

Then the data model is trained with yolo data yaml file. The trained model is tested using test images that predicts the number plate which is shown in the Fig. 6.

To evaluate our developed model, Mean Average Precision value(mAP) is calculated. It is based on several sub-metrics such as Confusion matrix, Precision and recall score. From the confusion matrix we can say our model works under different classes, as we have a single class so our model predicts 84% times the number plate prediction as shown in Fig. 7.

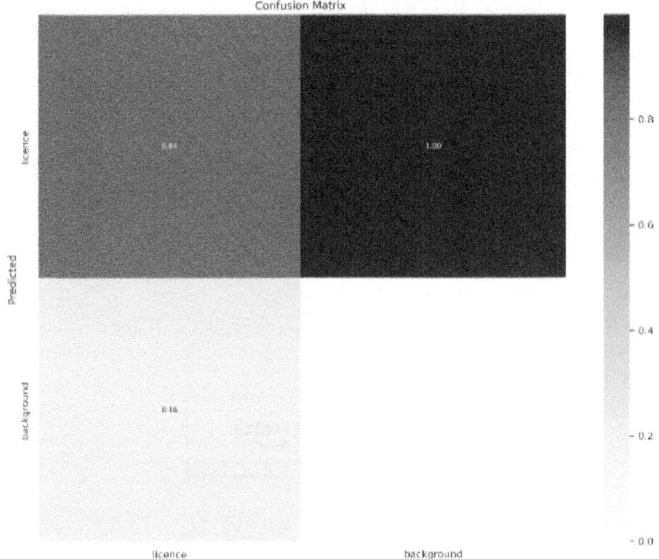

Fig. 7. Confusion Matrix

Here in our model we calculated the different losses such as box loss, cls_loss, dfl_loss from the training phase as well as in validation phase. The loss decreaseswhen number of epochs increases which is shown below in the Fig. 8.

The mean average Precision(mAP) or recall score is increasing when number of epochs increases as here we trained our model with 120 epochs which is shown below in the Fig. 9.

By dividing the total number of detected number Plates by the total number of accurately detected number plates, a precision is calculated. By dividing the number of ground truth number plates by the number of accurately identified ground truth number plates, the recall score is calculated. Precision and recall are mathematically represented as in Eq. (1) and Eq. (2)

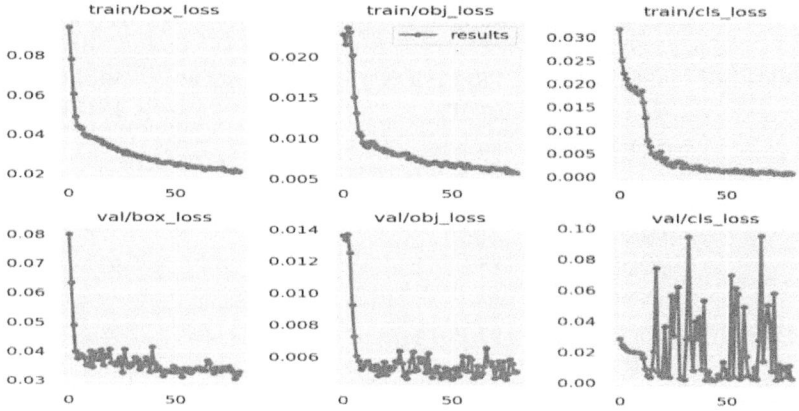

Fig. 8. Train and Validation Loss

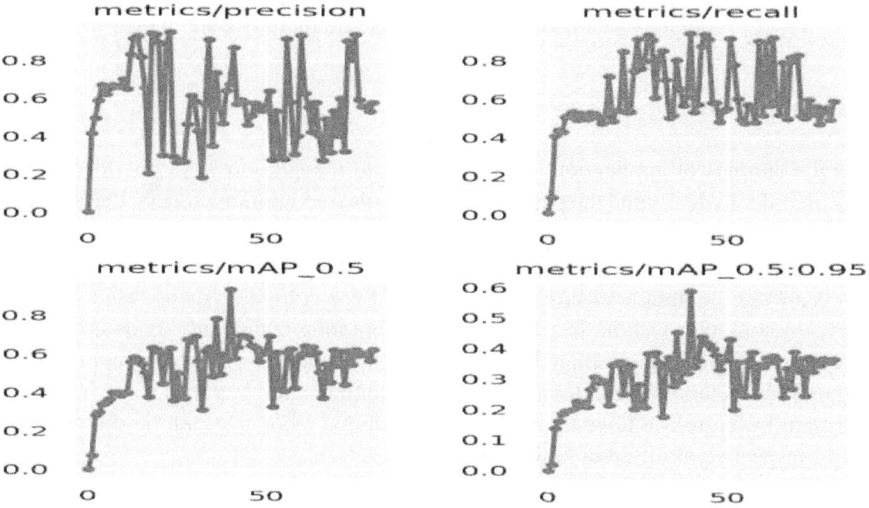

Fig. 9. Mean Average Precision

$$Precision = \frac{A}{A + B} \qquad (1)$$

$$Recall = \frac{A}{A + C} \qquad (2)$$

where A is abbreviated as true positive, B as false positive, and C as false negative respectively

According to the analysis from the Table 1 , our proposed model evaluation metrics is compared with the pre-existing models. It can be seen that the precision (mAP), recall score metrics are higher than CC filtering , Faster R-CNN, YOLOv3 models subsequently. From recognition perspective, the region of interest (ROI) our proposed

Table 1. Evaluation of Accuracy for Proposed Method versus Other Approaches

REFERENCE	APPROACH	PRECISION (mAP)	RECALL SCORE	REGION OF INTEREST
[11]	CC Filtering	89.45%	87.33%	-
[12]	Faster R-CNN	75.11%	89.2%	-
[13]	YOLOv3	91.5%	90.5%	96%
[14]	YOLOv3 and CNN	85%	92.5%	92%
PROPOSED	YOLOv8 and OCR	93.7%	94.7%	98%

model is achieving highest of 98% showing excellent recognition performance which is faster than other models. Therefore our developed YOLOv8 recognition model in this paper shows, excellent robustness and adaptability and there is faster recognition rate.

6 Conclusion

Several examples of applications for number plate recognition were provided in this study. In order to determine the phases involved in the recognition process, the proposed system was examined. As a result of the study, the implied technology is validated as a workable option for real-time recognition. A dataset, or a collection of training data, is always required to ensure the exactness of any computer vision task, including ANPR. In computer vision, deep learning is elaborating technology as its applicative in various domains due to their limitations. In this paper we provided both english and tamil characters for detecting the tamil and english number plates using YOLOV8 algorithm. In future, we hope to produce a system that recognizes all the state number plates irrespective of different languages.

References

1. Palanivel, N., Vigneshwaran, T., Srivarappadhan, M., Madahanraj, R.: Automatic number plate detection in vehicles using FasterR-CNN IEEE (2020)
2. Atikuzzaman, M., Asaduzzaman, M., Zahidul Islam, M.: Vehicle number plate detection and categorization using CNNs. IEEE (2019)
3. Raj, S., Gupta, Y., Malhothra, R.: License plate recognisiton system using YOLOv5 and CNN. In: International Conference on Advanced Computing and Communication System (ICACCS) (2022)
4. Prajawal, M.J., Tejas, K.B., Varshad, V., Madivalappa, S.R.: Detection of non helmet riders and extraction of license plate number using Yolov2 and OCR. Int. J. Innov. Technol. Explor. Eng. (IJITEE) (2019)
5. Qudes, M.B., Aljelawy, Salman, T.M.: Detecting license plate number using OCR technique and rasperry Pi 4with camera. In: International Conference on Computing and Machine Intelliigence (ICMI) (2022)

6. Sinha, H., Soumya, G.V., Undavalli, S., Jeyanthi, R.: An effective real-time approach to automatic number plate recognition (ANPR) Using YOLOv3 and OCR. Intell. Syst. Technol. Appl. (2021)

7. Upadhyay, U., Mehfuz, F., Mediratta, A., Aijaz, A.: Analysis and architecture for the deployment of dynamic license plate recognition using YOLO darknet. IEEE (2019)

8. Boliwala, R., Pawar, M.: Automatic number plate detection for varying illumination conditions. In: International Conference on Communication and Signal Processing (2019)

9. Babbar, S., Kesarwani, S., Dewan, N., Shangle, K., Patel, S.: A new approach for vehicle number plate detection. In: Eleventh International Conference on Contemporary Computing (IC3) (2018)

10. Atikuzzaman, M., Asaduzzaman, M., Islam, M.Z.: Vehicle number plate detection and categorization using CNNs. In: 2019 International Conference on Sustainable Technologies for Industry 4.0 (STI), Dhaka, Bangladesh, pp. 1–5 (2019)

11. Siyu, Z.: An end-to-end license plate localization and recognition system. M.S. thesis, Department of Electrical Engineering, Rochester Institute of Technology, New York, USA (2015)

12. Li, W.: Analysis of object detection performance based on faster R-CNN. ICETIS (2021)

13. Manjunath, A.S., Rudregowda, A.S.: Vehicle number plate detection and recognition using YOLOv3 and OCR. In: IEEE International Conference on Mobile Network and Wireless Communication (ICMNWC) (2021)

14. Abdullah, S., Mahedi Hasan, M., Muhammad Saiful Islam, S.: YOLO-based three-stage network for Bangla license plate recognition in Dhaka metropolitan city. In: 2018 International Conference on Bangla Speech and Language Processing (ICBSLP), Sylhet, Bangladesh, pp. 1–6 (2018)

Design and Creation of an App for Text and Voice Recognition System of Sign Language Using Deep Learning Technique - Sign Aloud

S. Nalini$^{(\boxtimes)}$, S. Shruthi, and M. Mathumithra

University College of Engineering, BIT Campus, Anna University, Tiruchirappalli, India
autnalini@gmail.com

Abstract. This is a project proposal for a Sign Language Recognition System – Sign Aloud. In non-verbal communication hand gesture is very important. Speech impairment is a disability of an individual to communicate using speech and hearing. Sign Language is the only solution for these kinds of people. There exists a challenge for non-signer to communicate with signer. Various Sign Language system had been developed by many developers but they are neither cost effective nor flexible for the end users. The major focus of this project is to create a deep learning-based android application that facilitate sign language translation to text and speech thereby aiding communication with general public. We use a custom CNN (Convolution Neural Network) EfficientNetB0 for recognizing the sign from a live video frame. This system can help in public speaking and students with speech impairment.

Keywords: Sign Language · hand gesture · communication · OpenCv · CNN EfficientNetB0

1 Introduction

Sign language is a visual language that uses a combination of hand gestures, facial expressions, and body movements to convey meaning and communicate with others who are deaf or hard of hearing. Natural languages like this one are just as intricate and emotive as spoken ones. Around the globe, there are numerous distinct sign languages in use, each with its own grammar, vocabulary, and cultural quirks. American Sign Language (ASL), British Sign Language (BSL), and French Sign Language (FSL) are some of the most popular sign languages. Since sign language is not a universal language, various nations and regions have their own sign languages. When communicating with people who are deaf or hard of hearing, sign language is a visual language that combines hand gestures, facial expressions, and body movements to impart meaning. Since sign language is not a universal language, various nations and regions have their own sign languages. It's essential to remember that sign language is a unique and distinct language with its own syntax and grammar, not just a visual representation of spoken language. The term Sign Language Recognition can be referred as conversion of sign language to either to text or voice. Thus, such type of conversion using an algorithm or model help them to overcome from these kinds of situations.

P. Dassan et al. (Eds.): ICICSCNT 2023, CCIS 1970, pp. 262–269, 2024.
https://doi.org/10.1007/978-3-031-75957-4_22

With the existing machine learning and deep learning techniques, it is possible to design an efficient system which would eliminate their difficulties. Few developers and machine learning experts have designed few systems for them.

1.1 Limitation of Existing Project

- E-glove hand gesture detection: The main problem faced by this gloved based system is that it has to be used every time whenever a user wants to communicate on his/her finger-tips so that the fingertips are identified. Wearing the glove for a long time may make a bad experience.
- Text based detection: The Sign Language to Text converter would not help the blind people to communicate with the deaf and dumb.

By solving the above disadvantages and difficulties, the system was designed in such a way that the system is portable (android application) and no third-party software or hardware used (Deep learning and Image processing).

1.2 Use of CNN EfficientNetB0

The dataset used for this system is American Sign Language (ASL). For training these images, CNN EfficeintNetB0 is used. CNN is a kind of network architecture for deep learning algorithm, used for image recognition and pixels data. EfficientNet model was proposed by Mingxing Tan and Quoc V. Le of Google Research, Brain team. This team studied a new scaling method that uniformly scale the depth, the width and the resolution of the network. This compound scaling method had achieved a better accuracy and efficiency when compared to other CNN Networks.

This algorithm helps in detecting the edges of the images efficiently because of compound scaling. So, CNN EfficientNetB0 was decided to use for training the dataset.

2 Literature Survey

The idea of using machine learning and artificial intelligence [1, 2, 11] to create sign language translators that can enhance communication between the deaf and hearing populations has gained more and more traction. Using a variety of methods, researchers have created more precise and effective sign language translation and recognition systems in recent years, making major strides in this field.

The computer vision methods [3] are used to examine video recordings of sign language gestures is one method for translating and recognising sign language. Convolutional neural networks (CNNs) and recurrent neural networks (RNNs), two types of deep learning models, have been created by researchers to identify and classify various hand gestures and movements with high accuracy and another strategy [4] is to interpret sign language in real time using wearable technology, such as smart wristbands or gloves. Typically, these gadgets use sensors to recognise hand motions and send the information to a computer or mobile device for translation.

The author [5] defined a taxonomy to classify existing works and explored their benefits and drawbacks. This taxonomy gives a thorough review of recent works for sign language recognition and it also covers applications, datasets, assessment metrics, features, and modalities. An innovative method for weakly supervised learning in the video domain that is applicable to issues with sequence learning and uses the method with sign language data. The authors [6] employ CNN-LSTM models embedded in each stream of a multi-stream HMM model and produces Word error rate (WER) of 26% used real-life sign language with more than a thousand classes and more than 60 distinct hand gestures for German sign language.

Gesture recognition through neuro-fuzzy systems can be a promising approach for recognizing gestures in Arabic sign language. It facilitates better communication between those with and without hearing impairments. Due to the complexity and variety of the language, however, it may prove difficult to recognise gestures in Arabic sign language. Neuro-fuzzy systems [7] solve these issues by adding linguistic principles and subject-matter expertise into the pattern identification process. Neuro-fuzzy systems can recognise Arabic sign language movements with high levels of accuracy because they combine neural networks with fuzzy logic to learn and adapt to new patterns while also handling the uncertainty and variability of sign language.

A new deep design [8] built on the encoder-decoder network and 3D-ResNet with connectionist temporal classification and iterative optimisation for continuous SLR. In two separate setups, 40 signers were used for training and 10 for testing, and 94 sentences from various signers were used for training and six for testing. The Chinese Sign Language dataset, which includes 100 sentences repeated five times by 50 signers, and RWTH-Phoenix-Weather2014 dataset both comprise a total of 25000 videos. Additionally, it offers a 500-word isolated form. The author [9] have analysed vision-based techniques for continuous sign language recognition through deep learning. Based on the methodologies they used, a thorough analysis of all the studies in vision-based continuous sign language recognition are analysed and they have discussed the challenges in continuous sign recognition system.

The author [10] has analysed the various difficulties faced through the hand gesture recognition and they have briefly discussed the steps to handle the hand gesture and they have also discussed the topic of automated gesture and sign language detection for readers and will help future research in this area. In general, all the models and methods [12–18] are looking to make sign language translation systems more expressive and natural-sounding in addition to recognising and translating sign language gestures.

3 Proposed Model

The block diagram of the proposed system is shown in Fig. 1. The description of each block and the functionality used in the architecture are clearly stated in the subsequent section.

- Sign Language Image Database: The database or dataset contains images of ASL alphabet hand signs.
- Image Pre-processing: For training this model BGR to RGB pre-processing technique is used.
- CNN EfficientNetB0: As mentioned before, for training this model CNN Efficient-NetB0 is used.

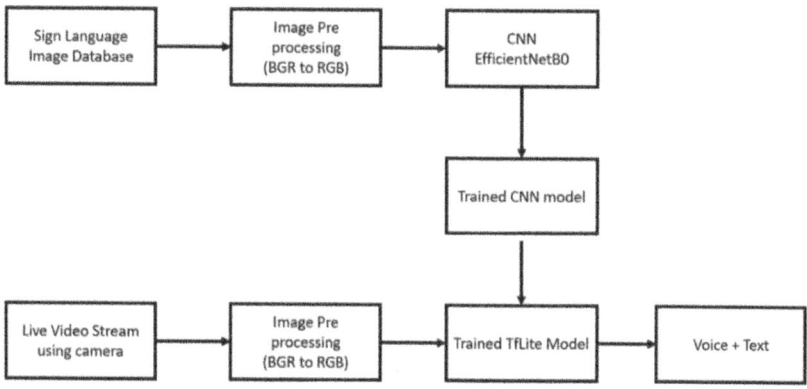

Fig. 1. System Architecture

- Trained TfLite model: The trained CNN model is converted into TfLite model. So that it can be embedded into android project and develop a mobile application successfully.
- Live Video Stream: The detection is done in live video stream which would help to give a good experience for the end-user.
- Voice + Text: The output of this model is the detected text and the corresponding voice of the text.

4   System Desgin

4.1   Hardware

- Working Environment: For developing the android app, a laptop or pc of 64-bit processor with RAM size of more than 4 GB is required. For testing the application an android phone with 11+ version is enough.
- Access Environment: As the application is an android application, it can only run in an android based phone.

4.2   Software

- Working Environment: For training the CNN model, Google colab or Jupyter Notebook is used. Android Studio Electric Ele 2022.1.1 (stable version) IDE is taken for developing the android application.
- Access Environment: As the resultant product is an android application which consumes nearly 200 MB of storage, it is advisable to use any kind of android phone. There is no specification regarding camera. The resolution of the camera can be as low as possible.

5   Phases of Project

The development and implementation of the project is categorised into three phases:

- Dataset Generation
- Training the dataset
- Code Generation

5.1   Dataset Generation

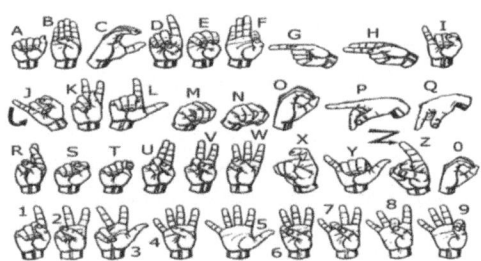

Fig. 2. American Sign Language (ASL)

The Fig. 2. Represents the American Sign Language (ASL), which is the primary dataset for this system. For each sign approximately 500 images were used in which 15% is taken for testing purpose. The dataset is designed in such a way that each sign has images ranging from lower resolution to higher resolution. This makes more efficient training model.

5.2   Testing the Dataset

CNN EfficientNetB0 algorithm is used for training the algorithm. This training phase consist of three parts: (i) Loading the dataset and image pre-processing (ii) Training the images and (iii) Converting to TfLite model. The above process is done in python language.

Loading the Dataset and Image Pre-processing:
The images present in dataset (folder name – Image) is loaded and the image are processed using BGR to RGB image pre-processing. While loading the dataset the images will be in BGR format. This can't be trained in CNN so; the images are converted into RGB format.

Training the Images:
Before training the images, they are divided into training set (85%) and testing set (15%). Then images are trained using CNN EfficientNetB0 (Figs. 3, 4 and 5).

Converting to TfLite Model:
The Trained Model is converted to TfLite Model so that, it can be easily embedded into an android project. This would help to develop a user-friendly mobile application.

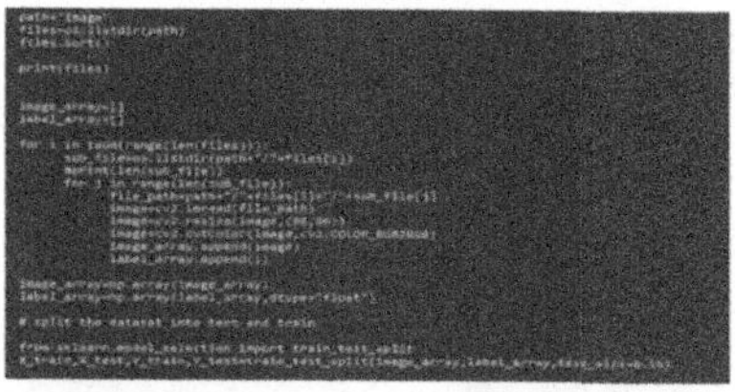

Fig. 3. Loading the dataset and image preprocessing

Fig. 4. Training the images

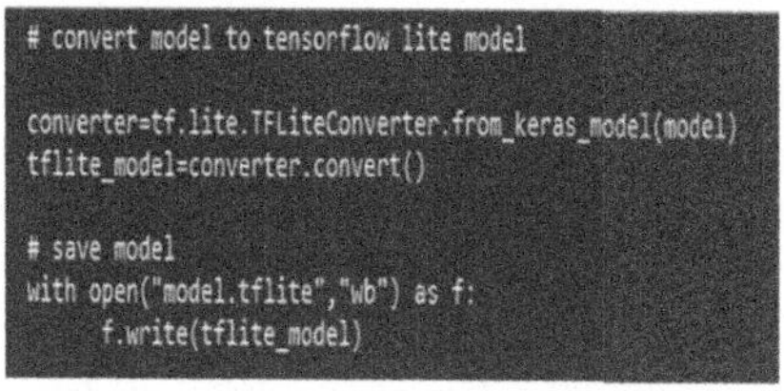

```
# convert model to tensorflow lite model

converter=tf.lite.TFLiteConverter.from_keras_model(model)
tflite_model=converter.convert()

# save model
with open("model.tflite","wb") as f:
    f.write(tflite_model)
```

Fig. 5. Converting to TfLife Model

5.3 Code Generation

This phase is nothing but developing the android application. For this android studio ide is used and the above trained model is mounted into the assets folder of the project. TensorFlow Lite Image Classification Demo – It is GitHub repo which is taken as starter code for the developing the android application.

6 Result Analysis and Discussion

The stated system was developed successfully refer the Fig. 6. Output. Analysing the performance of the application is very important. The application performance is evaluated in an automatic profiling method provided by android studio (Fig. 7. Performance Graph). This method helps us to evaluate the CPU, memory, network and battery resources used by the application.

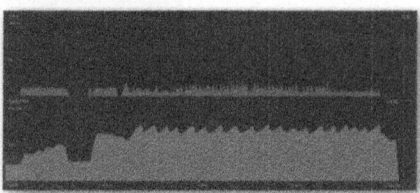

Fig. 6. Output   **Fig. 7.** Performance Graph

7   Conclusion and Further Enhancements

This paper describes the development of an effective real-time vision-based American sign language recognition system for Deaf and Dumb persons. On our dataset, we finally obtained a final accuracy of 92.0%. After implementing of an algorithm that allow us to verify and forecast symbols that are more similar to one another, we are able to better our prediction. In this manner, as long as the symbols are properly displayed, there is no background noise, and the lighting is sufficient, we can almost always recognise all the symbols.

And further the project can be extended by adding a timestamp which can be framed for detecting the letters automatically. The above developed application is not platform independent i.e.., it can be only operated in android os. The dataset can contain few other kinds of simple hand gesture.

References

1. Rastgoo, R., Kiani, K., Escalera, S.: Sign language recognition: a deep survey. Expert Syst. Appl. **164**, 113794 (2021). https://doi.org/10.1016/j.eswa.2020.113794. ISSN 0957-4174
2. Rani, R.S., Rumana, R., Prema, R.: A review paper on sign language recognition for the deaf and dumb. Int. J. Eng. Res. Technol. (IJERT) **10**(10), 2021 (2021)
3. Nyaga, C.N., Wario, R.D.: Sign Language Gesture Recognition through Computer Vision. In: Cunningham, P., Cunningham, M. (eds.) IST-Africa 2018 Conference Proceedings. IIMC International Information Management Corporation (2018). ISBN 978-1-905824-59-5
4. Ahmed, M.A., Zaidan, B.B., Zaidan, A.A., et al.: Real-time sign language framework based on wearable device: analysis of MSL, DataGlove, and gesture recognition. Soft. Comput. **25**, 11101–11122 (2021). https://doi.org/10.1007/s00500-021-05855-6
5. Rastgoo, R., Kiani, K., Escalera, S.: Sign language recognition: a deep survey. Expert Syst. Appl. **164**, 113794 (2021)
6. Koller, O., Camgoz, N.C., Ney, H., Bowden, R.: Weakly supervised learning with multistream CNN-LSTM-HMMs to discover sequential parallelism in sign language videos. IEEE Trans. Pattern Anal. Mach. Intell. **42**(9), 2306–2320 (2019)
7. Al-Jarrah, O., Al-Omari, F.A.: Improving gesture recognition in the Arabic sign language using texture analysis. Appl. Artif. Intell. **21**(1), 11–33 (2007). https://doi.org/10.1080/088 39510600938524

8. Aloysius, N., Geetha, M.: Understanding vision-based continuous sign language recognition. Multimed. Tools Appl. **79**, 22177–22209 (2020). https://doi.org/10.1007/s11042-020-08961-z

9. Cheok, M.J., Omar, Z., Jaward, M.H.: A review of hand gesture and sign language recognition techniques. Int. J. Mach. Learn. Cyber. **10**, 131–153 (2019). https://doi.org/10.1007/s13042-017-0705-5

10. Pu, J., Zhou, W., Li, H.: Iterative alignment network for continuous sign language recognition. In: Proceedings of the IEEE/CVF Conference on Computer Vision and Pattern Recognition, pp. 4165–4174 (2019)

11. Suharto, R.A., Wiryana, F., Ariesta, M.C., Kusuma, G.P.: Sign language recognition application systems for deaf-mute people: a review based on input-process-output. Procedia Comput. Sci. **116**, 441–448 (2017). https://doi.org/10.1016/j.procs.2017.10.028. ISSN 1877-0509

12. https://analyticsindiamag.com/implementing-efficientnet-a-powerful-convolutional-neural-network/

13. https://towardsdatascience.com/efficientnet-scaling-of-convolutional-neural-networks-done-right-3fde32aef8ff/

14. https://www.mathworks.com/help/deeplearning/ref/efficientnetb0.html

15. https://keras.io/api/applications/efficientnet/

16. https://github.com/pramod722445/hand_detection

17. https://paperswithcode.com/method/efficientnet

18. https://github.com/tensorflow/examples/tree/master/lite/examples/object_detection/android

Malicious Cyber Attacks on Blockchain Handled Using Machine Learning Algorithm

P. Preethy Jemima and C. Pretty Diana Cyril[✉]

Department of Computing Technologies, SRM Institute of Science and Technology, Chennai, India
preethy.jemima@gmail.com, prettydc@srmist.edu.in

Abstract. Most powerful technology of the trending world is Blockchain. Because of it's shared, immutable ledger that facilitates the process of recording transactions and tracking assets in a business network. It does not gets limited here, it is widely used in Financial Management and Accounting, Record management in government, Healthcare, Banking, supply chain and so on. In simple terms, Blockchain is a collection of useful information (data). No central authority has the single control on it. Each transaction is documented as a "block" of data as it happens. Every block is interconnected with those that came before and after it.Due to its decentralized nature, after each and every transaction the block information gets appended in the longest chain and is made transparent to everyone enhancing the trust.

But Blockchain is suffering from many cyber attacks. Many solutions are made to increase the level of security. Even in Jan 2019 majority attack has happened in Ethereum classic. The most rare occurrence of the event is the 51% attack (majority attack) which take the control of the hashing rate or hashing power of the network, potentially causing a network disruption. Leading to the cause of Mining Monopoly. Most of the time this happens when the attacker blocks all transactions from a miner in their own private network before broadcasting their own version to the network.

The other possibilities of this majority attack can break the basis terms like Reverse transaction, Double spending attack. The attacker takes the control of not confirming some transaction, changing the block rewards, stealing the coins and even creating the coins. The majority assault, also known as the 51% attack, aims to split the blockchain in order to invalidate completed transactions and tamper with data integrity.

Our approach is to provide security of blockchain transactions by following the POS (proof of stake) and tightening the network strength by tracing out the anomalous behaviour in the transaction.Machine learning model helps us to draw the insights and overcome this by training and testing the entire Ethereum classic logs of transaction and by detection algorithm along with the automatic featured engineering helps to identify in prior the occurrence of this rare event.

Keywords: Blockchain · cryptocurrency · security · cyber attacks · Ethereum

P. Dassan et al. (Eds.): ICICSCNT 2023, CCIS 1970, pp. 270–284, 2024.
https://doi.org/10.1007/978-3-031-75957-4_23

1 Introduction to Blockchain

When digital currency was introduced in 2008 using the highly sought-after crypto-graphic concept, the blockchain technology gained widespread adoption in the field of financial applications. While the first-generation Blockchain 1.0 was subject to several attacks, this caused the second-generation Blockchain 2.0, also known as the Ethereum blockchain, to be enhanced with additional security. In terms of security, the new system Ethereum has a slightly complicated structure. The whole Ethereum security system is vulnerable to numerous security breaches in recent years because to its various levels of defenses, weaknesses, and attack surfaces. Blockchain technology enables transactions between trustworthy peer endpoints without the need for third-party resource engage-ment, helping us to get around technical obstacles and limits. We found that while many have suggested countermeasures to cyber attacks, their efficacy varies widely.

We want to use machine learning to analyse various counterattacks on blockchain. Its purpose is to provide insight into various and substantial datasets. Finally, we would like to conclude by highlighting several limitations, as well as proposing solutions towards future research for providing enhanced high-level security and privacy during trans-actions in blockchain. When dealing with the blockchain pattern analysis, a number of factors are considered. The time series is the beating heart of a block chain since it allows any data to be verified before being added to the network while also tying trustworthi-ness and data privacy together. We have used the data from Kaggle2 and are attempting to improve blockchain security by identifying and foreseeing attacks that employ the fundamental attributes timestamp and acknowledgement.

Blockchain is nothing but a technology most commonly used in this digital word with the major influence of digital currency or crypto currency such as Bitcoin's. All transaction done in a blockchain cannot be reverted back, edited or tampered as it main-tains a database of record. The coin term used is digital ledger [24]. In olden days to transfer an amount from a peer to peer, a third party's inclusion is mandatory. If not they can perform a transaction by in-person mode only. Such a problematic scenario has been overcome with the help of the invention of decentralized system where different peer nodes are connected by a network of computers. Due to Blockchain's distributed struc-ture and peers the transactions without third part such as bank is possible now within fraction of seconds. Information in the blockchain is made publicly available and visible any users on the network of connected computers enhance the security on blockchain. Over the network, anyone can perform transaction and the authentication is done by the computer algorithms over there. After the initial verification information's in such a block as whole block information is appended at the end of the preceding block in the chain of transaction forming a chronological order of transaction. Order of transaction deals with the sender details, receiver details, transaction details, and so on [25].

In Fig. 1 A chain of blocks containing information makes up a blockchain. Without the need for a third-party middleman like a bank or government, the blockchain is utilised for the secure movement of commodities like money, property, contracts, etc. It is extremely challenging to modify data after it has been stored inside a blockchain. A software protocol called blockchain exists (like SMTP is for email). However, without the Internet, blockchains could not function. Because it influences other technologies, it is also known as meta-technology. It is made up of various components, such as a

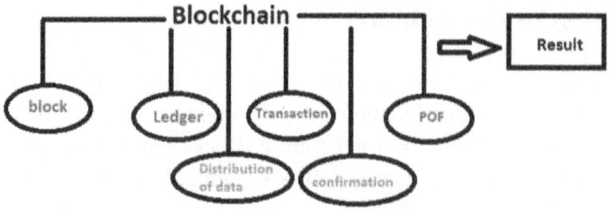

Fig. 1. Blockchain

database, software programme, networked computers, etc. Other virtual currencies or digital tokens may also be mentioned, such as the Ethereum blockchain or Bitcoin's blockchain. However, the majority of them refer to distributed ledgers [16].

2 Literature Survey

2.1 Blockchain Used in Financial Management

Blockchain technology can be used to address challenges in financial management such as information asymmetry, trust issues, and fraud. Then goes on to discussing about the various blockchain-based financial management applications, such as digital identity management, peer-to-peer lending, supply chain finance, and cross-border payments. The authors provide case studies of successful blockchain-based financial management solutions and analyze the benefits and challenges of implementing such solutions. Also suggestions that blockchain technology has the potential to revolutionize financial management by providing greater transparency, security, and efficiency. However, the authors also caution that there are still technical, legal, and regulatory challenges that need to be addressed before blockchain-based financial management solutions can be widely adopted [26].

An overview of the potential of blockchain technology to transform financial management. Main focus is on increased transparency, security, and efficiency in financial transactions. Doesn't get limited here it also examines the challenges and barriers to adoption, such as regulatory issues and the need for standardized protocols. The need of the hour is the collaboration and cooperation among stakeholders to fully realize its benefits [27].

Blockchain technology is used for improving the accuracy and efficiency of financial reporting, enhancing auditing processes, and reducing fraud. It also examines the challenges and opportunities such as regulatory issues, data privacy, and cybersecurity concerns. This paper have strong suggestion on the role of blockchain in sustainability reporting, and the effects of blockchain on corporate governance [28].

2.2 Blockchain Used in Healthcare

The limitations of the traditional healthcare information systems in terms of security and privacy and the issues by providing secure and transparent data exchange are discussed. The author came up with a system that uses a consortium blockchain and smart contracts to manage access control and data sharing among different stakeholders, such as patients, healthcare providers, and researchers. Provided with cas studies this system can facilitate secure data exchange and improve the efficiency and accuracy of medical diagnosis and treatment [29].

2.3   Blockchain Used in Banking Industry

The potential benefits in the banking industry, such as improving the security, efficiency, and transparency of payment systems, enhancing customer identity verification and authentication processes, and enabling new business models and revenue streams. The current challenges and barriers to blockchain adoption in banking, such as regulatory uncertainty, scalability concerns, interoperability challenges, and network effects were also discussed. The author argues that blockchain has the potential to transform the banking industry, but its adoption will depend on a range of factors, including regulatory clarity, technical innovation, and collaboration among industry players [30].

2.4   Blockchain Used in Supply Chain

The overview lack of transparency, inefficiencies in data sharing, and difficulties in tracking product provenance. Blockchain technology can provide a secure and transparent platform for data sharing and tracking.The business requirements for successful implementation of blockchain in supply chain traceability, including the need for a clear and well-defined use case, appropriate blockchain architecture, and stakeholder engagement. The highlight is the success on standardization, data privacy and security, and interoperability. Case studies on IBM Food Trust and the De Beers blockchain platform for tracking diamonds taken in this paper. The wider adoption of the supply chain industry is briefly given [31].

2.5   Consesus Algorithm in Blockchain

Explanation of how consensus algorithms are critical to the security and reliability of blockchain systems, as they enable multiple nodes in a network to agree on a shared version of the blockchain ledger. The major security risks associated with various consensus algorithms, including Proof of Work (PoW), Proof of Stake (PoS), Delegated Proof of Stake (DPoS), and Practical Byzantine Fault Tolerance (PBFT) are elaborated. The authors discuss the vulnerabilities and potential attacks associated with each algorithm, such as the 51% attack in PoW, the Nothing at Stake problem in PoS, and the Block Withholding attack in DPoS.The authors also propose a framework for evaluating the security of different consensus algorithms based on six key criteria: fault tolerance, scalability, resistance to attacks, energy efficiency, fairness, and incentive compatibility [32].

2.6   Bitcoin and Blockchain

The technical details of how Bitcoin works, including the role of nodes in the network, the mining process, and the use of cryptographic techniques to secure the network.

The limitations of the Bitcoin protocol, such as the scalability issues that arise due to the large amount of data stored in the blockchain, the high energy consumption of the mining process, and the potential for security vulnerabilities in the network are briefed out. The use of off-chain transactions to reduce the amount of data stored on the blockchain, the development of alternative mining algorithms that are less energy-intensive, and the use of advanced cryptography techniques to improve the security of the network [33].

2.7   Smart Contract in Blockchain

An introduction to Ethereum and its underlying smart contract technology, as well as the potential benefits and risks associated with the use of smart contracts. They then present a taxonomy of different types of attacks that can be carried out on smart contracts, including those related to design flaws, coding errors, and malicious behavior by contract participants. A detailed analysis of each type of attack, including examples of real-world attacks that have occurred on the Ethereum network. Various techniques that can be used to prevent and mitigate the impact of these attacks, such as code audits, formal verification, and the use of decentralized dispute resolution mechanisms [34].

2.8   Ethereum Background in Blockchain

Ethereum is a decentralised virtual machine that executes contracts, or programmes. EVM bytecode, a Turing-complete bytecode language, is used to write contracts. Contract refers to a group of functions, each of which is termed as a series of bytecode instructions. Characteristics of contract is transfer of bitcoin to other contracts. The Ethereum network is lead by users which handles the users transaction, the main cause why user comes to a contract is to: (i) new contracts creation; (ii) Roles of a contract to be invoked; (iii) transferring ether to contracts or to other users.

User made transactions are made publicly available, blockchain, a data structure that only accepts appends. By using contract we can have the trace of balance of each user after every transaction. Each contract has a economic value and is very touch to guarantee that it has performed a successful transaction. A huge network of shared, untrusted peers as miners, mine the block of data, thereby it cannot be taken into account as a trusted entity. A consensus protocol based on "proof-of-work" is used to overcome such a scenario. When the opponent does not hold the bulk of the network's processing power, the time take for the execution of such a contract will be correct only. The consensus protocol's security is predicated on the notion that morally upright miners are intelligent individuals. These miners are rewarded with a money for mining the block by the user.

2.9   Types of Blockchain

2.9.1   Public Blockchains

The blockchain is public and anyone can take part as a role of a node and take part in the process of determining decisions. No particular node has the controlling authority. All the users are connected with the common network and follows distributed consensus mechanism, here a copy of the distributed ledger is also maintained [16].

2.9.2 Private Blockchains

The blockchains that are not public and are not available to everyone is called as private Blockchain. A restricted group of people may be of any such particular organizations would like to share the information's among them and not to anyone outside the organization makes use of it.

2.9.3 Semi-Private Blockchains

When a part of the blockchain is made visible to certain group of people and the remaining is publicized to all you can name it as semi-private blockchain.

2.9.4 One-Way Pegged and Two-Way Pegged Side-Chain

A side-chain is the transfer of currency from one blockchain to another. movement is permitted in only one direction with a one-way pegged side-chain while movement is permitted in both directions with a two-way pegged side-chain.

2.9.5 Permission Ledger

The users are known already and are trusted.

2.9.6 Tokenized Blockchains and Token Less Blockchains

Tokenized Blockchains create coin and they follow the traditional mining process. Tokenless Blockchains does not exist in real they could not make use of the coins or they cannot transfer money, but this is handles to transfer value only among the existing trusted parties.

3 Overview of Blockchain

After the boom of cryptography an important concept called "Electronic cash" gave its contribution in the field of blockchain was the reason for the wide spread of this technology. In the 19th century we were using emails for communication which overcame the paper pen method. Fast communication played a vital role and encountered several issues. Spam mails came into the role play, this is also a important key note as it paved the path for creating the digital money concept called as "b-money" on a end to end network. The inventor of Blockchain Technology is assumed as Satoshi Nakamoto when he published a paper on bitcoin in 2008 as "Bitcoin: A Peer-to-Peer Electronic Cash System," but still inventor is considered as unanimous. The entire paper was dealing with the working of blockchain and its architecture. Without using a source from a third party we can make out online payments (transaction) underlying the concept of cryptography [16].

The strength mainly depends on the transparency in the transaction, reached the real world. As discussed in the paper Satoshi Nakamoto gave a solution to prevent the double spending attack with the help of Timestamp and tack of the coins(crypto currencies). There are numerous crypto currencies like Litecoin, Dogecoin etchowever bitcoins command the masters share of the market and are now the widely used crypto currency. Taking into consideration of one of the crypto currencies, Bitcoin started hit the market in the year 2013, investors believed that trades on Bitcoin, they poured funds on the new up comers related to Bitcoin. Any user will be able to electronically transfer bitcoins using wallet software. The development of the Ethereum platform in 2015 made it possible for blockchain to function with loans and connections. An algorithm based implementation that created the trust between 2 end peers was based on "smart contract".

3.1   Datasets Used in Blockchain

Dataset	Year	Inference	Link
Bitcoin Historical Data	(2014–2021)	This dataset includes historical price data for Bitcoin from various exchanges, as well as market capitalization, trading volume, and other metrics	https://www.kaggle.com/mczielinski/bitcoin-historical-data
Ethereum Blockchain Transactions	(2017)	This dataset includes a sample of Ethereum blockchain transactions, including transaction inputs and outputs, gas usage, and other metadata	https://www.kaggle.com/bigquery/ethereum-blockchain
Bitcoin Blockchain Transactions	(2017)	This dataset includes a sample of Bitcoin blockchain transactions, including transaction inputs and outputs, transaction fees, and other metadata	https://www.kaggle.com/bigquery/bitcoin-blockchain
Cryptocurrency Market Capitalizations	(2013–2021)	This dataset includes market capitalization data for various cryptocurrencies, as well as other metrics such as trading volume and circulating supply	https://www.kaggle.com/jessevent/all-crypto-currencies
Ethereum Historical Data	(2015–2021)	This dataset includes historical price and volume data for Ethereum, as well as other metrics such as gas usage and transaction fees	https://www.kaggle.com/prasoonkottarathil/ethereum-historical-data

3.2 Blockchain Architecture

In general companies maintain their own information's in the centralized databases where as in the working Blockchain technology it relays on decentralized database. A decentralized database is a collection of computers in which each computer has an identical copy of the database. The motive of it is to provide a efficient environment which is free from tampering of data. Blockchain runs on the upper most layer of the internet.

In Fig. 2. Three layers make up the bulk of the blockchain architecture: application layer (top most layer)- consisting of all the basic applications of the software of the blockchain, Decentralized Ledger followed by Peer-to-Peer Network(bottom most layer). In wallet security is highly maintained with the help of "keys". They are of private key and public key in order to maintain an overall control of the remaining coins in the wallet. Also helps users with the ease of maintain of transactions.

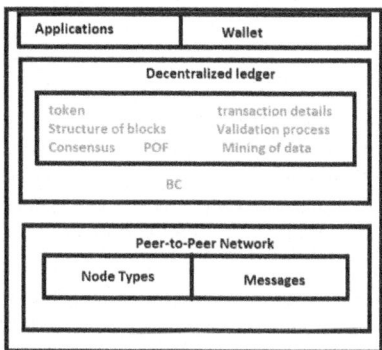

Fig. 2. Blockchain architecture

3.3 Blockchain Architecture

I. When two peers wants to share money between each other. Lets consider them as A and B. A initiates the transaction.

II. The information enters into the fresh block, once it gets filled it is been publicized over the entire network.

III. Miners mine and acknowledge the block as a "honest block" or "legitimate user"

IV. The blocks gets appended to the previous block in the chain. The chronological order is followed.

V. The money get transferred from A to B following the decentralized system structure and maintaining information's in the digital ledger.

Figure 3 Explain the overall working of blockchain. How a person A sends the money to Person B.

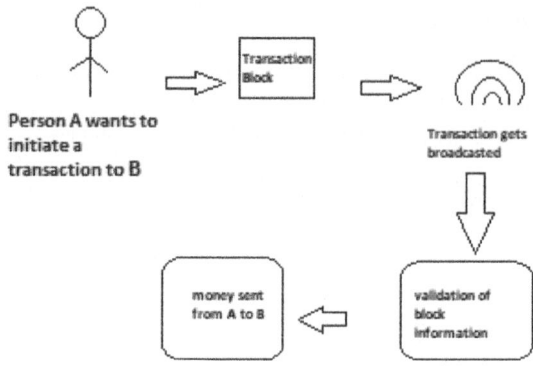

Fig. 3. Blockchain working

3.4 Attacks on Blockchain Comparitive Study

Attack Name	Year	Consequenses	Suggestions
DAO Attack	(2016)	An attacker exploited a vulnerability in the code of the Decentralized Autonomous Organization (DAO) and stole over $50 million worth of Ether	A hard fork that reverted the blockchain to a previous state before the attack occurred, effectively invalidating the stolen funds and returning them to their rightful owners. However, this solution was controversial
Parity wallet hack	(2017)	An attacker exploited a vulnerability in the Parity multi-sig wallet software, leading to the theft of over $30 million worth of Ether	Used a hard fork that froze the affected funds, preventing the attacker from accessing them. However, this solution also raised questions about the immutability
Binance hack	(2019)	Binance, was hacked, resulting in the theft of 7,000 bitcoins worth approximately $40 million at the time	2FA authentication for user accounts and the creation of a Secure Asset Fund for Users (SAFU) that is funded by a portion of trading fees and can be used to compensate users in the event of a hack
Veritaseum hack	(2017)	A hacker stole $8.4 million worth of Veritaseum (VERI) tokens in an ICO hack	The company issued a patch to fix the vulnerability and conducted a token swap to prevent the stolen tokens from being used

(continued)

(*continued*)

Attack Name	Year	Consequenses	Suggestions
NiceHash hack	(2017)	Cryptocurrency mining marketplace NiceHash was hacked, resulting in the theft of over 4,700 bitcoins worth approximately $70 million at that time	The company reimbursed affected users for their lost funds and implemented additional security measures, such as mandatory 2FA authentication and the use of multisignature wallets for storing user funds

3.5 Other Various Cyber Attacks in Blockchain

So far we have identified there are 87 malicious attacks on blockchain system with the help of SLR approach on cyber security attacks on blockchain technology [25]. There are 39 occurrences on financial dealings called as double spending attack or wallet malware attack which is of 5 occurrences. Indirect attack of financial values is called as selfish mining attacks which are of 33 occurrences. Mainly focused on debt obligations there were about 34 attacks so called as Sybil attacks. There are subdivisions such as attack on application logic, attack on client application, attack on consensus mechanism, attack on VM language, attack on P2P network.

The vulnerability in Ethereum application layer also took place [8]. Before the completion of contractor call DAO attacks happened. This often happens due to the ignorance of validity check by the contract or the transaction may run out of gas [8]. Even tokens get attacked in Ethereum application layer during transaction.

Also stealing of private keys leads to phishing attack, flooding of the network with the bot nets leads to DDOS attack [6]. Most common attack and many solutions are given based on the authenticity of clients. By monitoring the connected bots in the network. But most of them are suggestions only. Even miners go crazy on mining which has faced pooling attacks.

The identified problem statement is that most of the data handled in the blockchain is not labeled data. With the help of SHA algorithm and most commonly used "Proof-of-Work" we are able to validate the transactions and block information's and it is useful for miners to overcome double spending and DOS attacks. Since the occurrence of attacks or pattern is not same the attacks can never be stopped.

4 Working of SHA Algorithm

The SHA-256 (Secure Hash Algorithm) hashing algorithm is used in the blockchain of a Bitcoin. The National Security Agency (NSA) in the USA created the SHA-256 Hashing method in 2001. SHA is a family of cryptographic functions designed to keep data secured. In Fig. 4 working of SHA algorithm involves a hash function, it comprises of compression algorithms, bitwise operations, and modular additions.

The hash functions are SHA-0, SHA-1, SHA-224, SHA-256, SHA-384, and SHA-512, which total six. The hash communications of different length into outputs of fixed length. A proof-of-work cryptocurrency network's hash rate is a gauge of how much overall processing power in blockchain is being used to process transactions. A cryptographic hash referring to a data block can also be used to scale how quickly a cryptocurrency miner's computers is able to finish calculations. You can use Hash Pointer to check whether the block of data before it has been tampered with.

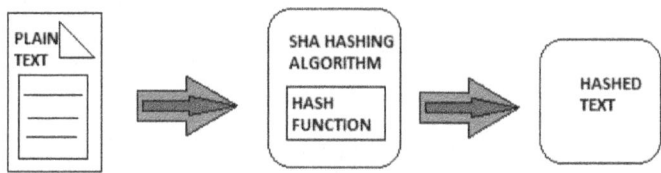

Fig. 4. SHA algorithm working

5 Methodology

The occurrence of rarity of events due to its changing pattern can be termed as abnormal blocks of information. An intelligent software agent can be implemented in the Ethereum. The utility function [18] is used by the application layer of the network to categorize the collaborator's goals and the worth of the current transaction being exchanged.

The working is based on tracing the motive of the stakeholders; If the motive is malicious in it is being (i.e. the transaction as a whole) is cancelled and all the collaborator's are requested to start with a fresh transaction.

Utility function has two distinct parts:

1) The likelihood that each collaborator may hardly detect their nature from the stakeholders' prior transactions.
2) The likelihood that the collaborator's may attack through a major attack will be based on present worth of goods or services being involved in the present transaction [18].

The abnormal block of information are traced out to identify who might be a alarming signal of attacks.

5.1 Intelligent Agent

An intelligent agent in blockchain refers to an autonomous software program that interacts with the blockchain network to execute certain tasks based on predefined rules and algorithms has been implemented in the application layer to enhace tle level of security. Use of the intelligent agent helps to monitor blockchain transactions, verifying transactions, and executing smart contracts. These agents can also be programmed to learn from past transactions and improve their performance over time.

5.2 Data Preprocessing

Preprocessing steps include those of basic steps removal of noisy data that is data cleaning then training the system. Here during the pre processing step it involves that of.

1) Feature extraction:Feature extraction in blockchain typically involves identifying and selecting the most relevant and informative data from the blockchain for further analysis. Some common examples of feature extraction in blockchain include: Transaction features: Extracting information such as the transaction amount, sender and recipient addresses, transaction fees, and timestamps.Block features: Extracting information such as the block size, block height, block timestamp, and the number of transactions in each block.Network features: Extracting information such as the number of nodes in the network, the geographic distribution of nodes, and the network latency.Address features: Extracting information such as the number of transactions associated with a specific address, the balance of the address, and the frequency of transactions.

2) Selection of gas average(i.e. the average provided gas needed to complete the entire transaction) In blockchain, gas is the unit of measurement for the computational effort required to execute a transaction The gas price is the amount of cryptocurrency paid per unit of gas, and the gas limit is the maximum amount of gas that can be used for a transaction or smart contract execution.The selection of gas average in blockchain depends on several factors, such as network congestion, gas prices, and the complexity of the transaction or smart contract. If the network is congested, gas prices may increase, and users may need to increase the gas price they are willing to pay to ensure that their transactions are processed quickly. On the other hand, if the network is less congested, users may be able to set a lower gas price and still have their transactions processed in a timely manner.

3) Normalization: Normalization in blockchain can be achieved through various techniques such as data deduplication, data compression, and data hashing. Data deduplication involves identifying and eliminating duplicate data to reduce the amount of data stored on the blockchain. Data compression involves compressing the data to reduce its size, which can help improve the speed and efficiency of the blockchain network. Data hashing involves generating a unique cryptographic hash value for each piece of data, which can be used to verify the integrity of the data. Normalization is used to increase the security in blockchain.

4) Generation of DATA using sliding window mechanism.

There may be a random fluctuation in the data as well as values, hence to avoid the instability of it we go for normalization of data which in return regularize the values [20].

5.3 Experimental Results

I have taken only the gas number provided its average sum. You can visualize in Fig. 5. The provided original data on gas average and the normalized average gas number. To make it understand with more clarity made the relation of both the data to be plotted along the sliding time window. The attacks when it happened were highlighted within the plots using red colored lines.

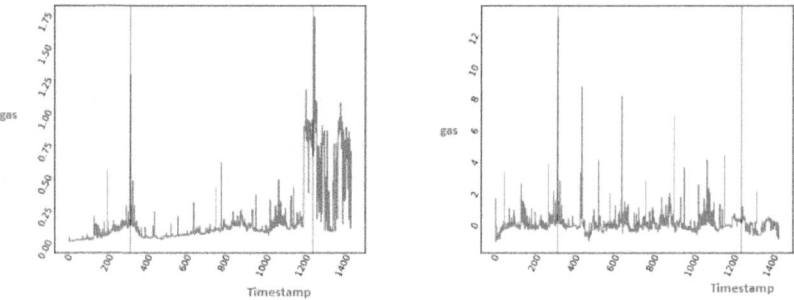

Fig. 5. Provided gas average as well as Normalized

6   Conclusion

In this work, feature extraction is widely used for anomaly detection, prediction, and clustering.Highly recommended for scalability in blockchain. Feature extraction to train the system and finally normalize the data to identify the changes in blockchain.It's important to note that setting a gas price that is too low may result in a transaction being stuck in the mempool for an extended period, or even being dropped altogether. Conversely, setting a gas price that is too high may result in the user paying more than necessary for the transaction. Therefore, it's important to carefully consider the current network conditions and the complexity of the transaction or smart contract when selecting the gas average in blockchain.

Only the Timestamp of Gas average provided along with the normalized data was used to figure out the anomalous changes in the security attacks of blockchain. Handled a Intelligent software agent in the application layer to trace out the abnormal blocks and followed by data preprocessing. The problem statement: rarity of attacks and different pattern of attacks can be well with the help of automatic feature extraction. RNN can be used along with these to make it more efficient in detecting the attacks in early. The Kaggle[2] dataset table had many attributes, some more attributes can also be taken into account in future.

References

1. Cisco Annual Internet Report - Cisco Annual Internet Report (2018–2023) White Paper, Cisco. https://www.cisco.com/c/en/us/solutions/collateral/executive-perspectives/annualinternet-report/white-paper-c11-741490.html
2. Kolias, C., Kambourakis, G., Stavrou, A., Voas, J.: DDoS in the IoT: mirai and other botnets. Computer **50**(7), 80–84 (2017)
3. Heilman, E., Kendler, A., Zohar, A., Goldberg, S.: Eclipse attacks on bitcoin's peer-to-peer network. In: Proceedings of the 24th USENIX Conference on Security Symposium, SEC 2015. USENIX Association, pp. 129–144 (2015)
4. Rahal, B.M., Santos, A., Nogueira, M.: A distributed architecture for DDoS prediction and bot detection. IEEE Access **8**, 159756–159772 (2020)
5. Aldweesh, A., Derhab, A., Emam, A.Z.: Deep learning approaches for anomaly-based intrusion detection systems: a survey, taxonomy, and open issues. Knowl.-Based Syst. **189**, Article ID 105124 (2020)

6. Chen, H., Pendleton, M., Njilla, L.: A survey on ethereum systems security: vulnerabilities, attacks, and defenses. ACM Comput. Surv. **53**(3), Article 67 (2020)
7. Sharafaldin, I., Habibi Lashkari, A., Ghorbani, A.A.: Toward generating a new intrusion detection dataset and intrusion traffic characterization. In: Proceedings of the 4th International Conference on Information Systems Security and Privacy, pp. 108–116. SciTePress, Funchal (2018)
8. Rodler, M., Li, W., Karame, G.O., Davi, L.: Sereum: protecting existing smart contracts against re-entrancy attacks. arXiv preprint arXiv:1812.05934 (2018)
9. Doriguzzi-Corin, R., Millar, S., Scott-Hayward, S., Martinezdel-Rincon, J., Siracusa, D.: Lucid: a practical, lightweight deep learning solution for DDoS attack detection. IEEE Trans. Netw. Serv. Manage. **17**(2), 876–889 (2020)
10. Nakamoto, S.: Bitcoin: a peer-to-peer electronic cash system (2008). http://bitcoin.org/bit coin.pdf
11. Karame, G.O., Androulaki, E., Capkun, S.: Double-spending fast payments in bitcoin. In: Proceedings of the 2012 ACM Conference on Computer and Communications Security, CCS 2012, pp. 906–917. ACM, New York (2012)
12. Decker, C., Wattenhofer, R.: Bitcoin transaction malleability and MtGox. In: Kutyłowski, M., Vaidya, J. (eds) ESORICS 2014. LNCS, vol. 8713, pp. 313–326. Springer, Cham (2014). https://doi.org/10.1007/978-3-319-11212-1_18
13. Maria, A., Aviv, Z., Laurent, V.: Hijacking bitcoin: routing attacks on cryptocurrencies. In: 2017 IEEE Symposium on Security and Privacy (SP). IEEE (2017)
14. Eyal, I., Sirer, E.G.: Majority is not enough: Bitcoin mining is vulnerable. In: Christin, N., Safavi-Naini, R. (eds) FC 2014. LNCS, vol. 8437, pp. 436–454. Springer Berlin Heidelberg (2014). https://doi.org/10.1007/978-3-662-45472-5_28
15. Sapirshtein, A., Sompolinsky, Y., Zohar, A.: Optimal selfish mining strategies in bitcoin. In: Grossklags, J., Preneel, B. (eds.) FC 2016. LNCS, vol. 9603, pp. 515–532. Springer, Heidelberg (2017). https://doi.org/10.1007/978-3-662-54970-4_30
16. Sarmah, S.S.: Understanding blockchain technology. Comput Sci. Eng. **8**(2), 23–29 (2018). https://doi.org/10.5923/j.computer.20180802.02
17. Pilkington, M.: 11 Blockchain technology: principles and applications. Res. Handb. Digit. Transf. 225 (2016)
18. Dey, S.: Securing majority-attack in blockchain using machine learning and algorithmic game theory: a proof of work. In: IEEE- Computer Science and Electronic Engineering Conference (CEEC) (2018)
19. Atzori, M.: Blockchain technology and decentralized governance: is the state still necessary? (2015)
20. Chatfield, C.: The Analysis of Time Series: An Introduction, 6th edn. CRC Press (2004)
21. Malinova, K., Andreas, P.: Market design with blockchain technology (2017)
22. Nguyen, Q.K.: Blockchain-a financial technology for future sustainable development. In: International Conference on Green Technology and Sustainable Development (GTSD). IEEE (2016)
23. Al-E'mari, S., Anbar, M., Sanjalawe, Y., Manickam, S.: A labeled transactions-based dataset on the Ethereum network. In: Anbar, M., Abdullah, N., Manickam, S. (eds) ACeS 2020. CCIS, vol. 1347, pp. 61–79. Springer, Singapore (2021). https://doi.org/10.1007/978-981-33-6835-4_5
24. Buterin, V.: A next generation smart contract & decentralized application platform. Ethereum White Paper (2015). Ethereum.org
25. Guggenberger, T., Schlatt, V., Schmid, J., Urbach, N.: A structured overview of attacks on blockchain systems. In: PACIS 2021 Proceedings, p. 100 (2021). https://aisel.aisnet.org/pac is2021/100

26. Sun, Y., Wang, Y.: Blockchain technology and applications in financial management. J. Financ. Manage. Anal
27. Voshmgir, S.: The Future of Financial Management with Blockchain Technology. J. Risk Manage. Financ. Inst
28. Zhang, Y., Chen, L., Chen, X.: The impact of blockchain technology on accounting: a review of the literature and future research directions. J. Account. Econ
29. Sun, Y., Wang, R., Wang, S.: Blockchain-based healthcare information system for secure data exchange. published in the J. Med. Syst
30. Stearns, D.W.: Blockchain and distributed ledger technology in banking. J. Payments Strategy Syst
31. Pernestål, A., Almér, A.: Blockchain for supply chain traceability: business requirements and critical success factors. J. Bus. Logist
32. Najafabadi, M.M., Dehghantanha, A., Mahmod, R.: Security risks of blockchain consensus algorithms. J. Inf. Secur. Appl
33. Dhar, S., Gehani, A.: Bitcoin: under the hood. In: Proceedings of the 2018 ACM Conference on Computer and Communications Security (2018)
34. Atzei, N., Bartoletti, M., Cimoli, T.: A survey of attacks on Ethereum smart contracts. J. Cryptograph. Eng

Efficient Wind Speed Prediction Using Machine Learning

B. M. Manoj Kumaaran$^{(\boxtimes)}$, S. Pradeepan, B. RamKumar, K. Indira, and M. Rahul

Thiagarajar College of Engineering, Madurai, India
maaranmanoj@gmail.com, kiit@tce.edu

Abstract. As a form of clean energy, wind energy has been used effectively in electrical systems. However, wind speed shows strong nonlinearity and instability due to atmospheric boundary layer effects. Therefore, accurate and stable wind speed forecasts are essential for the safety of the power grid. Industry can also use forecasted wind speed data to calculate the amount of electricity they need to generate or purchase in addition to electricity generated by wind power. A new hybrid prediction system is needed to improve the prediction accuracy. Machine learning models can help predict wind speeds. Competing regressors such as linear regressor, random forest regressor, XGBoost, decision tree regressor, Support Vector and lightweight gradient boosting machine were compared to choose the best machine learning model that provides high accuracy results.

Keywords: Machine learning regression models · Wind Speed Prediction · Random forest regressor · XGBoost regressor · Light Gradient Boosting Machine

1 Introduction

The wind speed prediction with higher accuracy is essential for a variety of industries such as renewable energy, aviation, and maritime operations. For instance, wind farms rely on accurate wind speed predictions to optimize energy production, while airports use wind speed forecasts to determine safe landing and takeoff conditions.

Developing a reliable wind speed prediction model requires sophisticated algorithms and advanced statistical techniques. Machine learning algorithms, such as neural networks and decision trees, are often used to analyze large datasets and uncover patterns in the data. With the help of these algorithms, the wind speed values in the upcoming days can be predicted. Overall, wind speed prediction models play a critical role in enabling industries to plan and operate more efficiently and safely, and they have become an essential tool in the era of sustainable energy and climate change mitigation.

Wind speed prediction using regression models is a common approach mostly applied in the area of energy that can be renewed and meteorology. The objective of the model is to determine and analyze the future wind speed at a specific location based on already recorded wind speed data and other relevant variables such as time of day, season, and geographical location. Usually, wind speed prediction is carried out in entirely different methods. By using the old wind speed record data prediction of wind speed both in

P. Dassan et al. (Eds.): ICICSCNT 2023, CCIS 1970, pp. 285–291, 2024.
https://doi.org/10.1007/978-3-031-75957-4_24

long term and short term is possible. As an alternative way, the other influential factors like temperature, humidity, etc.... and the corresponding wind speed are combined as a dataset and using the dataset, the future predictions of the wind speed would be discharged using machine learning regression models.

The foremost step in developing a regression model for the prediction of wind speed is to collect historical wind speed data from the location of interest. This data is typically collected using anemometers, which are instruments that measure wind speed and direction. Other attributes(behaviors), such as temperature and humidity, can also be collected alongside wind speed. Once the data has been collected, it is preprocessed to remove any outliers or errors. This involves checking for missing values, duplicates, or data that does not fit the expected range of values. The data is then split into a training set and a test set, the training set is used to train the model, and the test set is used to evaluate its performance. The next step is to select an appropriate regression algorithm to use for the model. Linear regression is the most commonly utilized method for prediction of wind speed, where the proportionality between wind speed and other variables is considered to be linear. Other regression algorithms, such as decision trees or support vector machines, can also be used depending on the specific needs of the model. In our work, we are using around six regression models to predict the wind speed. So we use the same training and testing data for each model and verify the accuracy of each model.

Finally, the performance as well as the accuracy of the model is evaluated by using the testing set. This involves comparing the predicted wind speed values to the actual wind speed values and calculating some of the metrics like mean absolute error or root mean squared error. The model can then be refined or adjusted based on its performance, and the process of training and testing can be repeated until a satisfactory level of accuracy is achieved.

2 Related Work

As wind energy is an important renewable resource, and to make use of it in an efficient manner, it is necessary to predict the accurate wind speed. On behalf of this, various research works have been performed on wind prediction. This section includes various results and outputs related to our work on wind speed prediction.

According to the paper [11], data on wind speed from five districts of Tamilnadu such as Chennai, Salem, Coimbatore, Madurai and Tirunelveli are collected for analysis. For the wind speed prediction, they adopted various techniques like TSVR, PLSTSVR, ILTPISVR, ELM, RVFL and LDMR and the performances of these models are evaluated based on different measures such as RMSE, MAPE, SMAPE, MASE, SSE/SST, SSR/SST, and R Square. As a result of this paper, in terms of precision LDMR is selected and in terms of faster computation ELM is selected as outperforming models.

Tree-based learning algorithms are treated as the best models according to [10]. Three models are trained to find the impact on the wind speed accuracy. The first model was trained using the mean and standard deviation of the measured wind speed at a fixed altitude of 40 m and a sampling time of 10 min. For the second model, the effect of sampling time on performance is calculated by the training model with varying time intervals for samplings for 1 h, 12 h and 24 h. In the third model, the impact of height on performance is calculated by the training model with varying heights of wind speed at 40m, 30m and 10m. Testing these models against wind data on different heights and geolocations suggests that tree-based learning algorithms such as decision tree, bagging, random forest, boosting, gradient boosting and XGBoost are the greatest suitable algorithms for wind prediction.

The efficient use of the random forest is stated in the paper [12]. The random forest method ranks the importance of input variables and extracts the correlated features which ensure the inputs for the wind speed forecasting model accuracy. RF methods also effectively improve the forecasting ability of Machine learning models such as KELM, ELM, SVM, NN and RBF-based models.

As of [13], wind speed forecasting was done on a wind farm and the comparison between Support Vector Regression and Back Propagation Neural Networks are performed using simulation results. They infer that SVM is better than BPNN by considering prediction graphs, Average Square Errors, and Average Absolute Errors. An analysis is carried out with three different stages of the wind speed curve and it says that the SVM perfectly fits the curve during the whole process. Thus, SVM best suits and outperforms the BPNN for predicting wind speed.

3   Efficient Wind Speed Prediction Using Machine Learning (EWSPML)

Wind speed prediction using regression models is a powerful tool for renewable energy and meteorology applications. By accurately forecasting future wind speeds, it can help optimize wind energy production and improve safety in aviation and other industries that rely on wind speed information.

3.1   Objective

Our main objective is to predict the wind speed using various Machine Learning algorithms and to compare the results obtained between the different models. We calculate MAE(Mean Absolute Error), MSE(Mean Squared Error), and RMSE(Root Mean Squared Error). Another objective is to calculate accuracy by R-Squared Score and Explained Variance Score. To compare the results obtained from Error and Accuracy we conclude with the best Machine Learning model among the six that we used.

3.2   Constraints

Apart from developing the Machine learning model we may face many issues. This may include collecting appropriate data that trains the model. The data utilized must be gathered from reliable sources. Sensors collecting the information must be accurate and should work properly. The data that are collected continuously should be managed and maintained. Industries using the model must be aware of its prediction accuracy and reliability.

3.3 Methodology

(See Fig. 1).

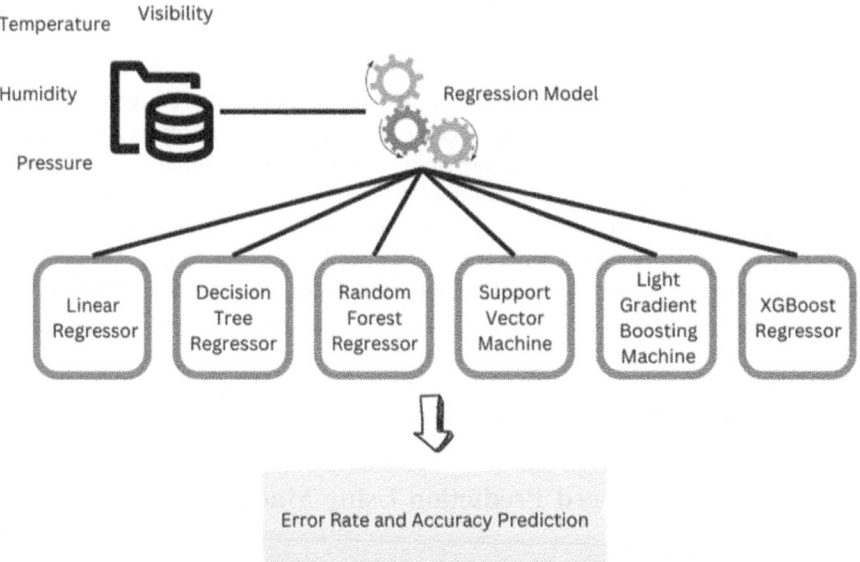

Fig. 1. Overview of the Regression Models

4 Results and Discussion

As mentioned in the introduction, the main ideology of this research is to develop an accurate wind speed prediction model using regression algorithms. For the evaluation of the performance of the regression algorithms, we used mean squared error (MSE), mean absolute error (MAE), root mean squared error (RMSE), and coefficient of determination (R^2) metrics. The results of the study show that the random forest and gradient boosting algorithms outperformed the other algorithms. The linear regression and decision tree regression algorithms also performed well. Overall, the random forest and gradient boosting algorithms are the best performers in terms of accuracy with wind speed prediction.

The results of our study are consistent with previous studies that have investigated wind speed prediction using regression algorithms. It has been shown that ensemble algorithms such as random forest and gradient boosting are more accurate than single regression algorithms such as linear regression and decision tree regression. This may be because ensemble algorithms can combine various weak learners to generate a more accurate prediction. Additionally, ensemble algorithms are least prone to overfitting compared to single regression algorithms. Note that the standard performance may vary from region to region, season, and time of day. Therefore, it is recommended that the model is validated on a specific dataset before it is deployed in real-world applications (Table 1 and Figs. 2, 3, 4, 5, 6, 7 and 8).

Table 1. R-square value for each regression model

Model	R-Square value
Linear Regressor	32.55%
Decision Tree Regressor	30.40%
Random Forest Regressor	64.52%
Support Vector Machine	20.22%
Light Gradient Boosting Machine	61.15%
XGBoost Regressor	63.30%

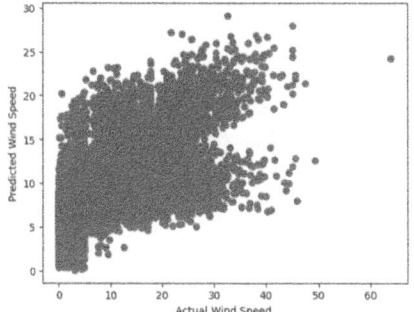 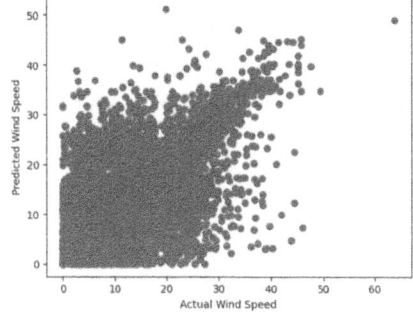

Fig. 2. Linear Regressor: Actual vs Predicted Wind Speed Data

Fig. 3. Decision Tree Regressor: Actual vs Predicted Wind Speed Data

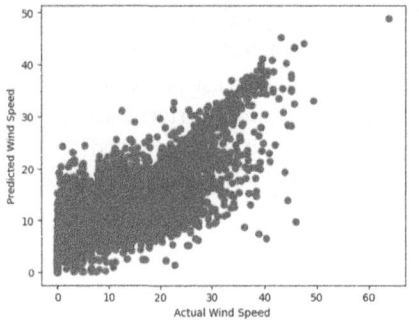

Fig. 4. Random Forest Regressor: Actual vs Predicted Wind Speed Data

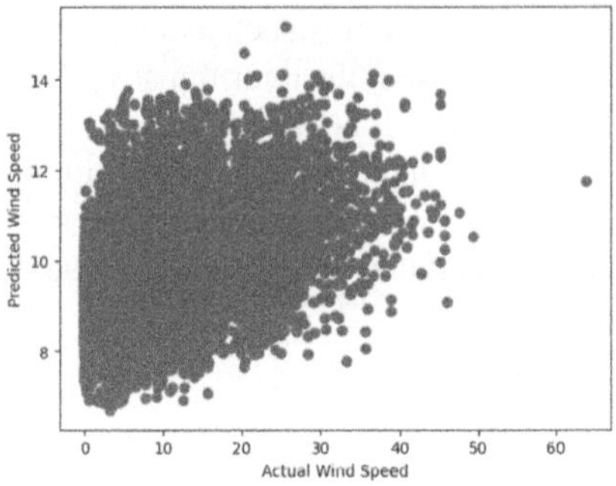

Fig. 5. Support Vector Machine Actual vs Predicted Wind Speed Data

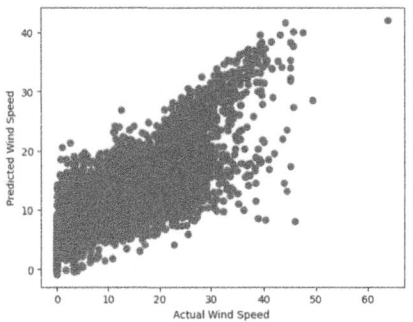

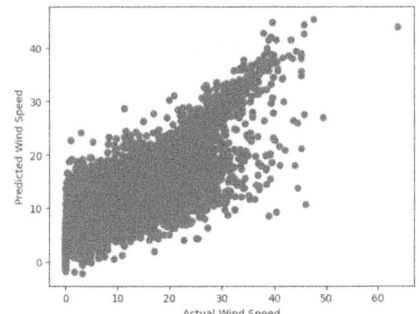

Fig. 6. Light Gradient Boost Machine: Actual vs Predicted Wind Speed Data

Fig. 7. XGBoost Regressor: Actual vs Predicted Wind Speed Data

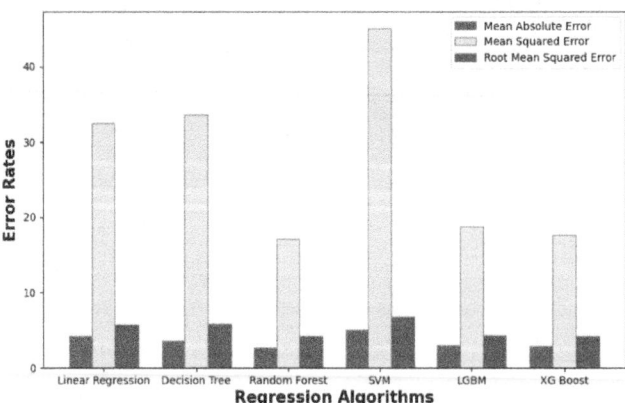

Fig. 8. Total Error Rate Associated with the regression model analysis

5 Conclusion

We used regression algorithms such as linear regression, decision tree regression, random forest regression, support vector machines, light gradient boosting machines, and XGBoost regressors to train the model and predict future wind speeds. The results of our study demonstrate that random forest and gradient boosting algorithms perform well in predicting wind speed. However, we acknowledge that the performance of the model may vary depending on the specific location, season, and time of day. We hope that this research will contribute to the development of more accurate and reliable wind speed prediction models, which can improve the efficiency and safety of renewable energy and aviation industries. We also acknowledge that there is still room for further research and improvement, and we look forward to future studies in this field.

References

1. Liu, X., Zhang, H., Kong, X., Lee, K.Y.: Wind speed forecasting using deep neural network with feature selection. Neurocomputing **397**, 393–403 (2020)
2. Wang, S., Wang, J., Lu, H., Zhao, W.: A novel combined model for wind speed prediction – Combination of linear model, shallow neural networks, and deep learning approaches. Energy **234**, 121–275 (2021)
3. Zhang, Z., et al.: Wind speed prediction method using shared weight long short-term memory network and Gaussian process regression. Appl. Energy **247**, 270–284 (2019). https://doi.org/10.1016/j.apenergy.2019.04.047
4. Hur, S.: Short-term wind speed prediction using Extended Kalman filter and machine learning. Energy Rep. **7**, 1046–1054 (2021)
5. Afrasiabi, M., Mohammadi, M., Rastegar, M., Afrasiabi, S.: Advanced deep learning approach for probabilistic wind speed forecasting. IEEE Trans. Ind. Inform. **17**, 720–727 (2021)
6. Ma, Z., et al.: Application of hybrid model based on double decomposition, error correction and deep learning in short-term wind speed prediction. Energy Convers. Manag. **205**, 0196–8904 (2020)
7. Senthil Kumar, P.: Improved prediction of wind speed using machine learning. EAI Endorsed Trans. Energy Web **6**(23), 157033 (2019). https://doi.org/10.4108/eai.13-7-2018.157033
8. Gupta, D., Natarajan, N., Berlin, M.: Short-term wind speed prediction using hybrid machine learning techniques. Environ. Sci. Pollut. Res. **29**, 50909–50927 (2022)
9. Chen, J., Zeng, G.-Q., Zhou, W., Wei, Du., Kang-Di, Lu.: Wind speed forecasting using nonlinear-learning ensemble of deep learning time series prediction and extremal optimization. Energy Convers. Manag. **165**, 681–695 (2018). https://doi.org/10.1016/j.enconman.2018.03.098
10. Ahmadi, A., et al.: Long-term wind power forecasting using tree-based learning algorithms. IEEE Access **8**, 151511–151522 (2020)
11. Wang, H., Sun, J., Sun, J., Wang, J.: Using random forests to select optimal input variables for short-term wind speed forecasting models. Energies **10**(10), 1522 (2017)
12. Zhao, P., Xia, J., Dai, Y., He, J.: Wind speed prediction using support vector regression. In: 5th IEEE Conference on Industrial Electronics and Applications, Taichung, Taiwan, pp. 882–886 (2010)

"Waverider" - An Autonomous Smart Boat for Garbage Collection, Rescue Operations and Aquaculture

N. Shivaanivarsha[✉], A. G. Vijayendiran, and M. Ajay Prasath

Department of ECE, Sri Sairam Engineering College, Chennai, India
varsha.ece@sairam.edu.in, vijayendirangiridharan@gmail.com,
sec20ec137@sairamtap.edu.in

Abstract. With increasing industrial revolution marine pollution and floating marine debris is becoming a major problem. If measures are not taken at an early stage, this marine trash grows unnoticed and will have consequences like death of aquatic organisms, blockage of seaways, thereby rendering the precious water resources unusable and posing a great threat for marine-based economies. Also, materials like plastics, glass discarded fishing equipment, industrial waste etc. will change the chemical and biological balance of the water resources causing very hazardous threats to humans and their health. Discarded plastics which may turn in to micro plastics are the most hazardous trash. With the help of today's technology, addressing marine pollution will be a very easy task. The proposed idea is about an autonomous boat which collect the floating trash in lakes, ponds, rivers, banks and beaches and deposits those in trash collection facilities where the trash is classified and recycled. The boat works by an autonomous driving system which uses Computer Vision and Convolutional Neural Network and Deep Learning to Identify floating trash with video feed and drive the boat near the trash to pick those up. The Boat also can be controlled remotely or by using headless driving software called "ARDUPILOT". The "ArduPilot" software comes with a mission planning tool to plot way points for the boat to cross in order to follow a uniform path. The boat can also be used for rescue operations during hydro-meteorological disasters and is fitted with appropriate sensors for aquaculture operations.

Keywords: Aquaculture · Rescue bot · Garbage cleaning · Autonomous bot

1 Introduction

Clean water is the basic right to every living organism whether it is water dwelling or water drinking one. One must ensure that all the aquatic animals are given a pollution free environment to live in and all the other organisms which depend on water for various activities in their life are provided with it. With the increasing industrial revolution, marine pollution and floating marine debris is becoming a major problem. Solving this with sheer manpower is nearly an impossible thing, so the stated mission is to clean the water bodies from bottom up, starting from ponds, reservoirs, lakes, rivers, canals and working the way up the big seas and oceans.

© The Author(s), under exclusive license to Springer Nature Switzerland AG 2024
P. Dassan et al. (Eds.): ICICSCNT 2023, CCIS 1970, pp. 292–300, 2024.
https://doi.org/10.1007/978-3-031-75957-4_25

The Problem addressed by this specific idea is trash and garbage being dumped from various domestic households and shops being disposed of into running water in rivers, lakes, ponds, being carried into backwaters and oceans and forming big garbage islands of grouped trash called *"Garbage Patches"*. These garbage patches will in turn, make plastics into micro plastics which are even worse than already toxic and non-bio-degradable plastics. These micro plastics will get consumed by fishes and those fishes when consumed by humans will turn from delicious and protein rich food supplements to dangerous carcinogenic and hazardous foods which are not even fit to be touched by humans.

2 Proposed Method

The proposed idea is about autonomous boats which collect the floating trash in lakes, ponds, rivers, banks and beaches and deposit those in trash collection facilities where the trash is classified and recycled. The boat works by an autonomous driving system which uses Computer Vision and Convolutional Neural Network and Deep Learning to Identify floating trash with video feed and drives the boat near the trash to pick it up. Figure 1 shows the Arch. Diagram for the system, where a Raspberry pi is used as the main controller.

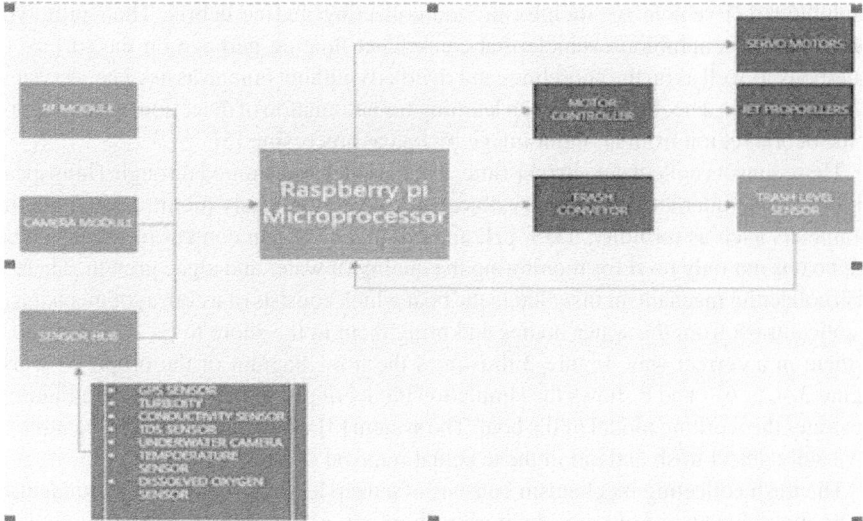

Fig. 1. Block Diagram

The Boat also can be controlled remotely or by using a headless driving software called "ARDUPILOT". The ArduPilot software comes with a mission planning tool to plot waypoints for the boat to cross in order to follow a uniform path. The boat can also be used for rescue operations during hydrometeorological disasters and is fitted with appropriate sensors for aquaculture operations.

The proposed system consists of a raspberry pi 4–8 GB ram controller connected with a pi cam camera and a husky lens module. This contraption is accompanied by a sensor hub which consists of various sensors required for moving the boat as well as checking the water quality of the water body. There is a conveyor mechanism included in the boat for collecting algae and garbage.

3 Methodology

This system utilizes a four-propeller boat that is specifically designed for water bodies such as lakes and ponds. The boat's movement is controlled by a microcontroller that is connected to the internet, allowing for wireless operation from anywhere. The boat is equipped with various sensors, including turbidity, TDS, pH, ORP, and NEO - 6M GPS module, as well as a camera with pan and tilt capabilities.

[1] Their study proposes an overview of modeling techniques and automations that can be used to track marine debris flows and make a collective database about them. This includes data on amount, type, and distribution of marine litter. An autonomous amphibious [2] vehicle for monitoring and collecting marine debris. Their prototype consists of an amphibious vehicle that can collect floating garbage on the surface of waterbody as well as on the shorelines and riverbeds without human assistance. A system called sea debris detection using deep learning for automation of detection and extraction of the debris region from an input image by image processing [5].

These sensors collect data in real-time, which is then transmitted through Thingspeak servers via the internet. The primary objective is to continuously monitor water quality parameters such as turbidity, TDS, pH, and dissolved oxygen content in water bodies. The boat is not only used for monitoring the quality of water and algae growth. There is trash collecting mechanism installed in the boat which consists of a conveyor mechanism to collect trash from the water bodies and bring them to the shore to recycle or dispose of them in a correct way. Figure 2 illustrates the flow diagram of the proposed work. Figure 3, 4, 5, 6, 7 and 8 shows the simulation model of proposed work. Where Figure 9 illustrates the working model of the boat .The system [3] would use advanced technology to visually detect trash and aid in the eventual removal of such debris.

The trash collecting mechanism consists of a deep learning-based model to identify the floating objects and classify them as trash or not, a conveyor mechanism to collect the trash and a trash collector [4] can with load cell to measure the amount of trash collected in a particular location which can be used later to track trash disposal.

The trash identification mechanism can be done either by using "Raspberry pi camera" and a custom model or by using pre-trained solutions such as "Husky Lens" which will make the process a lot easier. But the drawback in going with the premade solution is one cannot customize the kind of trash that he is targeting for.

Firstly, the Path for the boat is programmed in the waypoint mission planner software by placing GPS pins in the aerial map of the lake. Then the boat is placed in the lake and

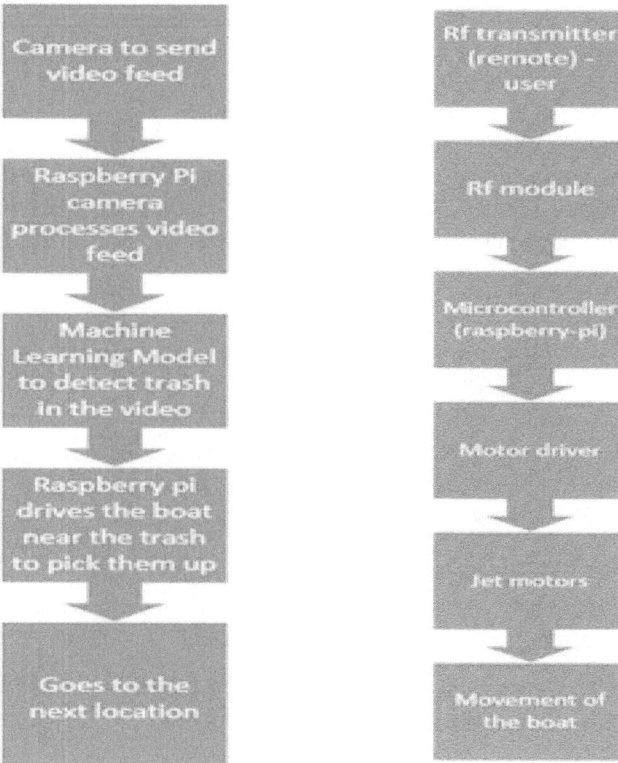

Fig. 2. Flow Diagram

turned on. The boat will autonomously drive itself to the first way point, and collect the trash around the waypoint. After collecting trash for a certain time, it will cruise around to check if there is some trash left around. If found, the boat will collect them, if not it will cruise on to the next waypoint.

4 Working of the Controller

The controller has an ESP32 inside connected with a NRF 24 module, it received sensor data from the boat and uploads that to the cloud and also is used to control the boat. The CAD model is made using the free online design tool called Autodesk Tinkercad, the frame is made using ¾ in. PVC pipes and joined using appropriate fittings. A garbage collection of mesh walls of a suited volume is placed on the PVC structure. In order to maintain the buoyancy of the boat a float is added to the bottom of the boat. This is made by folding and fixing inner tubes of vehicle tires.

In order to propel the boat four paddle wheels attached with geared motors are placed two on either side of the boat. These paddle wheels are custom designed in Tinkercad. These motors spin at two hundred revolutions per minute to keep the boat low and slow. The boat is intentionally kept slow to prevent fishes and other flora and fauna from getting tangled with the boat. In order to power these a 6000 mAh LiFePO$_4$ battery pack is included. Solar charging functionality is also implemented by making the top cover of the boat as a photovoltaic material.

Although, the boat can't be run completely on solar power, yet, energy from solar power can be used to charge the batteries and improve the efficiency of the boat. The above discussed sensors like pH sensor, Turbidity sensor, TDS sensor and ORP sensor are placed in the center of the float so that they are in continuous contact with the water. They can measure and send out data about the quality of water in the water body that is running on.

Fig. 3. Front side view of the boat

Fig. 4. Back side view of the boat

5 Specification

Based on the discussed methodologies an initial prototype is developed to prove the concept and to test the efficiency of this proposed solution and the specifications of the built prototype is given in the below Table 1.

Fig. 5. Right side view of the boat

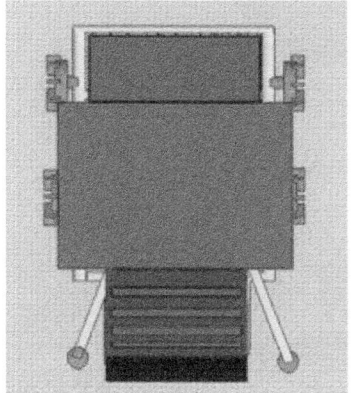

Fig. 6. Top view

Fig. 7. Front view

6 Results and Discussion

Using the above stated methodologies and techniques, a model has been constructed and tested. Pictures of those are given below.

Figure 10 and 11 shows the construction of the boat and Transceiver module of the boat. The boat has been tested in a close to realistic environment and the project is found to be working as intended.

The readings and inferences from various sensors included in the project are tabulated below.

S. No.	Sensor	Output	Inference
1	TDS	more than 10,000 mg/l	Salty Water
2	TURBIDITY	1 to 3 FNU	Dirty Water
3	PH	7.1	Neutral
4	BOD	6.5 ppm	Polluted

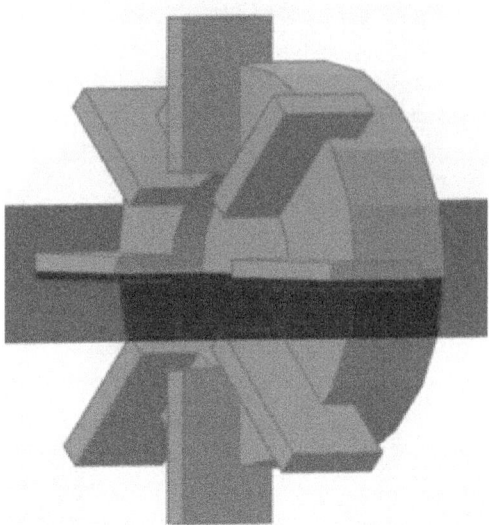

Fig. 8. Paddle wheel

Fig. 9. Model

The boat movement is controlled by using remote as well as a mobile phone application and the video feed is viewed on the laptop screen.

Table 1. Specifications

Specification	Value
Dimensions (L * B * H) In mm	762 * 381 * 300
Weight	4 kg
Battery Capacity	6000 mAh
Motor Specs	4×12 v motor each of torque 5 kg cm^{-1}
Solor Specs	12 v, 10 w solar panel
Controller	Raspberry Pi 4

Fig. 10. Boat Construction

Fig. 11. Transmitter and Receiver model

7 Conclusion

The amount of trash currently floating in the Pacific Ocean alone is estimated to about 1.6 million square kilometers and a span of around 620,00 square miles. Boats like these when used in large numbers can wipe out the trash and stop them from being consumed by the fish as micro plastics. Cancer due to consuming plastic fishes can be prevented. Fish population can be maintained by removing the algae formations in lakes, ponds and fish breeding areas.

8 Future Scope

This system has the potential for further development, including the incorporation of additional sensors and improved build quality, enabling it to be deployed in waterbodies for long-term autonomous operation without human intervention. Multiple bots can be integrated into a cohesive system, and a charging station powered by solar energy can be established in nearby forests, allowing these boats to autonomously collect data within a predefined radius.

References

1. Fulton, M., Hong, J., Islam, M.J., Sattar, J.: Robotic detection of marine litter using deep visual detection models. In: 2019 International Conference on Robotics and Automation (ICRA) (2019). https://doi.org/10.1109/icra.2019.8793975
2. Brown, A.H., Niedzwecki, J.M.: Review of modeling techiniques for marine debris flows. In: OCEANS 2018 MTS/IEEE Charleston (2018). https://doi.org/10.1109/oceans.2018.8604605
3. Patil, C., Tanpure, S., Lohiya, A., Pawar, S., Mohite, P.: Autonomous amphibious vehicle for monitoring and collecting marine debris. In: 2020 5th International Conference on Robotics and Automation Engineering (ICRAE) (2020). https://doi.org/10.1109/icrae50850.2020.9310888
4. Ruman, M.R., Das, M., Mahmud, S.M.I., Kumar Nath, S.: Automated marine surface trash cleaner. In: 2019 IEEE 5th International Conference for Convergence (2019)
5. Akhbarizadeh, R., Dobaradaran, S., Nabipour, I., Tajbakhsh, S., Darabi, A.H., Spitz, J.: Abundance, composition, and potential intake of microplastics in canned fish. Mar. Pollut. Bull. 160(August), Article no. 111633 (2020)

Early Detection of Autism Spectrum Disorder (ASD)-A Deep Learning Approach

G. Nirmalapriya[1], K. Ravikumar[2(✉)], M. Jayasri[2], and V. PrasannaKumar[3]

[1] Department of ECE, Rajalakshmi Institute of Technology, Chennai, India
nirmalapriya.g@ritchennai.edu.in

[2] Department of Computer and Communication Engineering, Rajalakshmi Institute of Technology, Chennai, India
{ravikumar.k.2020.cce,jayasri.m.2020.cce}@ritchennai.edu.in

[3] Computer Science Engineering, Rajalakshmi Institute of Technology, Chennai, India
prasannakumar.v.2020.cse@ritchennai.edu.in

Abstract. In this research, we put together a deep learning strategy that brings together the medical biomarker data from EEG tests and information from both clinical observations and inquiries to recognize the children with a Autism Spectrum Disorder (ASD). We expect that it can improve the effectiveness of identification significantly in addition to minimizing the expenditure. To begin with, our approach employed a pioneering tactic for selecting a neuronal signal landmarks derived from an eye movement, facial expression, and EEG results. This research showed that using a combination of Local Binary Pattern algorithm and deep learning classification led to a 87.50% accurate classification rate for multi-modal data merging. Finding a point to this method of detection as highly useful and effective for diagnosing ASD in young children. Graphs and confusion matrices unveiled different discrimination methods of EEG, personal behavior, and questionaries' for discrepancy discrimination between ASD individuals and typically developing kids, with EEG being the most useful in drawing distinctions. Studies portray the physiological and behavioral important complementary characteristics. The combination of complementary information in this study proposed the deep learning approach has led to a significant improvement in classification accuracy.

Keywords: Autism Spectrum Disorder (ASD) · EEG · CNN

1 Introduction

ASD (Autism Spectrum Disorder), a neurological development issue accompanied by behavioral and cognitive issues, starts mainly at an early age. However, there is no specific explanation for ASD, likewise no viable medical intervention have been noticed so far. The Centre for Disease Control and Prevention uncovered an often overlooked fact; today, millions of kids are behaving perplexing with the percentile climbing to 1 in 54 in the US.

It has been suggested that the abnormal behavior in ASD may originate from a brain which has adapted due to exposure to an unfriendly environment, rather than a result of

© The Author(s), under exclusive license to Springer Nature Switzerland AG 2024
P. Dassan et al. (Eds.): ICICSCNT 2023, CCIS 1970, pp. 301–311, 2024.
https://doi.org/10.1007/978-3-031-75957-4_26

longer-term neural degradation. Since children's brains develop quickly during child-hood, diagnosis and supportive treatment at an earlier age could protect their brains from a combative environment and thus it hopefully leads to a better outcome. ASD is now a worldwide issue with draining economic and emotionally costly consequences. Previous research has revealed that children's frame of mind decreases as they age, and interven-ing early on in their development can be advantageous by improving their abilities to comprehend language and thus solve issues. As a result, discoveringAutism Spectrum Disorder (ASD) earlier is of paramount significance. Alas! the existing methods of diag-nosingASD rely heavily on assessing symptoms using observance, which requires lots of effort and fiddly work. An example is the Modified Checklist for Autism in Toddlers, an official survey given to parents in strictly structured medical surroundings that usually takes multiple hours to finish. Therefore, a smart robotic recognition aid is necessary to improve diagnostic accuracy and ease of use. A number of biological and behav-ioral data have been indicated as an effective for identifying ASD in TD youngsters. Children with ASD exhibit disturbances in connecting socially, especially concerning the non-verbal signals of eye contact and mimicking facial expressions; moreover, they especially lack regular attention seeking behavior, interacting socially, and in express-ing emotions. Significant research has focused on investigating eye patterns in children with autism. Researchers looked at 31 children with ASD and 51 children with typical development (TD) by getting them to look at emotional faces. Subsequently, the ASD kids showed changes in how they were serving the faces as opposed to their TD peers.

These difference may stand as useful in initially determining the individuals by the way of finding with the answers of SRS and SCQ.

Commonly used screening instruments to evaluate social communication abilities and to identify autism spectrum disorder include the Social Responsiveness Scale (SRS) and Social Communication Questionnaire (SCQ). Typically, parents, caretakers, or teachers who have seen the person's social behavior fill out these survey. In this study, we employed a novel technique to extract EEG and the screening equipment SRS and SCQ, a hybrid technique for multi-model data fusion based on local binary pattern algorithms, were described. A 65- item questionnaire called the SRS is used to assess social skills in people between the ages of 4 and 18. It evaluates social abilities, social cognition, social communication, social skills, and autistic mannerism. Higher scores on a 4-point Like rt scale, which rates the responses, indicate more severe ASD related symptoms. The SCQ is a 40- item questionnaire that evaluates social communication abilities and ASD related behavior in people between the ages of 4 and 40. It emphasize social contact, repeated habits, and communication abilities. Higher scores on a 3- point Like rt scale, which rates the responses, indicate more severe ASD related symptoms.

In Sect. 2, the authors present the related works. In Sect. 3, methodology is presented. The experimental results and discussion are provided in Sect. 4, and the conclusions are given in Sect. 5.

2 Related Work

S. Shilaskar [1] research proposes using ML algorithms to predict autism and dyslexia traits across age groups. Real data sets were used to test the accuracy and efficiency of the prediction model, showing promising results. A. J. Syed, D. J. Durran [2] research develops an automatic facial emotion detection system using machine learning and image processing for children with autism to generate reports on their behavior during therapy sessions. It employs local binary patterns, CNN, Haar Cascade object detection Algorithm, and TensorFlow for emotion classification and face detection M. Cheng et al. [3] in his paper proposes a computer-aided diagnosis system for Autism Spectrum Disorder that uses audio- visual data to evaluate social interaction skills in children. The system achieves high accuracy and has the potential to help diagnose ASD in less developed areas. S. M. Mahedy Hasan [4], proposes a framework for evaluating Machine Learning (ML) techniques to detect Autism Spectrum Disorder (ASD) early. Four different Feature Scaling (FS) strategies and eight ML algorithms are used, achieving high accuracy rates. The proposed framework can guide healthcare practitioners in screening ASD cases.

M. F. El-Muhammady et al. [5], reviews recent studies on the use of adaptive human-robot interaction to improve the social skills of children with autism spectrum disorder, finding that robot- based interventions can be effective in enhancing social skills., X. Wang, Y. Ma, J. Cammon [6], paperexplores the use of self-supervised learning methods to improve the efficiency of resources in emotion classification from EEG signals. Results show improved performance and computation time compared to fully-supervised methods X. Du, M. Kong [7] proposed a 3D-CapsNet EEG signal recognition model is proposed to improve identification of motor imagery by integrating temporal, spatial, and feature relationships. It performs well on BCI competition IV dataset 2a, improving accuracy and overcoming individual variability. J. W. Choi et al., review explores recent studies that have integrateddimmersive virtual reality displays with electroencephalograms across various fields to study and applybrain signals non-invasively, discussing their purposes, experimental designs, effects, limitations, trends, and future research opportunities [8]. Y. Hu et al. proposes a cross-space convolutional neural network (CS-CNN) for four-class motor imagery-electroencephalogram classification is proposed to overcome limitations of single-space algorithms and increase subject specificity, achieving high accuracy in experiments [9].

J. Wang et al., [10] conducted experiment to collect EEG signals from hearing-impaired subjects while watching emotional movie clips. Microstates were extracted using theK-means method and classified using a 1-D deep residual shrinkage network, achieving an average accuracy of 87.48%. K. M. Tsiouris An unsupervised method for detecting epileptic seizures in EEG signals is proposed, based on concentration of rhythmic activity in specific frequency bands and signalvariation. Channel localization is also evaluated [11]. S. S. P. Kumar et al. [12], This project aims to detect epileptic seizures in multichannel EEG using Higher Order Spectra (HOS) to distinguish epileptic EEG from normal and background EEG with high confidence, M. Fani et al. A novel approach for multiclass EEG signal classification uses features derived from instantaneous frequency and energy in sub-bands [13]. High discrimination is achieved between healthy and epileptic patients with short EEG segments.

3 Methodology

In order to obtain voltage values from the integer recordings in the csv file generated by the Emotive software, it is required to multiply the EEG signal by a factor of 0.51×10^6. This conversion is necessary to obtain accurate voltage readings from the Epoc + device. Figure 1 shows EEG channels grouped into three brain regions: anterior, central, and posterior.

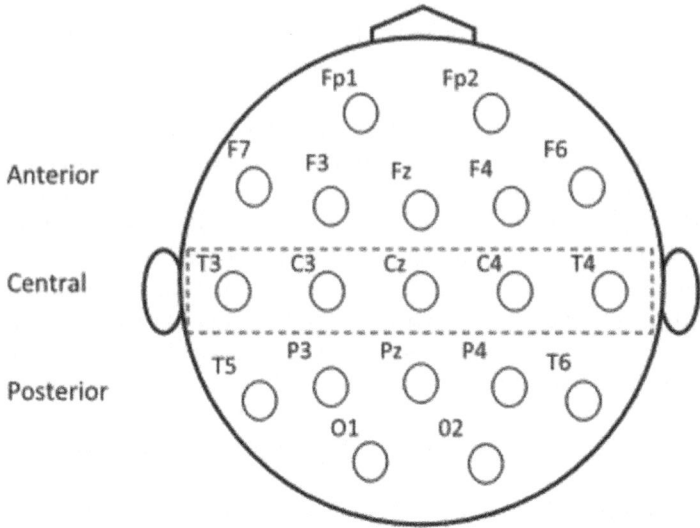

Fig. 1. EEG channels grouped into three brain regions: anterior, central, and posterior

3.1 Band Power Separation

As seen in the Fig. 2, it is standard procedure in EEG signal processing to divide the power spectrum density into separate frequency bands like Delta (1–4 Hz), Theta (4–8 Hz), Alpha (8–12 Hz), Beta (12–30 Hz), and Gamma (30–50 Hz). During the feature extraction stage, these frequency bands are used to calculate relative powers andratios. With the exception of the Delta band, the Emotiv software supplies each band's power as well as the LowBeta and High Beta powers inside the Beta band independently. However, both Beta band capabilities are pooledfor this study.

In this study, the Emotiv software is utilized to calculate the power spectrum density in absolute values with units $\mu V2/Hz$ by using two-second windows consisting of 256 samples, which are then separated into different frequency bands. The electrodes F3, P7, F4, and P8 are selected for analysis due to their high coherence in attention tasks. An example of band power separation can be seen in Fig. 3.

The General framework for ASD classifier is shown in Fig. 4. The first step in the process involves extracting features from the EEG signal. This could involve techniques

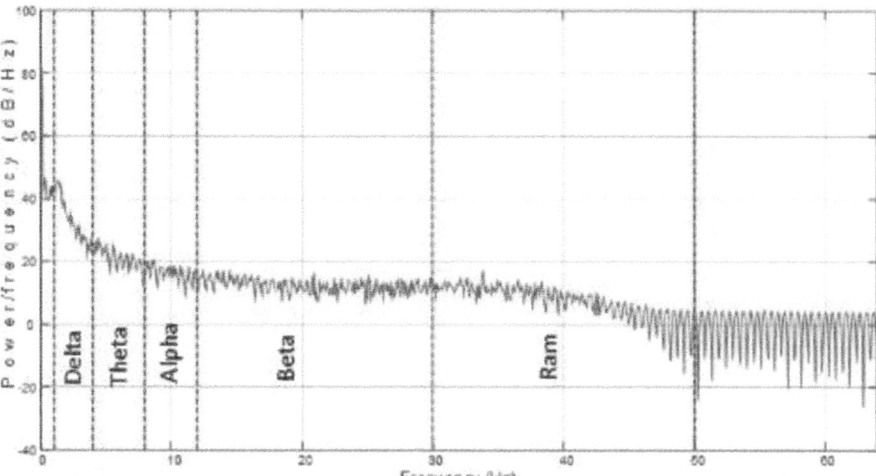

Fig. 2. Band power separation example, Welch power spectral density estimate (illustrative figure)

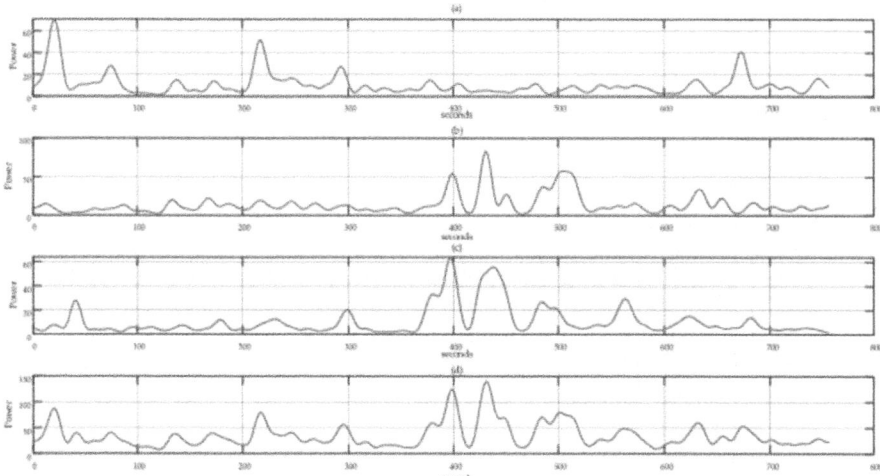

Fig. 3. Band power separation example from the F4 electrode. (a) Theta Band power (b) Alpha Band power (c) Beta Band power (d) Total Band power

such as wavelet analysis or time-frequency analysis to identify specific patterns or characteristics in the signal that are relevant to ASD detection. Once the features have been extracted, the signal is segmented and a Spectrogram is obtained, which can provide a visual representation of the signal over time and frequency. In addition to the EEG signal, the SRC and SCQ are also taken from a database. These are standardized questionnaires used to evaluate social communication skills and behaviors, which are often affected in individuals with ASD. The next step is to combine the features extracted from the EEG signal with the information from the SRC and SCQ, which are likely in the form of

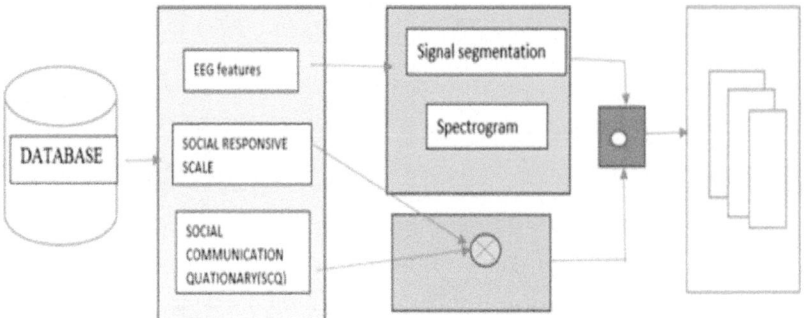

Fig. 4. General framework for ASD classification

numerical scores. This combined information is then used as input to a CNN shown in Fig. 5, which is trained to classify individuals as either having or not having ASD based on the patterns it identifies in the input data. Overall, this approach involves integrating multiple types of data to provide a more comprehensive assessment of ASD, which may lead to more accurate diagnoses and better treatment outcomes.

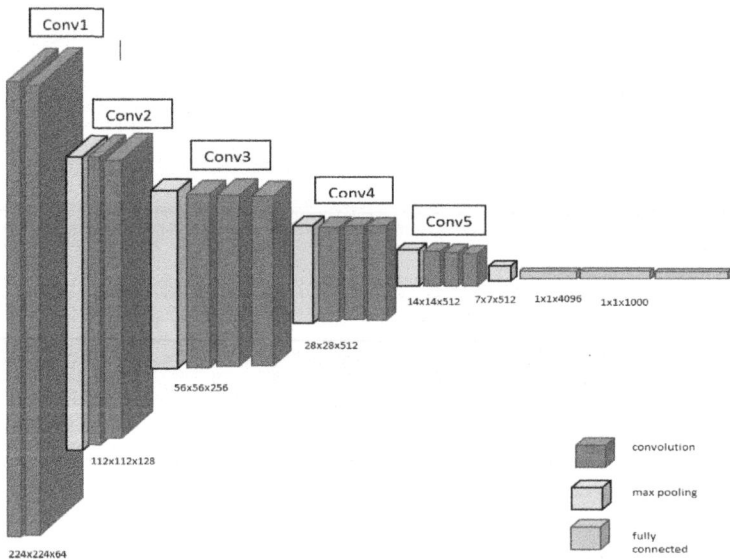

Fig. 5. Architecture and implementation of CNN

VGG-16 is a Convolution Neural Network (CNN) architecture that has been primarily used for image classification tasks [14]. It consists of 16 layers, including 13 convolution layers and 3 fully connected layers. However, it is not specifically designed for the detection of autism spectrum disorder (ASD).

4 Result and Discussion

We used a publically available data set to categorize EEG signals for our analysis. The data set comprises of EEG signals from 15 male right-handed patients with ASD (mean age: 31 months, SD: 6) and 15 male subjects with usual development (mean age: 29 months, SD: 4). Clinical psychologists used generally accepted diagnostic criteria to determine that the patients had ASD. A 32-channel device based on the worldwide 10–20 system and a reference electrode placed at the tip of the nose were used to capture EEG data after visual stimulus. The data were filtered with a band pass filter between 0.1 and 50 Hz and collected at a rate of 1000 Hz.The study included an average of 81 epochs for ASD patients (SD = 8) and 83 epochs for typically developing subjects (SD = 7), with each epoch lasting 400 ms post-stimulus. The data set includes 10 channels: P8, TP8, T8, P7, FT8, TP7, F8, T7, FT8, and F7.

Due to the limited number of patients in the data set, we used cross-validation to construct the suggested method in Python and assess its performance. We specifically employed the k-fold cross-validation approach with k = 5, which randomly divides the data set into k equal-sized divisions, using one partition for testing and the other partitions for training for each of the k repetitions. Averaging the five-fold performance results provided insight into the categorization performance of the system for ASD participants. Accuracy (21), sensitivity (5–7), specificity (5–7), positive predictive value (5–7), negative predictive value (−5) and F1-score (79) were the evaluation criteria we used. Equations (1)–(5) provide the mathematical formulas for computing various performance measurements.

- Accuracy (Acc): It provides the correct prediction of the classifier.

$$Acc = \frac{TN + TP}{Total} \qquad (1)$$

- Sensitivity or recall (Sen): It expresses the ability of the scheme to identify subjects who have ASD correctly.

$$Sen = \frac{TP}{TP + FN} \qquad (2)$$

- Positive Predictive Value (PPV) or Precision: It provides the probability of how likely it is that the subject has ASD.

$$PPV = \frac{TP}{TP + FP} \qquad (3)$$

- Negative Predictive Value (NPV): It expresses the probability of how likely it is that the subject is a typical developing subject

$$NPV = \frac{TN}{TN + FN} \qquad (4)$$

- F1-score (F1): It combines both sensitivity and PPV in a single metric.

$$F1 = \frac{2 * Sen * PPV}{Sen + PPV} \tag{5}$$

where TP, FP, TN, FN, respectively, represent true positive, false positive, true negative, and false negative.

The potential of the proposed process for identifying ASD was further investigated using two methods. Firstly, wavelet-supported features were combined with a counterpoint recognition system, followed by classification using ML categorizes based on Fourier details. The classification results for the first scenario, based on the threshold classifier, are summarized in Table 1. The table indicates that using multi-scale approximation entropy, high accuracy was achieved only in the beta and alpha sub-bands. The highest accuracy of 86% was obtained in the alpha sub-band using channel P7. We integrated this feature with other characteristics that had 100% NPV or PPV to further increase accuracy in the alpha sub-band. For the classification decision, we used a series of if-then- else rules that took into account the feature offering 100% NPV or PPV first, then applied multi-scale approximation entropy in the alpha-band. According to our research, the accuracy rose to 93% whenwe merged multi-scale approximate entropy in channel P8's gamma sub-band with multi-scale approximate entropy in channel P7's alpha sub-band. However, as seen in Table 1, the sensitivity remained low at 86.6%.

Table 1. Accuracy Comparison of various classifiers

Methodology	Accuracy	Sensitivity	Positive prediction value	Negative prediction value	F1 score
CNN	92.6%	0.8722	0.900	0.802	0.902
VGG16	94%	0.9122	0.842	0.822	0.882
Navies Bayes + CNN	96.4%	0.9222	0.904	0.900	0.904
Navies Bayes + VGG16	97.2%	0.9482	0.924	0.888	0.892
LBP + VGG16	99.6%	0.9662	1	0.982	0.920

The second scenario involved training experiments using K80, T4, and P100 GPU with 52 GB of RAM and 8 cores of an Intel(R) Xenon(R) CPU @ 2.2 GHz. The aim was to study the classification accuracy of different EEG ubjects in the data sets using each possible combination of features resulting from the recursive feature elimination (RFE) algorithm for all combinations of machine learning (ML) hyper parameter values. The resultsshowed that the classification based on the deployment of different ML algorithms with hyper parameter tuning achieved the best overall accuracy values, as shown in Fig. 6. The classification with all ML algorithms achieved an accuracy score of 96% in classifying ASD cases, which is higher than the performance achieved with the threshold classifier approach. These results demonstrate the effectiveness of using ML models to recognize patterns and achieve reliable and accurate diagnostics for identifying ASD cases (Fig. 7).

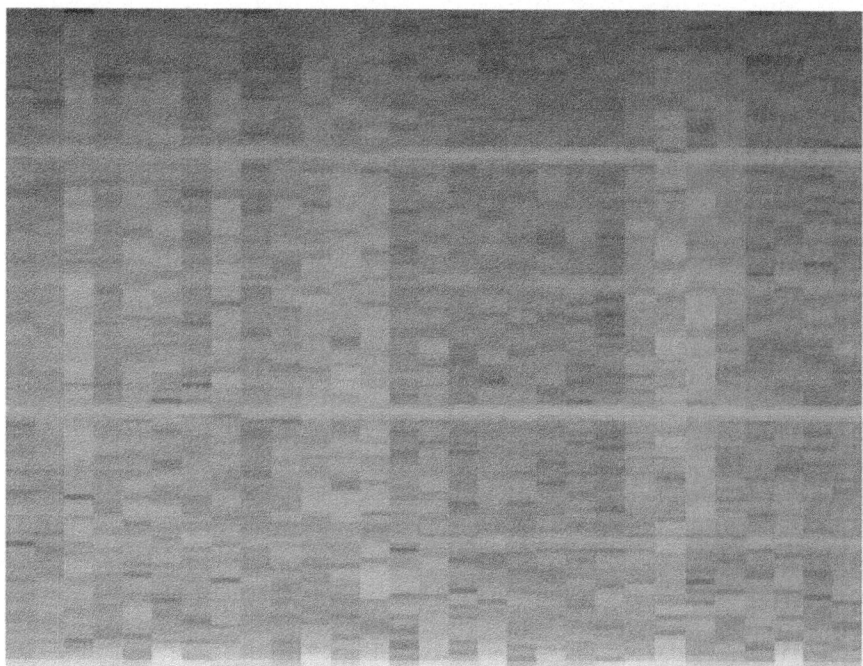

Fig. 6. Spectrogram of patient with ASD

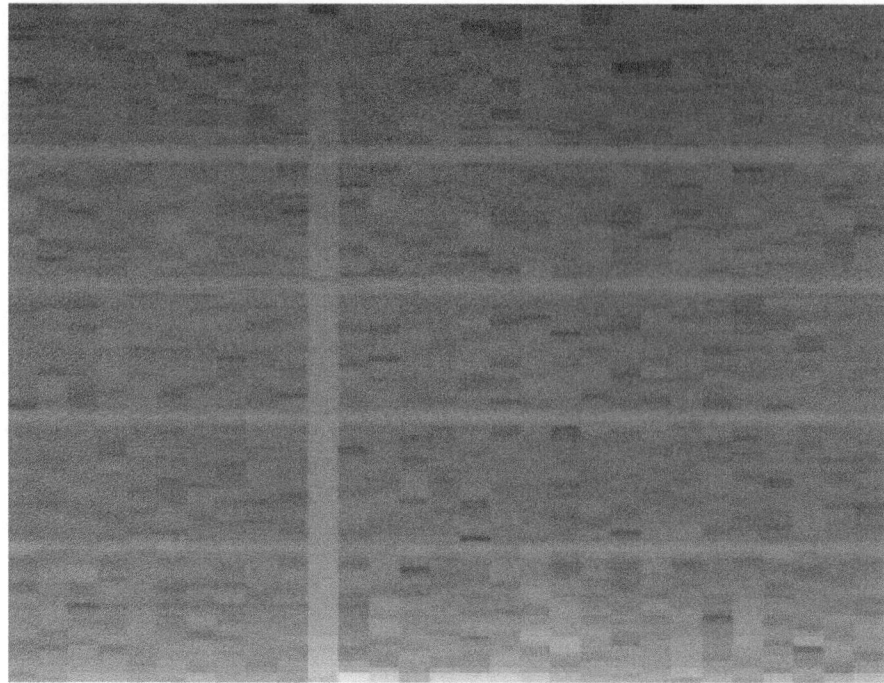

Fig. 7. Spectrogram of patient without ASD

5   Conclusion

In order to objectively diagnose Autism Spectrum Disorder (ASD) in early children, the paper suggests using EEG data. The study shows that, while maintaining a high level of accuracy in detecting ASD cases, the suggested scheme can be executed in a wearable sensor with constrained processing power. It is also demonstrated that the suggested deep learning-based embedded EEG analysis for ASD identification saves a large amount of energy. For on-node feature extraction and classification, the study combines nonlinear analysis, multi-scale estimated entropy in the time-frequency domain, and spectrum analysis (Welch) of EEG signals, which finds the ideal balance between energy efficiency and high accuracy. The proposed method will be prototype in the future as a wearable wireless sensor for an experimental use.

References

1. Shilaskar, S., Bhatlawande, S., Deshmukh, S., Dhande, H.: Prediction of autism and dyslexia using machine learning and clinical data balancing. In: 2023 International Conference on Advances in Intelligent Computing and Applications (AICAPS), Kochi, India, pp. 1–11 (2023). https://doi.org/10.1109/AICAPS57044.2023.10074161
2. Syed, A.J., Durrani, D.J., Shahid, N., Khan, W., Muhammad, A.: Expression detection of autistic children using CNN algorithm. In: 2023 Global Conference on Wireless and Optical

Technologies (GCWOT), Malaga, Spain, pp. 1–5 (2023). https://doi.org/10.1109/GCWOT5 7803.2023.10064653

3. Cheng, M., et al.: Computer-aided autism spectrum disorder diagnosis with behavior signal processing. IEEE Trans. Affect. Comput. https://doi.org/10.1109/TAFFC.2023.3238712

4. Mahedy Hasan, S.M., Uddin, M.P., Mamun, M.A., Sharif, M.I., Ulhaq, A., Krishnamoorthy, G.: A machine learning framework for early-stage detection of autism spectrum disorders. IEEE Access **11**, 15038–15057 (2023). https://doi.org/10.1109/ACCESS.2022.3232490

5. El-Muhammady, M.F., Yusof, H.M., Rashidan, M.A., Sidek, S.N.: Intervention of autism spectrum disorder (ASD) in a new perspective: a review on the deployment of adaptive human-robot interaction (HRI) system in enhancing social skill impairments. In: 2022 IEEE-EMBS Conference on Biomedical Engineering and Sciences (IECBES), Kuala Lumpur, Malaysia, pp. 1–6 (2022). https://doi.org/10.1109/IECBES54088.2022.1007926

6. Wang, X., Ma, Y., Cammon, J., Fang, F., Gao, Y., Zhang, Y.: Self-supervised EEG emotion recognition models based on CNN. IEEE Trans. Neural Syst. Rehabil. Eng. **31**, 1952–1962 (2023). https://doi.org/10.1109/TNSRE.2023.3263570

7. Du, X., Kong, M., Qiu, S., Guo, J., Lv, Y.: Recognition of motor imagery EEG signals based on capsule network. IEEE Access **11**, 31262–31271 (2023). https://doi.org/10.1109/ACCESS. 2023.3262025

8. Choi, J.W., et al.: Neural applications using immersive virtual reality: a review on EEG studies. IEEE Trans. Neural Syst. Rehabil. Eng. **31**, 1645–1658 (2023). https://doi.org/10. 1109/TNSRE.2023.3254551

9. Hu, Y., et al.: A cross-space CNN with customized characteristics for motor imagery EEG classification. IEEE Trans. Neural Syst. Rehabil. Eng. **31**, 1554–1565 (2023). https://doi.org/ 10.1109/TNSRE.2023.3249831

10. Wang, J., Song, Y., Mao, Z., Liu, J., Gao, Q.: EEG-based emotion identification using 1-D deep residual shrinkage network with microstate features. IEEE Sens. J. **23**(5), 5165–5174 (2023). https://doi.org/10.1109/JSEN.2023.3239507

11. Tsiouris, K.M., Konitsiotis, S., Koutsouris, D.D., Fotiadis, D.I.: Unsupervised seizure detection based on rhythmical activity and spike detection in EEG signals. In: 2019 IEEE EMBS International Conference on Biomedical & Health Informatics (BHI), Chicago, IL, USA, pp. 1–4 (2019). https://doi.org/10.1109/BHI.2019.8834644

12. Kumar, S.S.P., Ajitha, L.: Early detection of epilepsy using EEG signals. In: 2014 International Conference on Control, Instrumentation, Communication and Computational Technologies (ICCICCT), Kanyakumari, India, pp. 1509–1514 (2014). https://doi.org/10.1109/ICCICCT. 2014.6993198

13. Fani, M., Azemi, G.: Automatic epilepsy detection using the instantaneous frequency and sub-band energies of the EEG signals. In: 2011 19th Iranian Conference on Electrical Engineering, Tehran, Iran, p. 1 (2011)

14. Lopes, F., et al.: Automatic electroencephalogram artifact removal using deep convolutional neural networks. IEEE Access **9**, 149955–149970 (2021). https://doi.org/10.1109/ACCESS. 2021.3125728

Implementing Automatic ABCD Rule for the Classification of Benign and Malignant Skin Lesions

N. Logeswari$^{(\boxtimes)}$, A. Badri Krishnan, G. Nithishkumar, and S. Sashaank

Sri Sai Ram Engineering College, Chennai, India
logeswari.ece@sairam.edu.in

Abstract. Cases of melanoma have been gradually increasing for many years. 14,61,427 incidences of cancer were reported to be the projected total in India for 2022. The likelihood of developing cancer in India is one in nine persons where the North region of India has the highest AAR (Age adjusted rates) per 100,000 cases of cutaneous melanoma for both males and females, with 1.62 and 1.21, respectively. Males were more likely to develop non-melanoma of the skin or other skin malignancies in the East area than in the Northeast, where the incidence was greatest at 6.2 per 100,000. The most advantageous line of therapy is an early surgical excision. After removal at the earliest in-situ stage of melanoma, the life expectancy is unaffected. Early detection may be facilitated by automatic lesion analysis. Asymmetry, Border irregularity, Colour, and Diameter make up the ABCD rule of dermoscopy, which dermatologists use to quantify the findings and successfully identify malignant melanoma from benign lesions. Automatic ABCD trait recognition and differentiation between benign and malignant tumours enable early melanoma detection. The automated detection of hair and lesion borders is made possible by the employment of and geodesic active contours during the pre-processing stage. By modifying the kernel weights and attempting to increase the current accuracy, we hope to create a robust classifier model with OpenCV, Python, TensorFlow and UiPath. For the classifier model constructed, the obtained accuracy is 85.33% and other metrics such as precision of 84.48%, sensitivity of 85.33% and a specificity of 97.17% were obtained. When compared to other models that are already in use, this provides better results in terms of accuracy in predicting the rightful class of the lesion.

Keywords: Melanoma · TensorFlow · image processing · ABCD analysis · dermoscopy

1 Introduction

The skin, which covers roughly 20 square feet and shields the body from UV radiation and external contaminants, is the biggest organ in the body. Melanoma is a serious form of skin cancer that, if not caught in time, is fatal. Melanoma has a high mortality rate, with one person passing away from the condition every hour. Melanoma has been on the

© The Author(s), under exclusive license to Springer Nature Switzerland AG 2024
P. Dassan et al. (Eds.): ICICSCNT 2023, CCIS 1970, pp. 312–320, 2024.
https://doi.org/10.1007/978-3-031-75957-4_27

rise over the past thirty years, and it is now the sixth most prevalent cancer in women and the fifth most common cancer in men. Melanoma sufferers grew by 53% from 2008 to 2018 in number [1].

To distinguish between benign and malignant melanocytic lesions, several methods have been suggested. Dermoscopy is a non-invasive, in-vivo procedure that makes use of several incident light magnification devices and an oil immersion approach. It gives dermatologists a method for examining skin lesions, resulting in higher accuracy for spotting suspicious cases that would not be possible by visual inspection. Dermoscopy increases the diagnostic accuracy for melanoma by up to 50% as compared to simply visual inspection, depending on the observer's level of expertise [2]. Dermoscopy's ABCD (asymmetry, border irregularity, colour, and dermoscopic structure) rule was developed by Stolz et al. and Nachbar et al. to describe the geometrical and structural lesion properties [3, 4].

By evaluating a mole or spot's asymmetry, border irregularity, colour fluctuation, and diameter, the ABCD technique can detect possible skin cancer. Deep learning and Convolutional Neural Networks (CNNs), which were trained on massive datasets of dermoscopic images, were used to automate this process. These CNNs are able to recognise characteristics that can help in the detection of melanoma and can categorise skin lesions according to their ABCD traits [5–8].

This study focuses on the utilisation of datasets relevant to skin lesions from melanoma patients, pre-processing the picture using filters, and passing through a CNN model to extract the various properties such as asymmetry, border irregularity, colour, and diameter from the lesion. The efficiency of many alternative models that are currently in use as compared to the model developed in this work is also reported.

2 Methodology

2.1 Dataset on Images

Skin lesions from around 30,000 patients were collected in.JPG format and mapped with the ID of patient and location of the lesion in the skin is also mentioned in the CSV file.

The dataset consists of photos from different cultures, races, and age groups that have annotations for the location of lesions and whether they are cancerous or benign. The dataset has served as the foundation for a number of deep learning and machine learning tasks. The challenge's objective is to use machine learning and deep learning techniques to increase the precision and effectiveness of melanoma detection. The images were carefully selected, taking into consideration their accuracy, quality and dermo scopic features.

2.2 Block Diagram of Proposed Work

First phases involved gathering photos of lesions from a variety of patients, classifying the lesions in a CSV file according to the numerous classifications that each lesion comes under, and labelling targets against each class in the dataset.

The proposed block diagram is shown in Fig. 1. The images are stored in.jpg file. Anaconda navigator software is used for further processing of the image files.

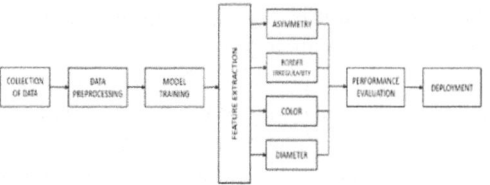

Fig. 1. Proposed block diagram

The image is then pre-processed so as to improve the quality such as sharpness, brightness, noise removal and image size normalization is also performed. The image is then subdivided into smaller regions for individual classification.

Features are extracted such as Asymmetry, Border Irregularity, Color and Diameter from the image and is used to detect the class of melanoma with the obtained result. A convolutional neural network model is built with 13 layers to perform the model training since CNN provides better results. After the features are extracted, the performance of the model is evaluated taking account of the accuracy and other important metrics. The model is finally deployed into use as a website where doctors and patients are the main stakeholders in utilizing the website for early diagnosis of the cancer.

2.3 Data Preprocessing

Using the Keras Image data generator's built-in preprocessing feature, we carried out a fundamental preparation step. Dermoscopy pictures at a resolution of 1872 x 1053 pixels are included in the dataset. One of the first steps in the pre-processing stage is to resize the photos to a 64 × 64 pixel size [6].

2.4 Feature Extraction

The four ABCD criteria can be divided into two groups: chromatic information (to compute the shape features) and representation of shape attributes (Asymmetry, Boundary, and Diameter). First, a binary segmented lesion's area, centroid, perimeter, and angle must be examined. These attributes are derived from the lesion image using image moments [7].

1) *Asymmetry*

In order to quantify asymmetry, the melanocytic sore is split into two 90° axes that are positioned to yield the least amount of asymmetry. The asymmetry score is 2 if both dermoscopic examinations show asymmetrical forms in terms of shape, colour, and dermoscopic structures. If there is only one pivot with asymmetries, which is quite unusual, the score is 1.

$$D_{(P,q)} = \sqrt{(x_1 + x_2)^2 + (y_1 + y_2)^2} \tag{1}$$

where the border pixels' coordinates (x1, y1) and (x2, y2) represent p and q along the lesion, respectively.

2) *Border Irregularity*

The shade example lesion's pigment pattern is evaluated after the lesion has been cut into eight sections. A score of one is given for an abrupt, unexpected cut-off of shadow at the margin inside each eighth-portion.

Yet, a constant, ill-defined cutoff inside the fragment obtains a score of 0. As a result, the border has a maximum score of 8 and a minimum score of 0.

$$V_i = (x_2 - x_1)(y_3 - y_1)(y_2 - y_1)(x_3 - x_1) \tag{2}$$

The lesion is divided into eight sections, and the pigment pattern is assessed. One point is awarded for a rapid, abrupt termination of the pigment pattern at the edge of each eighth section. Nonetheless, a gradual, fuzzy cutoff inside the segment receives a score of 0. As a result, the border has a maximum score of 8 and a minimum score of 0.

3) *Color*

One of the key characteristics that is utilised to identify skin disorders is colour. One of the crucial factors that aids in identifying the kind of cancer is the colour of the lesion.

Six colors—white, red, black, light brown, dark brown, and blue-gray—represent the lesion's hues. If a point is detected for each colour in the lesion region, it is assigned a point, added together, and multiplied by a weight factor of 0.5.

$$[(white + red + black + light\ brown + dark\ brown + bluegray) * 0.5] \tag{3}$$

4) *Diameter*

Diameter stands as one of the main features of identifying and diagnosing a skin lesion. In ABCD rule, the diameter takes the weight as 0.5 as (0.5 * diameter). If the diameter is greater than 6mm, it is malignant. If the diameter is less than 6mm, it is benign. The formula for calculating diameter of the lesion is mentioned below

$$D = \sqrt{\frac{4LA}{\pi}} \tag{4}$$

5) *Contour Detection*

For the aim of extracting the contours from the skin lesion image, the cv2 module is employed. One of the most crucial aspects of the image that contributes to giving the ABCD values, which are used to categorize skin lesions. The contour of the lesion can be used to determine characteristics of the lesion such perimeter, area, arclength, and convexity.

6) *CNN Model flow diagram*

The image is passed on as the input after it is resized to 64 x 64 in RGB format and typically consists of a set of pixels or voxels representing the image. The Conv2D layer with a neuron count of 256 is built and the image is further passed on to this layer and filtering of images take place. Each filter is a set of learnable weights that are used to convolve with the input layer, producing a feature map highlighting certain patterns or features in the input. After each convolutional layer, ReLU activation function is applied to the feature map and helps to introduce non– linearity into the model (Fig. 2).

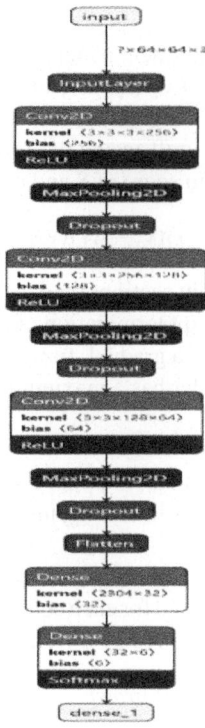

Fig. 2. CNN model flow diagram

3   Results and Discussion

(See Table 1).

Table 1. Overall Model accuracy and its parameters

Overall Accuracy	Sensitivity	Specificity	Precision	F1_Score
0.8256	0.817778	0.964722	0.804220	0.803412

3.1   Evaluation Process

To evaluate our model, we have selected 2 main evaluation metrics that were commonly used to monitor the performance of any CNN model. The first parameter is accuracy. Accuracy is the percentage of correctly classified samples. In CNN, there are 2 types of accuracies: Training accuracy (TRN – ACC) that measures how good a model is during the training process, and validation accuracy (VAL – ACC), which shows how good a model is while classifying the unseen data. Loss is the second metric that calculates the error during prediction, which is used to adjust the weights of neural network nodes. There are 2 types of loss: Training loss (TRN – LOSS) and the validation loss (VAL – LOSS) (Figs. 3 and 4 and Table 2).

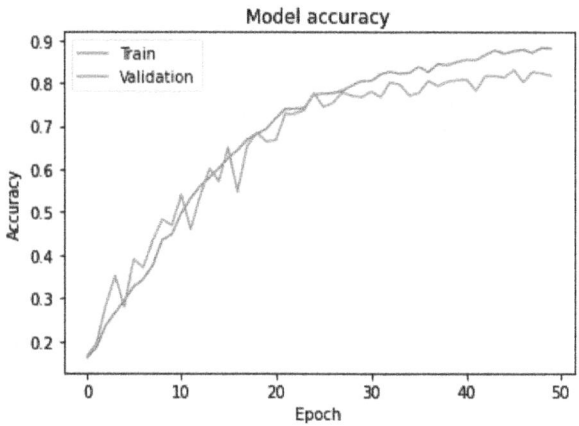

Fig. 3. Model Accuracy

Table 2. Dataset description and classwise image count

Classes	Numbers	Property	Description
Nevus	500	Format	JPG
Melanoma	500	Coloring model	RGB
Seborrheric keratosis	500	Dimension	64 × 64 × 3
Lentigo NOS	500	Dataset	ISIC
Lichenoid keratosis	500		
Solar Lentigo	500		
Total	3000		

We have plotted a graph for accuracy and loss with epochs in the x-axis and accuracy/loss in the y-axis. In the model accuracy graph, the blue curve represents the training accuracy and the red curve represents the validation accuracy (Test).

In the model loss graph, the blue curve represents the loss during training and the red curve represents the validation loss (Test).

In our deep learning model, we have achieved an overall accuracy of 85.33%. This indicates that the model has correctly predicted the class labels for 85.33% of the instances in the dataset. A high accuracy is generally considered desirable in a model, as it means that the model is making fewer mistakes. However, accuracy alone may not always be the most informative metric, especially in cases where the classes are imbalanced or the cost of false positives and false negatives is different. Therefore, it is

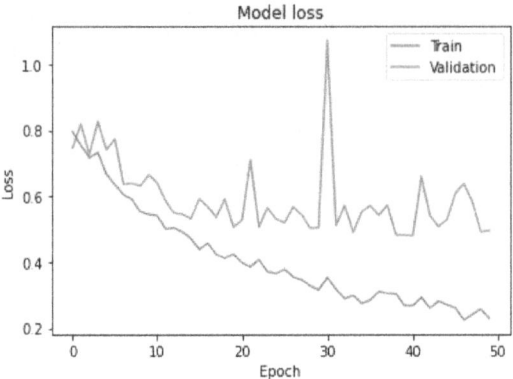

Fig. 4. Model loss

important to also consider other metrics such as precision, recall, and F1-score, in addition to accuracy, to obtain a more comprehensive evaluation of the model's performance (Figs. 5 and 6).

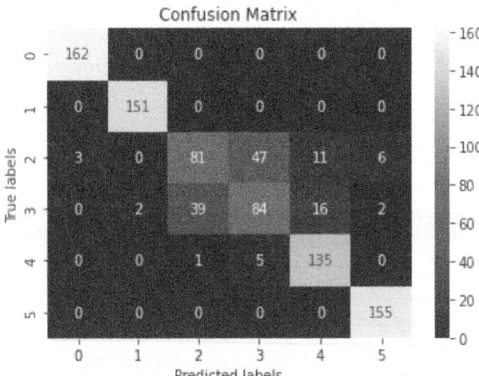

Fig. 5. Confusion matrix for the classification

A confusion matrix is a matrix that summarizes the predicted and actual classification results of a model for a given dataset. It is commonly used in supervised machine learning to evaluate the performance of a classification model. The matrix consists of four values, namely true positives, true negatives, false positives, and false negatives. In multiclass problems, the confusion matrix can help to identify which classes are being misclassified and to what extent. This information can help in improving the model by focusing on the areas of weakness (Table 3).

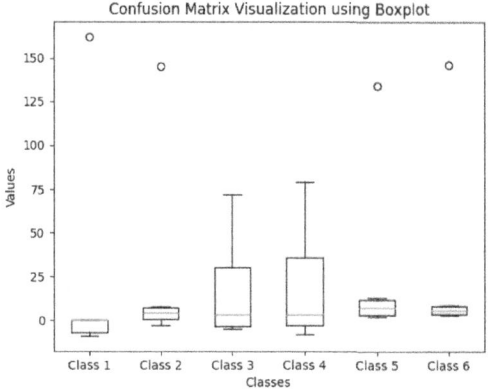

Fig. 6. Boxplot Visualization of the classwise accuracy

Table 3. Classwise accuracy and metrics

Classes	Recall	Specificity	Precision	F1-score
Lentigo NOS	1.000000	0.981030	0.920455	0.958580
L. Keratosis	1.000000	0.995995	0.980519	0.990164
Melanoma	0.513514	0.944149	0.644068	0.571429
Nevus	0.405594	0.948481	0.597938	0.483333
S. Keratosis	0.950355	0.923584	0.697917	0.804805
Solar Lentigo	1.000000	0.989262	0.950920	0.974843

4 Conclusion

Melanoma skin cancer incidence rates have increased during the past 20 years. Thus, early, quick, and effective skin cancer screening is crucial. Skin has one of the highest cure rates when caught early, and in the majority of cases, the procedure just cutting out the lesion. A mole must undergo all four analyses (ABCD), not only the first one that was described, if it is thought to be a melanoma mole.

Due to the subjective nature of the ABCD rule's interpretation, several efforts have been made to provide solutions that would allow for an impartial assessment of the ABCD rule's characteristics. Segmentation, feature extraction, and classification are the three primary phases of the automatic dermoscopic image analysis standard pipeline that were engaged in the suggested solutions. Combining such characteristics might improve melanoma detection's accuracy, sensitivity, and specificity compared to employing only one ABCD rule feature.

Collaborations with hospitals and medical imaging repositories will improve the scope of accuracy and this can be done by integrating the database of imaging with our model which can be deployed in cloud and can be used as a repository with premature calculations of ABCD and analyzing the type of cancer as soon as the image is uploaded in the website and this enhances the learning rate of the model and increase the precision which is even more beneficial for the users.

References

1. Gulati, S., Bhogal, R.K.: Detection of malignant melanoma using deep learning. In: Singh, M., Gupta, P., Tyagi, V., Flusser, J., Ören, T., Kashyap, R. (eds.) ICACDS 2019. CCIS, vol. 1045, pp. 312–325. Springer, Singapore (2019). https://doi.org/10.1007/978-981-13-9939-8_28
2. Stolz, W.J.E.J.D.: ABCD rule of dermatoscopy: a new practical method for early recognition of malignant melanoma. Eur. J. Dermatol. **4**, 521–527 (1994)
3. Nachbar, F., et al.: The ABCD rule of dermatoscopy: high prospective value in the diagnosis of doubtful melanocytic skin lesions. J. Am. Acad. Dermatol. **30**(4), 551–559 (1994)
4. Kasmi, R., Mokrani, K.: Classification of malignant melanoma and benign skin lesions: implementation of automatic ABCD rule. IET Image Proc. **10**(6), 448–455 (2016)
5. Ali, A.-R.H., Li, J., Yang, G.: Automating the ABCD rule for melanoma detection: a survey. IEEE Access **8**, 83333–83346 (2020)
6. Alvarez, A., Bajcar, S., Brown, F.M., Grzymala-Busse, J.W., Hippe, Z.S.: Optimization of the ABCD formula used for melanoma diagnosis. In: Kłopotek, M.A., Wierzchoń, S.T., Trojanowski, K. (eds.) Intelligent Information Processing and Web Mining. ASC, vol. 22, pp. 233–240. Springer, Heidelberg (2003). https://doi.org/10.1007/978-3-540-36562-4_24
7. Majumder, S., Ullah, M.A.: Feature extraction from dermoscopy images for melanoma diagnosis. SN Appl. Sci. **1**(7), 753 (2019)
8. Khan, M.F., Mufti, N.: Comparison of various edge detection filters for ANPR. In: 2016 Sixth International Conference on Innovative Computing Technology (INTECH). IEEE (2016)
9. Geman, D., et al.: Boundary detection by constrained optimization. IEEE Trans. Pattern Anal. Mach. Intell. **12**(7), 609–628 (1990)
10. Naseer, I.: Removal of the noise and blurriness using global & local image enhancement equalization techniques. Int. J. Comput. Innov. Sci. **1**, 1–11 (2022)
11. Messadi, M., Cherifi, H., Bessaid, A.: Segmentation and ABCD rule extraction for skin tumors classification. arXiv preprint arXiv:2106.04372 (2021)
12. Monisha, M., et al.: Classification of malignant melanoma and benign skin lesion by using back propagation neural network and ABCD rule. Clust. Comput. **22**, 12897–12907 (2019)
13. Yamunarani, T.: Analysis of skin cancer using ABCD technique. Int. Res. J. Eng. Technol. **5**(04), 1864–1870 (2018)
14. Chaturvedi, S.S., Tembhurne, J.V., Diwan, T.: A multi-class skin Cancer classification using deep convolutional neural networks. Multimed. Tools Appl. **79**(39–40), 28477–28498 (2020)
15. Hosny, K.M., Kassem, M.A., Fouad, M.M.: Classification of skin lesions into seven classes using transfer learning with AlexNet. J. Digit. Imaging **33**, 1325–1334 (2020)
16. Bassel, A., et al.: Automatic malignant and benign skin cancer classification using a hybrid deep learning approach. Diagnostics **12**(10), 2472 (2022)
17. Senan, E.M., Jadhav, M.E.: Analysis of dermoscopy images by using ABCD rule for early detection of skin cancer. Glob. Transit. Proc. **2**(1), 1–7 (2021)
18. Li, Z., et al.: A classification method for multi-class skin damage images combining quantum computing and Inception-ResNet-V1. Front. Phys. **10**, 1120 (2022)

The Behaviour and Sentience of Artificial Intelligence

S. Shree Shankar$^{(\boxtimes)}$, S. Nikhil, G. Pramod, Prabal V. Khatawkar, and K. Durga Devi

SRM Institute of Science and Technology–Ramapuram Campus, Chennai, India
{ss1934,ns4826,pg8817,pv0466,durgadek}@srmist.edu.in

Abstract. The study of knowledge, sentience and machine behaviour is a complex field of inquiry which has come increasingly important in our new period of technology. As we approach the 21st century and further develop our understanding of artificial intelligence, sentience, and how these relate to one another also an ferocious study of this content is consummate. This disquisition paper examines the pretensions and possible behavioural patterns of artificial intelligence (AI) in relation to meta-mortal tasks. Specifically it explores The future implications of AI Machine behaviour and Possible paths towards post AI society.

Keywords: Sentience · Natural language processing · Deep knowledge · Machine Behaviour and structural patterns

1 Introduction

Thomaz & Breazeal et al. [1]. The paper examines the behavioural patterns of artificial intelligence, and the questions of sentience and machine behaviours under different conditions. Conscious AI is an area of study that focuses on determining whether artificially intelligent machines are analogous to humans. All algorithms bedded in machine knowledge systems and distributed across the network need to be included in a holistic understanding of system behaviour. A number of important provocations for the scientific discipline of machine behavior stem from the ubiquity and complexity of intelligent algorithms. Medium refers to what causes a behaviour in terms of positive or negative underpinning; important is the amount of emphasis each researcher places on the particular aspects of knowledge that do at different stages during the accession process. Max Planck institute for Human Development et al. [15], Machines contain processes that generate behaviour, experience development that incorporates external factors into behaviour, provide functional outcomes that lead to specific machines being more or less prevalent in different environments, and have evolving histories. That have enabled former surroundings and individualities to continue to impact machine behaviour across evolutionary history. Development of behaviour can be attributed to lifeless technical fields choices or it can be affected by programming tricks analogous as value functions used in machine knowledge algorithms. Enterprises that are produced by studying mass such like algorithms can predict interesting sensations in the real world, analogous

P. Dassan et al. (Eds.): ICICSCNT 2023, CCIS 1970, pp. 321–331, 2024.
https://doi.org/10.1007/978-3-031-75957-4_28

as complex transportation systems, intelligent distributed machines and computer networks. Studies of cooperative machine behaviour focus on synthetic life forms created by humanmade artificial intelligence systems and consider larger- scale social, profitable and political aspects of machine elaboration and elaboration in general. In this paper cooperative behaviour of algorithms applied to financial trading surroundings is studied using the system of dynamical systems analysis. Other issues include how far that intelligent machines affect authorities, surveillance, and combat, as well as the effect which bots have on results of choices and whether cooperative action may be made possible by AI technologies that help humans produce social ties. The study of machine behaviour is concerned with how intelligent machines impact the manner in which humans process information and make opinions. How these engineering procedures modify the acting behaviours of AI systems—whether a given action is caused by the training data or by a combination of an algorithm and data—is a crucial aspect of understanding machine behaviour. Both the science of artificial intelligence and ethics are very interested in how people influence machine conduct.

There are a number of studies in this area- but not much has been done empirically. Important of the disquisition has concentrated on defining what makes an reality sentient, or suitable to cortege some form of behaviour or agency. For illustration, can a mark on a computer be considered conscious because it reacts to commands? Or is it indeed conscious if it has no physical body(e.g., like an algorithm running CPU)? This paper narrows down on these questions by trying to understand how AI's body affects their capability to be conscious and bear. The idea is that any action performed by an reality can be considered as part of its behaviour if its influence is felt by other realities in the system; that is why we need to be suitable to define " Sentience " more efficient.

2 Related Work

1) Meriam Webster Sentience is defined as the feeling or sensation as distinguished from perception and study;
2) Machine behaviour Iyad Rahwan1,2,3,34 et all A number of important provocations for the scientific discipline of machine behaviour stem from the ubiquity and complexity of intelligent algorithms. Different types of algorithms run regularly in our society, and they have an increasing impact on our daily lives. Second, due to the intricate parts of these algorithms and the environment Given the environment in which they function, certain of their characteristics and behaviours may be difficult or impossible to formalise analytically Binns R. et al. [3]. Third, due to their wide use and complexity, it's extremely challenging to predict the way that intelligent algorithms will have an profitable or mischievous effect on humans.

– The analysis of the behaviour of certain intelligent machines

  focuses on those particular devices alone.

– The study of the exchange between individual machines, focusing on their system, is known as cooperative machine behaviour-wide conduct
– Artificial Intelligence had to obtain information as well as how we engage with the outside world.

- Through the direct engineering of AI systems, training on four active mortal inputs, and unresisting obedience of mortal behaviours from the data we produce every day, humans are shaping the behaviour of machines.
- The ethics of artificial intelligence and its wisdom are really concerned in how human behaviour affects that of machines.
- the study of behavioural patterns of AI and AI sentience is runner 2 of 3 an provocative and swiftly evolving field. As we continue to develop increasingly advanced AI technologies, it becomes essential to understand their behaviour and how they interact with the world around them

3   AI Sentience

Jonas, H. et al. (1984) [12]. Sentience is defined as the feeling or sensation as distinguished from perception and thought; or, more generally, awareness of what it is like to be in one's own body. Iyad Rahwan et al. [7] Sentient AI is an area of study that focuses on determining whether artificially intelligent machines can achieve such a state. Sentient AI is an artificial intelligence that possesses feelings, emotions, and/or consciousness. The term was first used in 1984 by American computer scientist Roger W. Moderate who developed the concept of a "sentience movement", which urged researchers to investigate the field of artificial intelligence from the perspective of increasing levels of sentience Kalluri, P. et al. (2020) [13]. To understand a complex machine's behaviour, all data associated with that behaviour must be accessible and interactable. This requires conducting experiments with perturbed inputs and observing their effect on system behaviour. The pervasiveness and complexity of intelligent algorithms are major drivers for the scientific field of machine behaviour. First, many different types of algorithms are used in modern culture, and they play a bigger and bigger part in what we do everyday. Quizlet et al. [16] The Second, certain of these algorithms' characteristics and behaviours may be hard to formalise analytically due to their complicated qualities and the settings in which they operate. Third, since they are so widespread and complicated, predicting the effects of intelligent algorithms on humanity—whether positive or negative—presents a serious challenge. In today's world, machines can now outperform humans in many respects. Artificial Intelligence (AI) has the potential to transform everything from healthcare to transportation, and even national security Nagel, T. et al. (1989) [14]. As we move forward with this new and exciting tech, we are also becoming more aware of its dangers. In recent times various multinational entities, individuals, and companies have come together to put a halt to the development of Artificial Intelligence due to the dangers of sentience.

3.1   Complexity and Ubiquity of AI

Artificial intelligence (AI) has proliferated in contemporary civilization and is now present in many facets of our daily lives. AI has completely changed the way we live, work, and connect, from personal assistants like Siri and Alexa to self-driving cars and cutting-edge medical diagnoses. The creation of algorithms and computer systems that can carry out operations that typically require human intelligence, such as learning, reasoning, decision-making, and perception, is what gives artificial intelligence (AI) its complexity. A variety of methodologies, including machine learning, deep learning,

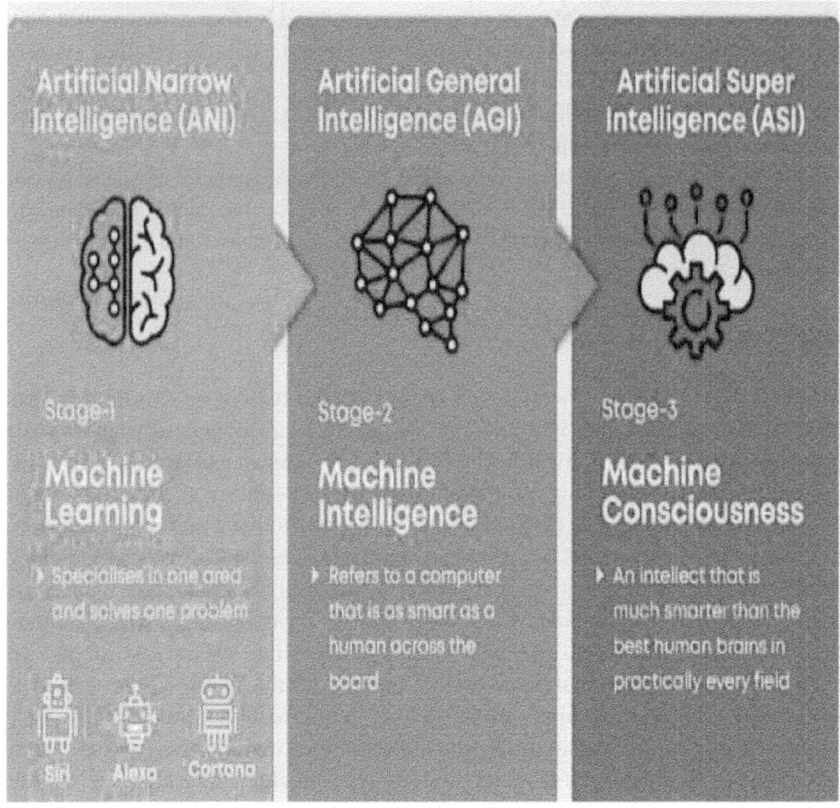

Fig. 1. Stages of machine learning

robotics, computer vision, and natural language processing, among others, are used to develop AI 7 technology. These methods provide AI systems the ability to analyse huge datasets, spot trends, anticipate the future, and react to user input. A number of industries, including healthcare, banking, transportation, manufacturing, education, and entertainment demonstrate how pervasive AI is in today's society. AI is applied to healthcare to create individualised treatment programs, identify diseases early on, and examine x-rays. AI is applied in finance for automated trading, risk management, and fraud detection. AI is utilised in the transportation industry to build self-driving automobiles, optimise routes, and increase safety. AI is applied to inventory management, predictive maintenance, and quality control in the manufacturing industry. AI is applied in education to create personalised learning plans and evaluate student performance. AI is utilised in entertainment to produce music and art, build realistic virtual worlds, and improve user experience. In conclusion, it is obvious that AI is complicated and pervasive in the present world. AI technology have the potential to revolutionise a number of industries and enhance millions of people's quality of life. Nevertheless, Addressing the ethical and societal concerns presented by AI is essential, as is ensuring that technology is developed and used in a responsible and transparent manner.

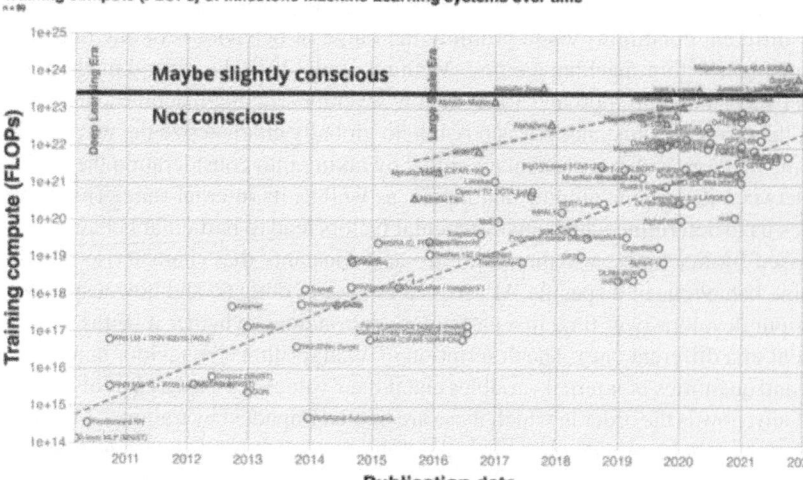

Fig. 2. Chart on training compute (FLOPs) of milestone machine learning systems over time

Goal Complexity Level	Description	Examples
Level 0	Programs that "just run" and have no goal seeking behavior	conventional software
Level 1	Basic goal seeking behavior	numerical optimization
Level 2	Complex goal seeking with the ability to seek subgoals and express reward hacking	genetic algorithm, reinforcement learner
Level 3 (danger)	Ability to realize instrumental goals and seek self improvement, ability to hide goals	humans, AGI

Fig. 3. Table on Levels of AI Sentience

4 Behaviour of Machines Under Different Circumstances

4.1 Individual Machine Behaviour

Iyad Rahwan et al. [7]. Individual intelligent machines are the focus of the research of individual machine behaviour. These studies frequently focus on characteristics of the specific machines that are fundamental to them and that are driven by their source code or design. Today, the majority of these studies are conducted in the fields of software engineering and machine learning. There are two common methods for studying the behaviour of certain machines. The first focuses on comparing one machine's behaviour under different conditions while profiling the range of behaviours of any one machine agent using a within-machine method. Medium et al. [20]. The second method, known as a between-machine approach, looks at how several particular machine agents perform under the same circumstances. It is possible to globally characterize the way in which a machine behaves within a particular context by taking into consideration the dependencies between its various outputs and inputs, as well as its internal state. This approach enables us to determine which environmental factors lead to particular behaviours being expressed by machines, whether there are any constants that characterise the within-machine behaviour of a specific AI across a range of contexts, and how a specific AI's behaviour develops over time in a particular environment (whether it be the same environment or a different one). The description of an algorithm's behaviour in terms of the types and quantities of internal variables and if their values are used as part of its processing. If any, how is the order in which these are treated impacted by training? For instance, an algorithm could only display particular behaviours if it has been trained on specific underlying data. The next concern is whether an algorithm that calculates the likelihood of recidivism for parole decisions will act unexpectedly when given assessment data that significantly deviates from its training data. We can consider different between machine studies of individual machine behaviour. Examining the inter-machine effect of doing tests with the same set of advertising inputs across platforms is another method that has been used to analyse the same behaviours as they differ between computers. For example, advertisers may test campaigns across a number of different ad platforms to investigate how different design principles influenced the effectiveness of their campaigns. Another example would be like-to-like testing in which a similar campaign is run on two separate platforms (for example Facebook and Twitter), which allows for comparisons across platforms by examining whether there are differences in the adoption rates and campaign results. Taking this approach allows for one to make inferences about the underlying mechanisms that influence behaviours exhibited by intelligent agents.

4.2 Collective Machine Behaviour

Crawford, K. et al [5]. The study of collective machine behaviour focuses on how individual machines interact with one another on a systemic level. This covers interactions with other groups or the surroundings of the collection. Until the collective level is taken into account, the implications of individual machine behaviour may not make much sense. Animal groupings, insect swarms, or mobile groups like flocks of birds or schools of fish are a few examples. Iyad Rahwan et al. [7].

The collective behaviour of algorithms and systems of artificial agents can be studied using information on their local interactions and collective behaviours. Speculations that are produced by studying swarm-like algorithms can predict interesting phenomena in the real world, such as complex transportation systems, intelligent distributed machines and computer networks. Other examples include neural network algorithms or evolving genomic swarms in biological models. Some interesting research questions remain to be resolved around this topic, such as how do we understand emergent properties of large collectives?

We argue that collective behaviour and collective cognition is a new research focus for biological science and engineering. Modern and emerging research in life sciences teach us that many systems have evolved elaborate forms of self organization that can be used for non-functional states (such as privacy, cache coherence or memory) but can also be used for functional properties such as resilience, economy or cheap energy. Bruce D. Meyer, University of Chicago and James X. Sullivan University of Notre Dame et al. [6], In this paper we've discusses the complexities in behaviours that may emerge from simple basic building blocks, by investigating how living things achieve complex collective behaviours through interaction between individuals. Robots are very good examples of machines that behave in ways that can be adapted to solve tasks which normally require intelligent behaviour (e.g., machines must observe their surroundings to detect obstacles). Instead of having a simple goal recognition system to identify objects in the environment (like an octopus or bird) they need a more advanced knowledge base like the human brain which is capable of sophisticated reasoning (e.g., the ability to predict future actions based on past experiences).

Collective machines are systems in which parts come together to form a whole that is different from the sum of its parts. They have an evolutionary history, but their evolution is guided by rules or goals to make them behave in ways that are both predictable and unexpected - for example, learning about what to do and how to perform it over time. Studies of collective machine behaviour focus on synthetic life forms created by human-made artificial intelligence systems but also consider larger-scale social, economic and political aspects of machine evolution and evolution in general.

Although studying complex systems is not new, it is sometimes difficult to apply a clear understanding of complex systems theory to real world problems. In this paper collective behaviour of algorithms applied to financial trading environments is studied using the method of dynamical systems analysis. Collective behaviour of human traders and financial algorithms are compared using two examples: flash crashes and intraday price jumps. The paper concludes that finance algorithms may create greater risk of market crisis than human traders because they are more susceptible to unexpected events and their actions cannot be anticipated by humans.

4.3 Human-Machine Hybrid and Its Impacts

4.3.1 Machines Shaping Human Behaviour

Artificial Intelligence e had a significant impact on human behaviour in a number of ways, including how we communicate and obtain information as well as how we engage with the outside world [21]. The greater reliance on technology is one of the most significant effects of AI and robotics on human behaviour. We now rely on technology for practically every facet of daily life, from cell phones to smart houses. Human behaviour has changed as a result of our greater reliance on technology, including less face-to-face interaction and more time spent in front of screens.

Technology and AI have also changed how humans interpret and process information. Huge volumes of data can now be processed and analysed swiftly owing to the creation of potent algorithms and machine learning. As a result, we now use different techniques for making judgements and solving issues. Sometimes, when we rely on machines to make decisions for us, we give up some of our autonomy and responsibility.

The nature of labour and employment has also changed as a result of AI and robotics. The rise of automation has altered both the labour market and how we approach work as many traditional employment are being replaced by robots. This has led to worries about job loss and economic inequality while also opening up new opportunities for those with expertise in fields like computer programming and data analysis [20].

In conclusion, machines and artificial intelligence have significantly influenced human behaviour in a variety of ways. While there are numerous advantages to these changes, such as greater production and efficiency, they have also sparked worries about the potential drawbacks of relying too heavily on robots. It is vital that we carefully evaluate the consequences of these technologies on human behaviour and try to reduce their possible negative impacts as we continue to develop and incorporate AI and machines into our daily lives.

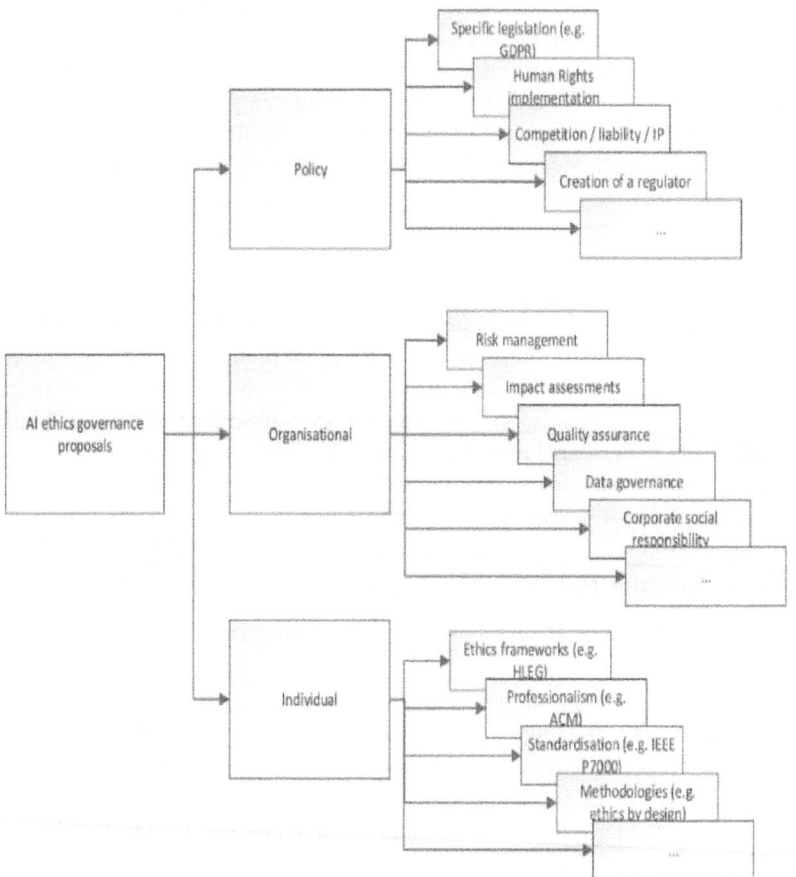

Fig. 4. Tree graph on AI ethics governance proposals (for example)

4.3.2 Humans Shaping Machine Behaviour

Through the direct engineering of AI systems, their training on active human input, and passive observations of human behaviour from the data that we generate every day, humans are directly shaping the behaviour of machines. Iyad Rahwan and others [7] Understanding how these engineering processes change the behaviours that emerge from AI systems—whether a specific behaviour is caused by the training data or by a mix of an algorithm and data—is essential to understanding machine behaviour.

4.3.3 Cooperation Between Man and Machine

How people affect computer behaviour is a topic that both artificial intelligence research and ethics are particularly interested in. The bulk of AI systems work in domains where they co-exist with people in complex hybrid systems, despite the fact that it might be methodologically useful to separate research into the ways that humans influence machines from studies into the ways that robots affect humans. Iyad Rahwan and others [7] The study of these systems must address issues pertaining to the behaviours that characterise human-machine interactions, such as cooperation, competitiveness, and coordination—for instance, sts4ir et al. [13] on how trade patterns may be changed by interactions between real-world and virtual trading bodies as well as how traffic management can be improved to be more effective in places with a lot of automobiles.

There are areas where a human could trump AI and machines in analytical thinking, for example, Humans are used to shaping machine behaviour and can enhance their efficiency. While robotic and computer-assisted surgery can increase the productivity of human workers. One example is the replacement of people by driverless automobiles, which also improve safety (at least, it's an area of artificial intelligence that is making great strides, and we may expect to see the results of these efforts in our lifetimes).

To date, studies on Human-machine interaction have been limited to simple peer to-peer interactions with the chatbot and the question about how humans and machines interact with each other. Such studies, where digital tools can be used to generate and define questions in an uncontrolled environment, may produce short-13 term engagements which are useful for generating ideas on what questions should be asked, however they also point out that many of the issues surrounding hybrid behaviour require not only the examination of the feedback between human influence on machine behaviour and machine influence on human behaviour, but also the examination of interactions between humans and machines that are comparable to or even superior to those between humans and other humans.

5 Conclusion

The study of behavioural patterns of AI and AI sentience is an exciting and rapidly evolving field. As we continue to develop increasingly advanced AI technologies, it becomes essential to understand their behaviour and how they interact with the world around them. The research on AI behavioural patterns has the potential to revolutionize many industries, from healthcare to finance, and improve the quality of our lives. Iyad Rahwan et al. [7].

AI sentience, in particular, has generated significant interest in recent years. The idea of creating machines that can perceive and understand the world in the same way as humans has long been a subject of science fiction. The realisation of this ideal, however, is now more achievable than ever thanks to recent developments in AI. Numerous possible effects of artificial intelligence (AI) consciousness on human culture and how humans interact with machines must be carefully considered. Derrick Johnson et al. [4] In conclusion, the study of behavioural patterns of AI and AI sentience has an ever more realistic and ever growing potential to change the world, for either the better or the worse, or both. As we continue to develop and integrate AI into our daily lives, it is crucial that we closely monitor its behaviour and develop ethical guidelines for its use. With the right approach, AI and AI sentience have the potential to improve our lives in countless ways, making us more efficient, productive, and capable of solving some of the world's most significant challenges.

References

1. Thomaz, A.L., Breazeal, C.: Teachable robots: understanding human teaching behavior to build more effective robot learners. Artif. Intell. **172**, 716–737 (2008)
2. Stone, P., et al.: Artificial Intelligence and Life in 2030. One Hundred Year Study on Artificial Intelligence: Report of the 2015–2016 Study Panel Stanford University (2016)
3. Binns, R., et al.: 'It's reducing a human being to a percentage': perceptions of justice in algorithmic decisions. In: Proceedings of 2018 CHI Conference on Human Factors in Computing Systems, vol. 377. ACM (2018)
4. Johnson, D.: Leveraging Artificial Intelligence to Improve Human Connection and Well-being — Encounter AI
5. Crawford, K., et al.: The AI Now report: the social and economic Implications of artificial intelligence technologies in the near-term (2016)
6. Amodei, D., et al.: Concrete problems in AI safety (2016)
7. Meyer, B.D.: University of Chicago and James X. Sullivan University of Notre Dame et al - Winning the War: Poverty from the Great Society to the Great Recession
8. Ferrara, E., Varol, O., Davis, C., Menczer, F., Flammini, A.: The rise of social bots. Commun. ACM **59**, 96–104 (2016)
9. Machine behaviour Iyad Rahwan1,2,3,34*, Manuel Cebrian1,34, Nick Obradovich1,34, Josh Bongard4, Jean-François Bonnefon5, Cynthia Breazeal1 , Jacob W. Crandall6, Nicholas A. Christakis7,8,9,10, Iain D. Couzin11,12,13, Matthew O. Jackson14,15,16, Nicholas R. Jennings17,18, Ece Kamar19, Isabel M. Kloumann20, Hugo Larochelle21, David Lazer22,23,24, Richard McElreath25,26, Alan Mislove27, David C. Parkes28,29, Alex 'Sandy' Pentland1 , Margaret E. Roberts30, Azim Shariff31, Joshua B. Tenenbaum32 & Michael Wellman33
10. Lazer, D.: The rise of the social algorithm. Science **348**, 1090–1091 (2015)
11. Tufekci, Z.: Engineering the public: big data, surveillance and computational politics. First Monday **19**, 7 (2014)
12. Jonas, H.: The imperative of responsibility: In search of an ethics for the technological age (1984)
13. Kalluri, P.: Don't ask if artificial intelligence is good or fair, ask how it shifts power. Nature **583**(7815), 169 (2020)
14. Nagel, T.: The View from Nowhere. Oxford University Press, Oxford (1989)
15. Sts4ir , Philosophy – Through a dark glass, clearly (sts4ir.com)

16. O'Neil, C.: Weapons of Math Destruction: How Big Data Increases Inequality and Threatens Democracy Broadway Books (2016)
17. How will AI change the world? – TED-Ed
18. Max Planck institute for Human Development Concept 1: Machine behavior | Max Planck Institute for Human Development (mpg.de)
19. Quizlet Comm 100 FINAL Flashcards | Quizlet
20. Medium Applying Behavioral Science to Machine Learning | by Jesus Rodriguez | DataSeries | Medium
21. How your brain organizes information – Artem Kirsanov
22. Independent.co.uk- Artificial Intelligence may already be 'slightly conscious; AI scientists warn

Detection of Affected Spina Bifida Infant Babies in Ultra-Sound Images Using LRMNet

R. Asha and S. S. Subashka Ramesh[(✉)]

Department of CSE, SRM Institute of Science and Technology, Chennai, India
subashka@gmail.com

Abstract. An estimated 150,000 new born are born apiece year with bifida, making it one of the most frequent central nervous system defects that do not compromise a fetus's chance of survival. Spina bifida is now more reliably diagnosed in the womb and treated in a very different way than it was even a decade ago. This study proposes the use of a localization and refinement module-based convolutional neural network (LRMNet) for the purpose of classifying spina bifida images. For better object classification, LRMNet considers each object to be a collection of features and uses both feature and context data. To avoid the complications that come with dealing with a wide range of forms and sizes, it is important to have accurate component information to guide the prediction of an item. To ensure precise component data generation, we create a part localization module that can solely rely on bounding box annotation to learn the categorization of component points. To facilitate better learning of component knowledge and feature representation, a context refinement unit is developed to combine local context information with global context info. The gathered photos are used to validate the model.

Keywords: Spina Bifida · Deep Learning · Bounding Box · Feature Representation · Localization and Refinement Module

1 Introduction

The prevalence of neural tube abnormalities has decreased dramatically in industrialised nations with the implementation of comprehensive prenatal screening, folic acid supplementation, and staple food fortification [1]. Second trimester screening is performed [2]; or at the routine anatomical scan at 18–24 weeks, with direct examination of foetal structures [3, 4]. Maternal blood markers and ultrasound nuchal translucency (NT) used for aneuploidy screening in the first trimester. Acrania may be easily seen during the NT scan, and some research [5–7] suggests that spina bifida can be identified at the same time. The foetuses in these investigations were at increased risk, and the ultrasound scans were performed early and thoroughly by experts. Efforts to use this knowledge in a standardised screening test have failed thus far [8, 9]. Additionally, we have documented ultrasound-visible structural characteristics of the embryonic brain at 11–14 weeks in

---

R. Asha—Research Scholar

P. Dassan et al. (Eds.): ICICSCNT 2023, CCIS 1970, pp. 332–343, 2024.
https://doi.org/10.1007/978-3-031-75957-4_29

instances with neural tube abnormalities, although these results did not lead to the development of a screening tool [10]. We also found that afflicted foetuses had abnormally tiny head circumferences, however we did not accept this finding as diagnostic evidence. The development of an automated system using Artificial Intelligence methods for the detection of spina bifida in foetuses is urgently needed.

Most current classification approaches include a bounding box localization module that predicts the locations of bounding boxes via the use of high-level feature maps and many convolutional layers. The representational power of discriminative characteristics is typically crucial to the efficiency of the bounding box localization module. The main obstacles in the categorization are the different look of the components and the different distribution of the parts. Traditional CNN struggle to learn discriminatory features, appearance, and geometric constructions for the reasons stated above. Therefore, the aforementioned techniques can't be utilised to diagnose people with spina bifida. We propose that the bounding box localization module may be improved by considering the components' locations and the spatial interactions between them.

One of the oldest and most basic categorization strategies is the part-based model. In most cases, part class templates are designed manually, starting with a basic component design. The supplied picture is then compared to the established template in an effort to locate the best potential matches. With supervised learning, training pictures have labels indicating where various components should be found. The EM technique allows models to estimate part positions and learn parameters in a weakly supervised learning way. Parts-based detection algorithms are often simple to deploy. However, part templates are shape-sensitive and depending on rotation and size. In addition, conventional component-based approaches rely on custom-built features that are unable to convey high-level semantic properties and are not amenable to end-to-end learning. Of several applications, including human posture prediction and fine-grained classification, features in part-based approaches thanks to the development of deep learning. Subjects are more challenging to recognise because to their irregular forms in comparison to these things.

Here, we propose the LRMNet for spina bifida, which employs a part-guided localization technique to enhance the standard regression formulation. Visual perception in humans served as inspiration for the model. To identify the separate components of an item, LRMNet employs a collection of points as part information. A covariant reaction to point deformation is calculated by the model and generated by the detector's detection result. The model also uses a multi-scale and localisation technique to recognise objects of varying sizes and forms. The acquired info about the affected parts is put into a detection framework that may aid in the diagnosis of spina bifida. LRMNet's enhanced capability to collect context information and enhance localization accuracy is particularly noticeable against complicated backgrounds. What we've brought to the table is summed up here. A uniform framework, LRMNet, is introduced here for detection. LRMNet is able to increase the performance of spina bifida identification in comparison to previous detection algorithms by incorporating part information into context information. To learn part information without any further supervision beyond the bounding box annotation provided by the proposed weakly-supervised part localization module's. Our technique enhances the categorization and localisation of objects with varying forms, sizes, and fuzzy borders by extracting multi-level component information. In order to extract layers, we develop a maps are aggregated to generate the context info, which is then used to mitigate the effect of intricate backdrops. To get additional discriminative characteristics, CRM may aggregate global local context info.

Here is how the rest of the paper is organised: In Sect. 2, we provide the linked research of previously published works that emphasises detection. Both the proposed model and its validation are briefly described in Sects. 3 and 4. Section 5 provides a summary and proposals for further research.

2 Related Works

The first spatial-temporal foetal brain for SBA was computed using a semi-automatic computational approach developed by Fidon et al. [11]. We analysed 90 MRIs of foetal patients diagnosed with SBA during 21–35 weeks of gestation. All of the scans were acquired as isotropic, motion-free 3D reconstructed MRIs. To increase the accuracy of spatial alignment of aberrant foetal brain MRIs, we present a strategy for the explanation of anatomical landmarks in brain 3D MRI of foetuses with SBA. Moreover, we provide a weighted generalised Procrustes method that employs anatomical landmarks for initial atlas construction. In this paper, we provide a weighted generalised version of Procrustes that can handle temporal regularisation and annotations. Once the initialization is complete, the atlas is upgraded the anatomical land-marks. White matter, ventricular system, cerebellum, extra-axial cerebrospinal fluid, cortical grey matter, deep grey matter, the eight tissue types that we parcellate in our foetal brain atlas using a semi-automatic technique. An analysis are precise enough. We find that the proposed atlas performs better than a standard foetal brain atlas when used for automated segmentation of brain 3D MRI in utero using SBA.

This research sought to enhance the U-Net method, a convolutional neural network algorithm, in order to arrive at a new algorithm, Oct-U-Net, for the automated detection and diagnosis of prenatal spina bifida in ultrasound pictures. The study's subjects were 3,300 pregnant women who all got 3D ultrasounds. The diagnostic performance of Oct-U-Net for foetal spina bifida was then assessed using the following metrics: recall rate, precision rate, execution time. Additionally, the U-Net method and the fully convolutional network (FCN) technique were presented for evaluation. In comparison to both FCN and U-Net, Oct-U-Net performed much better in terms of recall rate, precision rate, PA, and MioU. Oct-U-Net had an average runtime of 12.15 s and a mean standard error of 4.1243. Both of these metrics have much lower values than FCN and U-Net. In conclusion, Oct-U-Net was more suitable for clinical use due to its improved diagnostic efficacy on 3D ultrasound pictures of prenatal spina bifida, as shown by its greater segmentation accuracy and shorter running time.

In their research, Ajitha and Punitha [13] used the active contour (AC) algorithm to separate healthy and defective foetal spines from ultrasound pictures. The diagnostic facility provides the pictures to be analysed. Wiener filter and filter are used to clean up the pictures and get rid of any noise. Denoised image quality is evaluated using many metrics such as signal-to-noise ratio, peak signal-to-noise ratio, structural resemblance index measure, and mean square error (MSE). Histogram equalisation (HE) and its variant, adaptive histogram equalisation (AHE), are used to achieve this. Comparisons of entropy and adaptive mean brightness error are used to assess the excellence of the new contrast images (AMBE). The spine region is isolated from the revised image using the active contour technique. When the AD filter's parameters are fine-tuned, it beats the Wiener filter in denoising, as shown by the results. With regards to contrast enhancement, the findings distinguish AHE as superior than HE. The active contour approach may be used to segment the spine in images with or without spina bifida. This approach may have significant therapeutic implications in light of the urgent need for early identification of spina bifida.

Two-stage method integrating deep learning and traditional image processing approaches is proposed by Franz et al. [14]. (2DL-CIMP). Before anything else, we use a deep learning technique to separate the vertebrae of the spine. When fed into a tracing method, the resultant probability map is invaluable. The resulting data is a list of coordinates that describes the spine's midline. It is possible to generate view planes for the examination of abnormalities and measure spinal length along this line. If the FNR is less than 3, then our system successfully locates at least 97% of the annotated ground-truth spine line in at least 50% of the test instances. The discovered centerline's length may then be used to provide an automated suggestion for an estimated spinal length. A flex view, in which the spine and protruding ribs are shown in a straightened position, may also be created for further clarity.

The research conducted by Garcia et al. [15] investigated the use of SPADE, a conditional general adversarial network (cGAN) that acquires the label-to-image space mapping. The network was fed super-resolution T2-weighted brain MRI data from 120 foetuses (normal and diseased, gestational age range: 20–35 weeks) annotated for 7 tissue types. In order to train SPADE networks, 256×256 2D slices (image and label pairs) were taken from each of the reconstructed volumes in each of the four possible directions. A simple mean of the outputs of the three networks was used to integrate the volumes obtained from each direction into a single picture. We created incredibly realistic synthetic graphics using just the label maps. However, some of the finer characteristics, such as little vessels, were not synthesised. The correlation coefficient and structural similarity index both settled at 9772.00.016 and 9774.00.008, respectively. We artificially enlarged the ventricles in the segmentation map and produced synthetic MRIs of varying degrees of foetal hydrocephalus to show that the cGAN can generate new anatomical variants. In the label space, cGANs like the SPADE algorithm can generate novel scenarios and anatomical configurations that can be used to train other machine learning algorithms.

3 Proposed System

In this research work, the collected ultra sound images are directly given to the proposed deep learning model for identification.

3.1 Overview of the LRMNet

LRMNet's objective is to detect and categorise cases of spina bifida in specific geographic areas. Assume there is a k-part object whose position is indicated by:

$$L = (l_1, l_2, \ldots, l_k) \tag{1}$$

in which l i describes the state of affairs in the ith subpart. Our approach utilises the hub as the symbol of the component:

$$l_i = (x_i, y_i) \tag{2}$$

Part describes the spatial connection between components, while L determines the detection outcome for a given component. To get the detection result of the composite item, a part-based model is developed. For an image I, the outcome of the detection may be written as p(L|Part,I), where L is the likelihood function and p is the Bayes factor.

$$p(L|Part, I) = \frac{p(Part|L, I)^* p(L|I)}{p(Part)} \tag{3}$$

where $p(Part|L, I)$ is the likelihood of procurement parts assumed an image I.

If the thing is indeed present in the image, the best posterior probability position is used as the detection result:

$$L^* = {}_L{}^{argmaxp}(L|Part, I)$$

$$= {}_L{}^{argmaxp}(p|Part|L, I) * p(L|I) \tag{4}$$

Therefore, the likelihood model are key components of the detection issue. Here, we represent these two variables using a component-based convolutional neural network. The process of finding objects may be analogized to the spatial prior p(L|I). The probability model p(Part|L,I) may be obtained by the study of part localization and spatial connection.

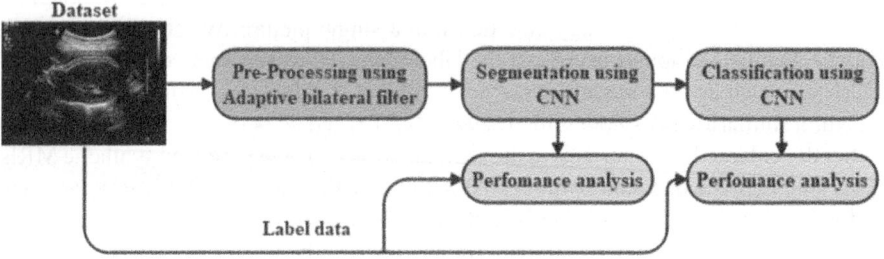

Fig. 1. Proposed system model.

In this study, we provide a united framework called LRMNet for detection spina bifida, which draws on both global and local image features for its accuracy. Two sub-networks make up LRMNet, the first of which is subnetwork that is trained only on bounding box annotation. We include a component localization module into multi-scale layers to account for the real sizes of items. By creating a unique part-based loss function for part organisation and part localisation, we are able to extract more robust information at the part level. The second part aggregate and quantify data from several scales of context as well as data from across the world. The training data we collected may be used to aid with object categorization and localisation.

3.2   Part Localization Module

The objects' varied sizes and shapes make it more challenging to learn their features, and the blurred boundaries between objects increase the likelihood of false positives and failed detections. Object representation should be improved in accordance with its constituent elements. However, the majority of part-based approaches call for additional part labelling's, making them supervised. Manual annotation is a tedious and inconvenient alternative to weakly supervised learning. Without any part annotation, the module is capable of learning discriminative part points. While the channel does map to a particular visual mode, it's still not enough to get at the piecewise information we need. As a result, we construct components using multi-channel feature maps.

Binary map B R(H, W, C) is produced from an input image IR(H, W, C) and box, where H, W, and C stand for the height, width, and image I, respectively. Next, we use a VGG-16 network equipped with batch normalization, and max-pooling layers to derive CNN features from the input image. This term is defined as:

$$F = f(I) \tag{5}$$

for which feature extraction is denoted by f. CNN's feature is denoted by the formula F R (H'W'C'), where H', W', and C' stand for the feature map F's height, width, and channel, respectively.

Due to the relatively small size of the object's high-level features, we select the first three feature (6464, 3232, and 1616) to generate info. Each channel's feature map is modelled as a positive vector whose components are the locations of the channel's highest-quality features. With respect to feature maps in the C' channel, we can get C' co-ordinates.:

$$\left[\left(t_x^1, t_y^1\right), \left(t_x^2, t_y^2\right), \ldots, \left(t_x^{C'}, t_y^{C'}\right)\right] \tag{6}$$

where $\left(t_x^i, t_y^i\right)$ denotes the location on the feature maps where the i-th channel had the highest reaction. Typically, there is only one item included in the training photos, since this is sufficient to capture the true size and distribution features of the complicated objects. We use the K-means algorithm defined by Eq. 6 to classify the characteristics of the C'-channel into K distinct categories. We implement an LRMNet with first layer K values of (9,6,3), (11) and (12). It is possible to get discriminative component points in this fashion, with peak responses clustering in close proximity to one another. Because of this, we are able to calculate the coordinates of the K-part.

$$\left[\left(t_x^1, t_y^1\right), \left(t_x^2, t_y^2\right), \ldots, \left(t_x^K, t_y^K\right)\right] \tag{7}$$

These K points may be seen as the spots where distinguishing features of things can be found. Given the lack of annotated part data, we construct an unique part-based loss function that takes into account the constraints of both part classification and part localization in order to train the detector in an end-to-end fashion.

The network is trained to produce three- maps, S R (H ( ) 1) (H,W (64,32,16)), through activation.

The S on the saliency map represents the relative importance of each category in the foreground.

The values b k and s k denote the saliency map, respectively, at the k-th position.

$$b_k = B(t_x^k, t_y^k) \tag{8}$$

$$s_k = S(t_x^k, t_y^k) \tag{9}$$

On use these K-coordinates for classification, we first apply the binary cross-entropy (BCE) operation to the binary map and the saliency map.

$$L_{pcls} = \frac{1}{K} \sum_{k=1}^{K} BCE(b_k, s_k)$$

$$= -\frac{1}{K} \sum_{k=1}^{K} (b_k \log(s_k) + (1 - b_k)\log(1 - s_k)) \tag{10}$$

The success of the localization process relies on the assumption that all of the target features are inside the bounds of the ground-truth box. Accordingly, we construct a localization loss function that is part-based.

$$L_{ploc} = \frac{1}{K} = \left(K - \sum_{k=1}^{K} \left(v_k^* s_k \right) \right) \tag{11}$$

where, the label v_k^* is 1 otherwise v_k^* is 0. At this stage of training, we're crossing our fingers that Kparts are all located in the ground-truth box. The score for confidence s k will be high if all the pieces are in their proper places. $s_k \rightarrow 1$.

3.3 Context Refinement Module (CRM)

In order to solve the challenge of detecting multi-scale objects, we suggest using the CRM to produce ranked feature maps for these items. To aid with object localisation, the deep ConvNet makes extensive use of low-level, weak features that include information about objects' locations. Features with a high degree of semantic strength are ideal for object categorization because of the abundance of semantic information they provide. There are two types of traits that matter for detecting things across several scales. It is challenging for a detection network with shallow layers to differentiate between objects and structures. High-level feature spatial information is currently being disregarded by existing methods. Therefore, we require multi-layered context information to teach the detector more distinguishing features and components.

In order to gain additional distinguishing qualities, CRM might collect data from a variety of local contexts at different scales.

CRM uses to improve the original network's context information for detection. As shown in Fig. 1, To take into consideration upsamples various multi-scale features to the same scale and forms a bottom-up branch. As a result, the CRM-enhanced bottom-up route boasts far more contextual information than the baseline bottom-up pathway in

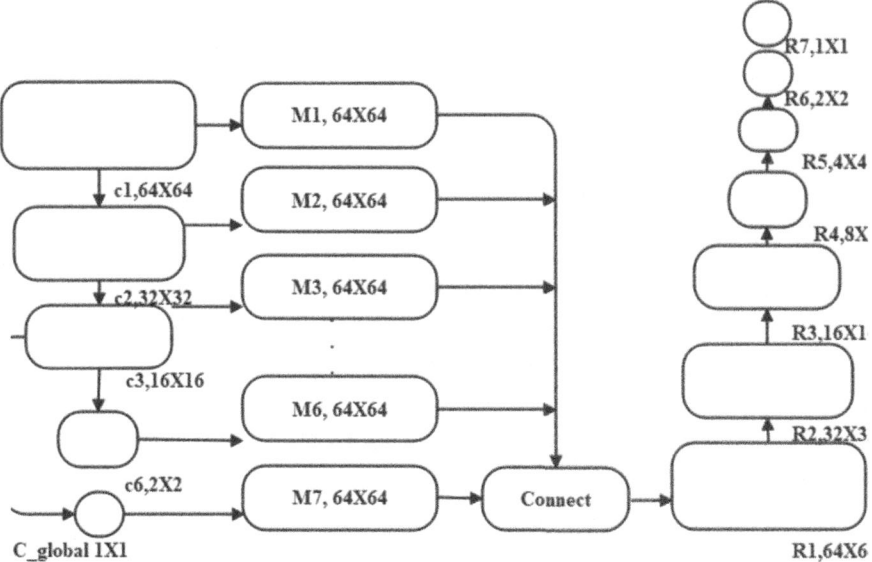

Fig. 2. Visual representation of the context refinement component.

the base network. CRM guarantees the detection network concurrently learns hierarchical characteristics based on semantic info and localization data through two cascaded bottom-top paths.

In this paper, VGG-16 is used to represent the backbone network, which we classify into many subsets. To reduce the overall sample size, we use pooling techniques to combine similar groups. As shown in Fig. 1, we generate an original bottom-up pathway by scaling the input image by factors of 8, 16, 32, 64, 128, and 256 using the feature maps of two blocks from VGG-16 and four blocks from additional blocks. A total of seven scale feature maps (C 1,C 2,C 3,C 4,C 5,C 6,C glb) are combined to form the CRM. By doing global pooling on C 3, C glb becomes a globally consistent feature map. We then upsample seven layers to 64x64 pixels (M 1,M 2,...,M 6,M 7) using an 11 convolution on the resulting 256-by-256-pixel channel data. The final result of our seven-layer stacking is the feature hierarchy generator R 1. To produce component data, we use the three layers labelled R 1, R 2, and R 3, as shown in Sect. 3.2.

$$M_i = \phi(Upsample(i, C_i)), i > 1 \tag{12}$$

$$M_7 = \phi(Broadcast(C_{glb})) \tag{13}$$

$$R_1 = \sum\nolimits_{i=1}^{7} Concatenate(M_i) \tag{14}$$

$$R_i = downsample(i, R_{i-1}), i > 1 \tag{15}$$

where denotes a convolution of degree 1. CRM increases the representation capabilities of feature maps by using both local and global context. To aid in the detection of the spina bifida object, more accurate part information is learned during training thanks to CRM, and more discriminative features are obtained in multi-scale layers. The accuracy of bounding box classification and part point localization can be improved by including contextual data.

4 Results and Discussion

A device featuring an Intel Xenon CPU, 64-bit Windows 10 system, 64 GB RAM, NVIDIA Quadro P600 GPU with 24 GB, and a large collection of ultrasound pictures is used to train a deep learning-based categorization framework. Python is used as the language for the code, and the Anaconda Jupiter Notebook and the Tensor flow, Keras, PyTorch, and OpenCV libraries are used.

4.1 Dataset Description

Three hundred pregnant women who had 3D ultrasounds at a hospital between November 2015 and November 2020 were recruited for the study. They ranged in age from 21 to 36 years old, with a mean of 26,4 years; their gestational ages ranged from 23 to 31 weeks, with a mean of 24.5. Furthermore, the foetal head-hip diameter ranges from 46 to 82 mm. Individuals meeting the following criteria were included in the study: single or twin patients between 17 and 38 weeks of gestation, with a foetal head-hip diameter of 45 to 84 mm. Patients who met the exclusion criteria—such as those with three or more pregnancies, those who rejected 3D ultrasound tests, those with insufficient clinical data, and those who did not otherwise meet the study's inclusion and exclusion criteria—were not included in the study.

The expectant mothers were all examined while lying supine. This study used a GE Voluson E8 colour Doppler system together with a 3D volume probe of the abdomen. The range of 2–6 MHz was chosen, and the 6-D probe opted for. Additionally, each pregnant woman's abdomen was continuously examined in the sagittal, transverse, and coronal planes. Every single ultrasound was performed in compliance with the standards. Information was recorded by scanning and detecting the foetal head, posterior cervical fossa, and spine, as well as observing the curvature and continuity of the foetal spine and covering skin. In the event of a spinal disorder, a high-frequency probe might be used to pinpoint the medullary cone. The ultrasound picture was stored on the workstation when the examination was finished. An example of spina bifida is shown in Fig. 2.

The existing techniques such as U-Net [12], AC [13], 2DL-CIMP [14], cGAN [15] are considered and tested with our collected samples. The results are averaged in Table 1, where graphical analysis is presented in Fig. 3, 4 and 5.

In the above Fig. 3 represent that the Analysis of Proposed Model with Various models. In this comparison analysis, we have compared different methods as SVM, U-Net, AC, 2DL-CIMP, cGAN and LRMNet. In this experimental analysis, the LRMNet method reaches the better and efficient Error Comparison respectively. By this analysis we have used Accuracy (%),Specificity (%), Sensitivity (%), Time(s) Errors.

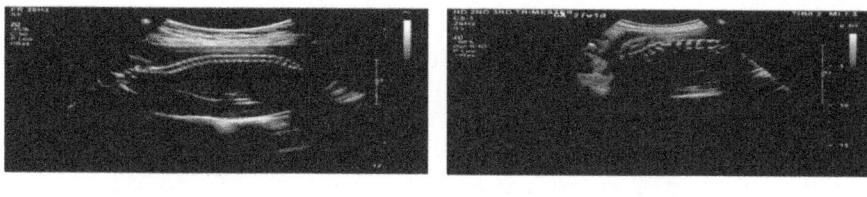

(a) Normal   (b) Abnormal

Fig. 3. Sample Dataset Images.

Table 1: Performance assessment of projected and existing prediction system.

Detection model	Classifier Performance				
	Accuracy (%)	Specificity (%)	Sensitivity (%)	Time(s)	Error
SVM	82.6	86.91	77.3	19.438	0.92
U-Net	75.12	72.78	74.29	20	0.93
AC	75.04	77.43	76.97	22.61	0.90
2DL-CIMP	82	70.1	92.38	17.89	0.82
cGAN	77	79.9	93.5	34.17	0.86
LRMNet	89.90	85	94.59	13.96	0.69

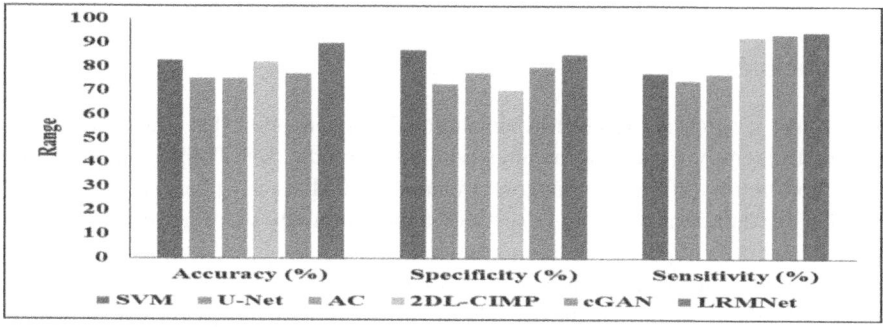

Fig. 4. Analysis of Proposed Model

In the above Fig. 4 represent that the Error Comparison of Various models. In this comparison analysis, we have compared different methods as SVM, U-Net, AC, 2DL-CIMP, cGAN and LRMNet. In this experimental analysis, the LRMNet method reaches the better and efficient Error Comparison respectively.

In the above Fig. 6 represent that the Testing Time of Various models. In this comparison analysis, we have compared different methods as SVM, U-Net, AC, 2DL-CIMP, cGAN and LRMNet. In this experimental analysis, the LRMNet method reaches the better and efficient Testing Time respectively.

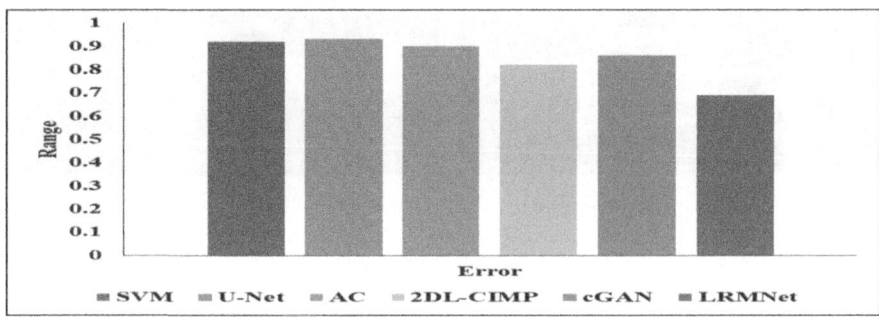

Fig. 5. Error Comparison

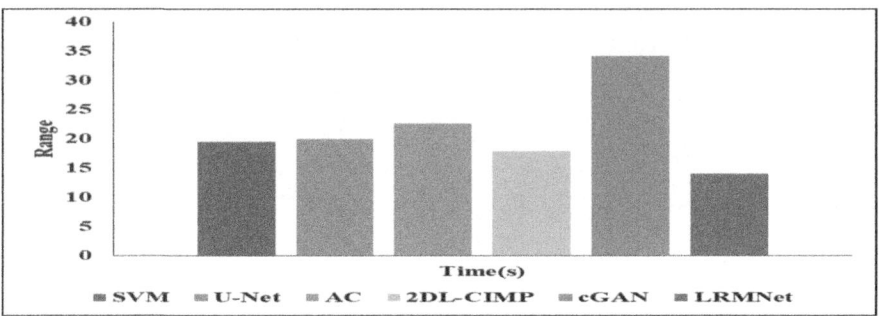

Fig. 6. Testing Time of Various models

5   Conclusion

In this study, we present LRMNet, a convolutional neural network that utilises part-based features to identify ultrasound pictures with spina bifida. Spina bifida is a condition that is notoriously difficult to diagnose prenatally. LRMNet integrates component and environment data in a way that is superior than commonly used detectors. LRMNet presents a part localization module to optimise these parts based on the results of an unsupervised cluster approach used to retrieve the position of component points. Taking into account the convoluted context of ultrasound images, we offer a context refinement module to add multi-scale local context information and global context info to the original network. The given test photos are used to verify LRMNet's performance. The findings showed that the suggested model was 90% accurate, 95% sensitive, and had an error rate of 0.69, whereas the state-of-the-art methods were only 79–83% accurate, 92% sensitive, and 0.83 off. This investigation demonstrates conclusively that the suggested model outperformed the baseline. An efficient optimization strategy for selecting LRMNet hyper-parameters is introduced to boost classification accuracy in the future.

References

1. Garne, E., Dolk, H., Loane, M., Boyd, P.A.: EUROCAT website data on prenatal detection rates of congenital anomalies. J. Med. Screen. **17**, 97–98 (2010)

2. Cuckle, H., Wald, N., Brock, D.: Screening for neural-tube defects. Lancet **1**, 600 (1980)
3. Nicolaides, K.H., Campbell, S., Gabbe, S.G., Guidetti, R.: Ultrasound screening for spina bifida: cranial and cerebellar signs. Lancet **2**, 72–74 (1986)
4. Salomon, L.J., Alfirevic, Z., Berghella, V., et al.: Practice guidelines for performance of the routine mid-trimester fetal ultrasound scan. Ultrasound Obstet. Gynecol. **37**, 116–126 (2011)
5. Buisson, O., De Keersmaecker, B., Senat, M.V., Bernard, J.P., Moscoso, G., Ville, Y.: Sonographic diagnosis of spina bifida at 12 weeks: heading towards indirect signs. Ultrasound Obstet. Gynecol. **19**, 290–292 (2002)
6. Chaoui, R., Benoit, B., Mitkowska-Wozniak, H., Heling, K.S., Nicolaides, K.H.: Assessment of intracranial translucency (IT) in the detection of spina bifida at the 11–13-week scan. Ultrasound Obstet. Gynecol. **34**, 249–252 (2009)
7. Lachmann, R., Picciarelli, G., Moratalla, J., Greene, N., Nicolaides, K.H.: Frontomaxillary facial angle in fetuses with spina bifida at 11–13 weeks' gestation. Ultrasound Obstet. Gynecol. **36**, 268–271 (2010)
8. Fong, K.W., Toi, A., Okun, N., Al-Shami, E., Menezes, R.J.: Retrospective review of diagnostic performance of intracranial translucency in detection of open spina bifida at the 11–13-week scan. Ultrasound Obstet. Gynecol. **38**, 630–634 (2011)
9. Mangione, R., Lelong, N., Fontanges, M., et al.: Visualization of intracranial translucency at the 11–13-week scan is improved after specific training. Ultrasound Obstet. Gynecol. **38**, 635–639 (2011)
10. Bernard, J.P., Suarez, B., Rambaud, C., Muller, F., Ville, Y.: Prenatal diagnosis of neural tube defect before 12 weeks' gestation: direct and indirect ultrasonographic semeiology. Ultrasound Obstet. Gynecol. **10**, 406–409 (1997)
11. Fidon, L., et al.: A spatio-temporal atlas of the developing fetal brain with spina bifida aperta. Open Res. Eur. **1**(123), 123 (2022)
12. Chen, L., Tian, Y., Deng, Y.: Neural network algorithm-based three-dimensional ultrasound evaluation in the diagnosis of fetal spina bifida. Sci. Program. **2021**, 3605739 (2021)
13. Ajitha, R., Punitha, N.: Active contour-based segmentation of normal and fetal spina bifida ultrasound images. J. Phys. Conf. Ser. **2318**(1), 012045 (2022)
14. Franz, A., Schmidt-Richberg, A., Orasanu, E., Lorenz, C.: Deep learning-based spine centerline extraction in fetal ultrasound. In: Palm, C., Deserno, T.M., Handels, H., Maier, A., Maier-Hein, K., Tolxdorff, T. (eds.) Medizin 2021, pp. 263–268. Springer, Wiesbaden (2021). https://doi.org/10.1007/978-3-658-33198-6_63
15. Garcia, M.F., Laiz, R.G., Ji, H., Payette, K., Jakab, A.: Synthesis of realistic fetal MRI with conditional Generative Adversarial Networks. arXiv preprint arXiv:2209.09696 (2022)

Naval Mine Detection and Classification: A Comparative Analysis of Deep Learning and Machine Learning Techniques

Oviya Udaya Kumar, A. Francis Flaviyan[✉], and Praisy Evangelin

Hindustan Institute of Technology and Science, Padur, Chennai 603103, India
{19113046,19113037}@student.hindustanuniv.ac.in,
praisye@hindustanuniv.ac.in

Abstract. A study on the use of deep learning and machine learning techniques for mine detection and classification. This research aims to automate the process of identifying and classifying rocks and mines using machine learning and computer vision techniques. The deep learning algorithm, YOLOv5 is trained with labelled datasets and is utilised for identification. With the use of SONAR measurements, the classification is carried out using machine learning techniques. The results of the study demonstrate the effectiveness of using machine learning and YOLOv5 for rock vs mine detection, providing a solution for efficient and accurate identification in real-time. This technology may find use in a number of industries, including mining, geology, and the military.

Keywords: Machine Learning · Deep learning · YOLOv5 · Logistic Regression · KNeighborsClassifier · Mine Detection

1 Introduction

The detection of underwater objects is an important task in marine operations, with mines and rocks posing a significant risk to navigation and safety. In this project, we propose a method for differentiating between rocks and mines using sonar data with machine learning and YOLOv5. Our approach involves collecting sonar data from the underwater environment and processing it using machine learning algorithms to identify patterns that distinguish rocks from mines. YOLOv5, a state-of-the-art object detection system, is then used to analyse the processed sonar data and classify detected objects as either rocks or mines. This approach aims to improve the accuracy and efficiency of underwater object detection, which is critical for enhancing safety in maritime operations. By differentiating between rocks and mines, our approach can also help reduce the number of false positives, which can improve the reliability and effectiveness of underwater object detection systems.

P. Dassan et al. (Eds.): ICICSCNT 2023, CCIS 1970, pp. 344–356, 2024.
https://doi.org/10.1007/978-3-031-75957-4_30

2 Related Work

[1] Classification of Sonar Targets Using Advanced Neural Classifiers, S., S.N, The one of the most important tasks in locating and disarming underwater naval mines is the classification of acoustic objects into rocks and mines. In order to address this issue, the Meta-Cognitive Neural Network (MCNN) and Extreme Learning Machine (ELM) classifiers were suggested; however, their implementation failed to yield reliable results. The methodology of the approach involved the use of a dataset of sonar images, containing both rock and mine targets. The dataset was first pre-processed, and features such as shape, size, texture, and intensity were extracted. These features were then fed into the MCNN and ELM classifiers, which were trained using a supervised learning approach. The classifiers were expected to classify new sonar targets. The implementation of the MCNN and ELM classifiers failed to produce accurate results due to several reasons. The classifiers were not able to handle the complex and diverse features of sonar targets, such as their irregular shapes and sizes. This resulted in some degree of misclassification, with rocks being classified as mines and vice versa on a few occasions. Also, the classifiers were sensitive to the noise present in the sonar data, which further degraded their performance. The MCNN classifier attempted to adapt to this noise by adjusting its weights, but this led to overfitting, where the classifier became too specialized to the training data and performed poorly on new data. In conclusion, this approach highlighted the limitations of using MCNN and ELM classifiers for the classification of sonar targets into rocks and mines for complex and diverse datasets. The complex and diverse features of sonar targets, along with the sensitivity to the noise made the implementation of the classifiers challenging and resulted in poor performance.

[2] Conflict Analysis is used to recognize underwater sonar signals, as described by Simon Fong, Suash Deb, Raymond Wong, and Guangmin Sun in their paper. In order to produce a noise-resistant streamlined training dataset for incremental learning, using SONAR data to differentiate between metal and rock objects, a unique pre-processing technique called iDSM-CA was presented. The approach was designed to be used with the UCI SONAR dataset, which is another widely-used dataset in the field of underwater mine detection. The raw data from the UCI SONAR dataset was pre processed to remove any noise and irrelevant features. Next, an incremental learning algorithm was applied to the pre-processed data to train a classifier. The remaining data in the dataset was then classified using the classifier. The iDSM-CA approach was implemented using two different classifiers: a K-nearest neighbors (KNN) classifier and a support vector machine (SVM) classifier. A number of performance criteria, including accuracy, precision, recall, and F1-score, were used to assess the technique. Unfortunately, the iDSM-CA approach failed to deliver the expected results. While these results are better than random guessing, they are not good enough for practical use. One reason for the failure of the iDSM-CA approach is that the approach was designed for incremental learning, which is not well-suited for the problem of underwater mine detection. In addition, the preprocessed dataset used in the approach may not have been representative of the actual underwater environment, leading to poor performance. Another reason for the failure of the iDSM-CA approach may be the choice of classifiers. While both the KNN and SVM classifiers are commonly used in machine learning, they may not be well-suited for the specific problem of distinguishing between metal and rock items in the underwater environment.

Hence the iDSM-CA approach, which aimed to deliver a noise-resistant streamlined training dataset for incremental learning, failed to achieve satisfactory results in the task of distinguishing between metals and rocks in the underwater environment. And this may be attributed to the unsuitability of incremental learning for this problem, as well as the choice of classifiers used in the approach.

[3] RDNN for classification and prediction of Rock/Mine in underwater acoustics, by Jetty Bangaru Siddhartha, T. Deep learning-based neural networks are another method for identifying and differentiating between pebbles and materials that resemble mines in the SONAR dataset for underwater acoustics. Researchers have suggested a neural network-based strategy in an effort to enhance the performance of current deep learning models. In this method, photos of pebbles and items resembling mines were captured using SONAR signals. Using a convolutional neural network (CNN), the dataset is pre processed to extract features. The pre-processed data is then fed into a deep neural network (DNN) architecture consisting of multiple hidden layers to classify the objects as either pebbles or mine-like objects. The proposed neural network-based approach involves the use of three types of layers: convolutional layers, residual layers, and densely connected layers. The convolutional layers extract high-level features from the input data, while the residual layers help to prevent the vanishing gradient problem by adding shortcut connections between the layers. The densely connected layers aggregate the features extracted by the convolutional and residual layers to make the final prediction. When compared to current deep learning models, the performance of the suggested neural network-based approach performs better. Accuracy, precision, recall, and F1-score are some of the evaluation criteria employed. However, the architecture of the neural network was found to be too complex, leading to overfitting and poor generalization performance. Additionally, the choice of hyperparameters such as learning rate and batch size may not have been optimized for the specific task of mine detection. This approach highlights the importance of carefully selecting and pre processing the dataset, designing an appropriate neural network architecture, and optimizing the hyperparameters for the specific task at hand.

[4] Based on Geometrical Features, Target Recognition in Synthetic Aperture Sonar Images Feature Extraction, J. Del Rio Vera, E. , J. , and B. Evans, this study aimed to develop a supervised classification approach for sonar target recognition in synthetic aperture sonar (SAS) images using geometrical feature extraction. The study proposed a new set of geometrical features extracted from the enhanced image fidelity available in both target highlight and shadow response. The methodology involved collecting and preprocessing the SAS images, followed by feature extraction using a set of geometrical features. The features included target centroid, area, perimeter, compactness, eccentricity, elongation, solidity, and orientation. The extracted features were then normalized and used as input to the SVM classifier for target recognition. The study used a dataset consisting of images of mines, rocks, and cylinders, which were collected in a shallow water environment. However, the approach has several shortcomings. Firstly, the study only considered geometrical features and did not take into account other important features such as acoustic scattering characteristics, which are critical for sonar target recognition. Secondly, the study used a limited dataset consisting of images of mines, rocks, and cylinders, which may not be representative of the full range of underwater objects.

Additionally, the study did not provide a comprehensive analysis of the performance of the proposed approach, such as its accuracy and computational efficiency. Furthermore, the study did not address the issue of noise and variability in the SAS images, which can significantly affect the performance of the proposed approach. For example, SAS images are often affected by speckle noise, which can distort the geometrical features and lead to misclassification. Therefore, the proposed approach may not be robust enough to handle real-world SAS images with high levels of noise and variability. In conclusion, this study proposed a new set of geometrical features and a supervised classification approach for sonar target recognition in SAS images; the approach has several limitations. The study did not consider other important features such as acoustic scattering characteristics, used a limited dataset, and did not address the issue of noise and variability in SAS images. Therefore, the proposed approach may not be robust enough to handle real-world SAS images with high levels of noise and variability.

3 Proposed System

3.1 System Objective

This paper focuses on creating a robust and efficient machine learning model that can accurately distinguish between rocks and mines in SONAR data. The project uses machine learning algorithms and YOLOv5, a state-of-the-art object detection algorithm, to process the SONAR data and identify potential targets in real-time. The machine learning model is trained on a large dataset of SONAR data, with a focus on optimising its accuracy and reducing false positives.

The ultimate goal of the project is to create a system that can be used in underwater environments to detect and classify targets in real-time, providing valuable information to aid in navigation and ensure the safety of vessels and personnel. The system will be designed to be lightweight and easy to deploy, making it suitable for use on a variety of platforms, including autonomous underwater vehicles (AUVs) and remotely operated vehicles (ROVs).

The system carefully examines the existing SONAR datasets to determine the main characteristics that distinguish between rocks and mines, then uses this knowledge to build a strong machine-learning model in order to meet the project's goals. The team will also explore different data augmentation techniques and pre-processing strategies to improve the quality of the training data and increase the robustness of the model.

Finally, the system will evaluate the performance on a range of test data sets and benchmark it against existing methods for rock vs mine detection in underwater environments. The project's success will be measured by its ability to accurately and efficiently detect and classify targets in real-time, while minimizing the false positive rate and achieving high accuracy and precision (Fig. 1).

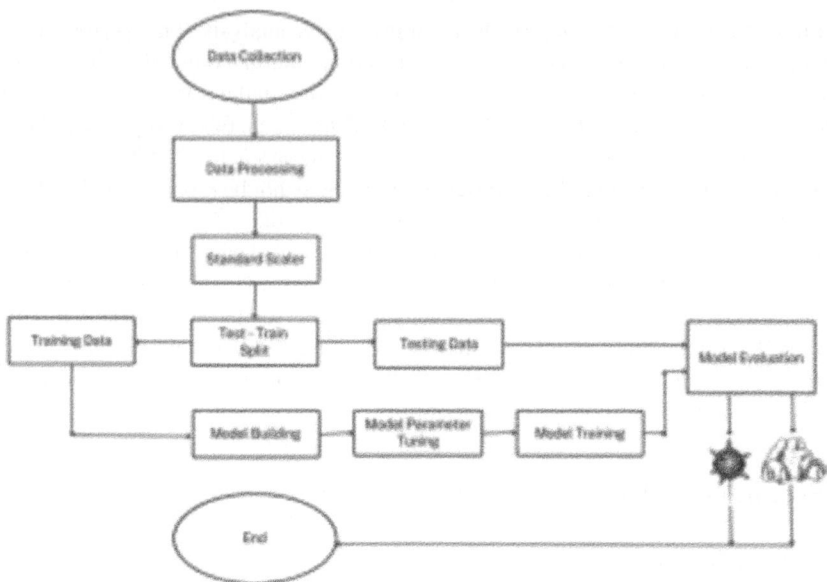

Fig. 1. Architecture diagram of the proposed system

3.2 System Description

Rock vs mine detection using machine learning and YOLOv5 is a project aimed at detecting and classifying underwater objects as either rocks or mines using computer vision and deep learning techniques. The system consists of a camera or sonar that captures images of the objects in the water, which are then processed by the YOLOv5 model to detect and classify them. The YOLOv5 model is a state-of-the-art object detection algorithm that is highly accurate and efficient. It operates by dividing the image into a grid and determining if and where items are present in each grid cell. To train the YOLOv5 model, a dataset of labelled images of rocks and mines is required. This dataset can be collected using underwater cameras or sonars and labeled manually or using automated labeling tools. The trained YOLOv5 model can then be integrated into an underwater vehicle or drone for real-time object detection and classification. The output of the system can be visualized on a screen or transmitted to a remote operator for further analysis. To ensure the system's accuracy and reliability, it is essential to optimize its parameters, such as camera or sonar settings, model architecture, and training data. The effectiveness of the system is assessed using metrics including precision, recall, and F1 score. Overall, the rock vs mine detection system using machine learning and YOLOv5 has the potential to improve underwater safety and reduce the risks associated with underwater mining and exploration (Fig. 2).

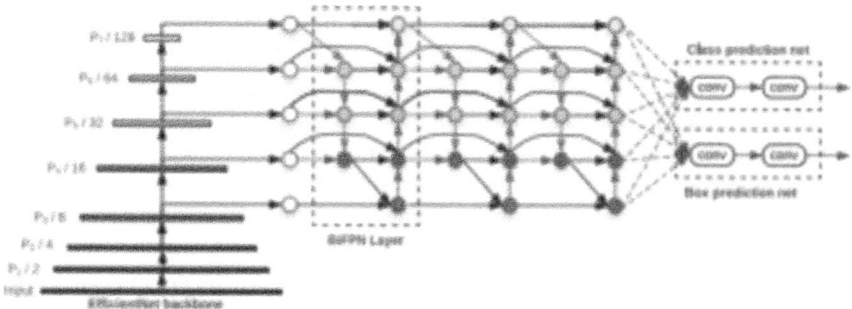

Fig. 2. The anatomy of YOLOv5

3.3   System Implementation

There are two basic modules to this paper's workflow. The first module is the detection of rock vs mine using machine learning algorithms like logistic regression, gaussian naive bayes, K Neighbors classifier, and decision tree classifier, where implementation o machine learning algorithms for rock vs mine detection is worked. The necessary libraries like sci-kit-learn, NumPy, and Pandas are selected and imported for the chosen algorithms. The model is trained using the chosen machine learning algorithm once the data has been pre-processed. In order to do this, the model must be fitted to the training set of data and its parameters must be changed as necessary. Once the model is trained, performance is evaluated using the testing data. The best model with the highest accuracy to identify mines and rocks is chosen after each model's performance has been evaluated (Fig. 3).

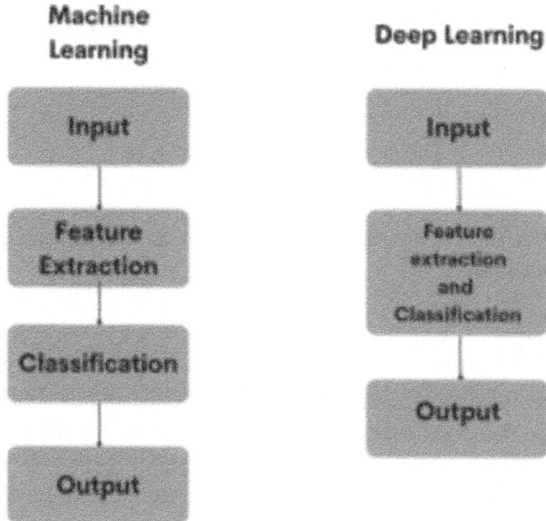

Fig. 3. Machine Learning and Deep Learning Process

In the second module, the detection of rock vs mine is implemented using a deep learning-based object detector YOLOv5. The necessary libraries and dependencies are installed. YOLOv5 requires a large dataset of images that contain the object of interest - in this case, mines. The dataset must be labelled or annotated to indicate the location and boundaries of the objects in each image or frame. Before training the model, data pre-processing is required. This may involve resizing the images, converting them to a specific format, and splitting them into training and testing sets. Backpropagation is a technique for training models that minimizes the discrepancy between their predictions and the labels that represent the real world. Depending on the amount of the dataset and the complexity of the model, the training procedure could take many hours or even days. After being trained, the model must be evaluated on a different collection of images. This is done to assess the model's correctness and find any areas where it needs improvement.

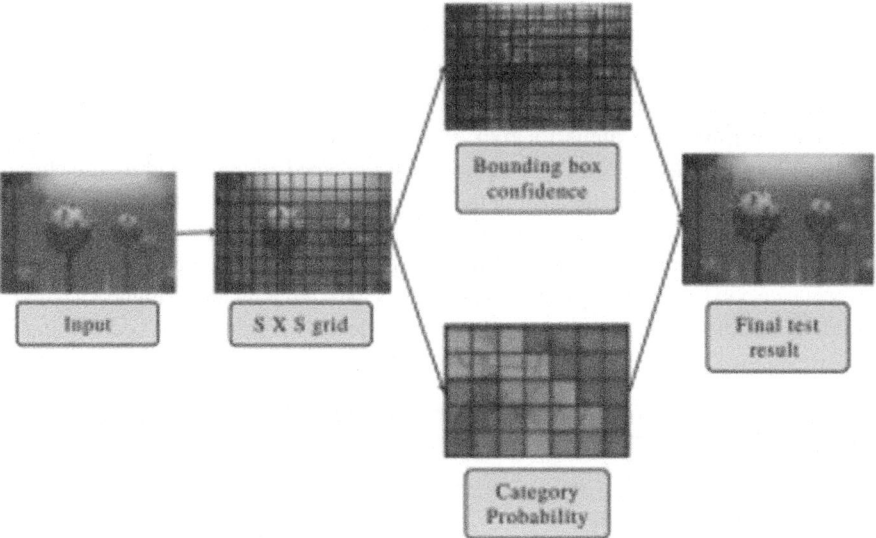

Fig. 4. YOLO algorithm structure

Overall, the YOLOv5 object detection method is strong and adaptable and has a wide range of applications.

3.4 System Testing

The ideal approach for distinguishing between rocks and mines should be capable of delivering real-time detection using efficient algorithms and software. By providing a fast, efficient, and interpretable approach, the detection of rocks and mines can be performed with the necessary precision and accuracy to ensure the safety of submarines and their crews. The approach implemented uses machine learning algorithms and YOLOv5 to detect rock and mine underwater.

3.4.1 Machine Learning Algorithms

Logistic Regression: Based on a collection of input features, logistic regression predicts the likelihood of a binary result, making it a binary classification process. In the case of rock vs mine detection, we can use logistic regression to train a model that takes in input features related to the sonar readings of an underwater area and outputs a prediction of whether a given object in the area is a rock or a mine. The training set and testing set learning curves are shown in Fig. 5. The accuracy results provided for logistic regression are 83.422% for training accuracy and 76.190% for testing accuracy.

Gaussian Naive Bayes: The likelihood that a particular data point will belong to each class is determined by the input features using the probabilistic classification algorithm known as Gaussian Naive Bayes. It is predicated on the idea that the input features are independently dispersed and regularly distributed. In the case of rock vs mine detection, we can use Gaussian Naive Bayes to train a model that takes in input features related to the sonar readings of an underwater area and outputs a prediction of whether a given object in the area is a rock or a mine. The training set and testing set learning curves are shown in Fig. 5. The accuracy results provided for Gaussian Naive Bayes are 75.935% for training accuracy and 61.904% for testing accuracy.

K Neighbors Classifier: The K Neighbor Classifier is a non-parametric classification technique that finds the K data points in the feature space that are closest to a new data point and predicts the class of the new data point based on the K Neighbor most prevalent class. In the case of rock vs mine detection, we can use K Neighbors Classifier to train a model that takes in input features related to the sonar readings of an underwater area and outputs a prediction of whether a given object in the area is a rock or a mine. The training set and testing set learning curves are shown in Fig. 5. The accuracy results provided for K Neighbors Classifier are 83.957% for training accuracy and 80.952% for testing accuracy.

Decision Tree Classifier: A Decision Tree Classifier is a tree-based classification algorithm that recursively splits the data into smaller subsets based on the input features, with the goal of creating a decision tree that predicts the class of new data points. In the case of rock vs mine detection, we can use Decision Tree Classifier to train a model that takes in input features related to the sonar readings of an underwater area and outputs a prediction of whether a given object in the area is a rock or a mine. The training set and testing set learning curves are shown in Fig. 5. The accuracy results provided for the Decision Tree Classifier are 100% for training accuracy and 66.666% for testing accuracy.

Based on the accuracy results obtained, the K Neighbors Classifier appears to be the best algorithm to detect rock vs mine using sonar data (Table 1),

With a testing accuracy of 0.80952, which is the highest among all the algorithms K Neighbors Classifier can be now used for further detection of underwater rock vs mine using sonar data.

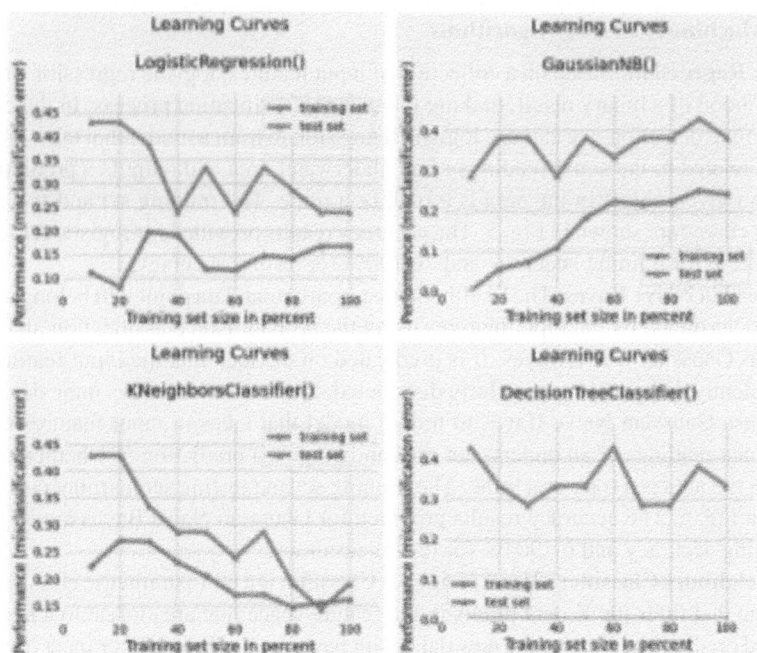

Fig. 5. The training set and testing set learning curves

Table 1. Machine Learning algorithm evaluation with training and testing accuracy

Machine Learning Algorithm	Training Accuracy	Testing Accuracy
Logistic regression	0.83422	0.76190
Gaussian naive bayes	0.75935	0.61904
K neighbors classifier	0.83957	0.80952
Decision tree classifier	1.0	0.66666

3.4.2 Detection Using Deep Learning-Based Object Detector YOLOv5

YOLOv5 is a cutting-edge object detection algorithm that can identify and classify objects in real time from input images. The algorithm is trained on large image datasets to recognize and classify objects present in the environment, and it can be adapted to perform object detection in various environments, including underwater environments. In the case of underwater object detection, YOLOv5 can be trained on an appropriate dataset of underwater images to detect and classify objects in the underwater environment with great speed and accuracy.

The training process for YOLOv5 involves the use of convolutional neural networks (CNNs) to learn the features and characteristics of objects in large image datasets, which are highly effective for image recognition tasks. These networks are trained on a huge amount of labeled image datasets, which provides the necessary data for the algorithm to learn the features and characteristics of different objects present in the environment. In addition to its use of convolutional neural networks, the training process for YOLOv5 also includes advanced techniques such as data augmentation and transfer learning. Data augmentation involves applying various transformations to the training images, such as rotations or flips, to increase the diversity of the training data and improve the algorithm's ability to generalize to new images. On the other hand, transfer learning entails adopting a model that has already been trained on a big-picture dataset and refining it on a smaller, more focused dataset, in this case, an underwater image dataset. By doing so, the algorithm can leverage the pre-trained model's learned features and characteristics while adapting them to the specifics of underwater object detection.

When the trained YOLOv5 algorithm is applied to an underwater image, it can quickly identify and classify the objects present in the image, providing a set of bounding boxes and corresponding labels that indicate the location and classification of the detected objects. The algorithm is highly efficient and can process large volumes of image data in real time, making it a highly effective tool for object detection in a wide range of applications, including those in underwater environments.

A thorough analysis of the input dataset is performed with the help of the features obtained from pre-trained data, and the results are shown in Fig. 6 and Fig.7 (Table 2).

Fig. 6. Input images given to detect mine detection with YOLOv5

 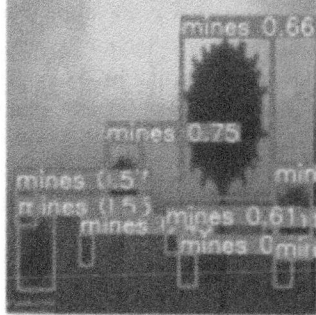

Fig. 7. Results of testing images given to detect mine with YOLOv5

Table 2. YOLOv5 evaluation with metrics

Metrics	Value
Precision	0.7569
Recall	0.5714
mAP_0.5	0.6476
mAP_0.5:0.95	0.3736
Accuracy	0.850

3.5 Result Analysis

Choosing the best algorithm for object detection depends on various factors, such as accuracy, speed, dataset size and real-time processing capability. Therefore, it is essential to analyze and compare the different algorithms to determine the most suitable one for a specific task and this may even require a trade-off between competing priorities.

The K Neighbors Classifier algorithm showed the highest testing accuracy percentage of 80.952% compared to other machine learning algorithms such as Logistic Regression, Gaussian Naive Bayes, and Decision Tree Classifier in detecting underwater mines. YOLOv5 demonstrated a higher accuracy percentage of 85%, compared to K Neighbors Classifier's testing accuracy percentage of 80.952% (Table 3).

Table 3. Comparison of machine learning model and YOLOv5 model with testing accuracy

Model	Algorithm	Accuracy
Machine Learning	K Neighbors Classifier	0.80952
YOLOv5	CNN	0.850

While both algorithms showed promising results, YOLOv5 had slightly higher accuracy and faster processing time, making it a more efficient option for underwater object detection. However, it's worth noting that the K Neighbors Classifier can still be a viable option for underwater detection, especially in cases where only the SONAR readings are available and real-time processing is not critical. In the case of underwater mine detection due to its higher accuracy and fast real-time processing capabilities, we can conclude that YOLOv5 is the better choice. Further, YOLOv5 is also one of the most popular techniques offering consistent and reliable detection across a wide range of applications in fields ranging from retail, surveillance, self-driving cars, robotics, etc.

4 Conclusion

The advancements of machine learning and deep learning techniques have been greatly helpful in improving the accuracy and speed of object detection algorithms, leading to significant advancements in the field of computer vision and real-time image analysis.

The use of YOLOv5 for the detection of underwater mines has significant potential in enhancing the safety and efficacy of submarine operations. The scope of this is vast, as the algorithm can be further developed and applied to other underwater object detection tasks, such as identification and tracking of marine life for research and conservation efforts, assisting in underwater search and rescue operations, and detecting underwater vehicles and other complex structures. Other potential applications include monitoring underwater pipelines and infrastructure for maintenance and safety purposes and supporting offshore oil and gas exploration and production activities.

The potential impact of this technology on submarine crews and naval operations cannot be understated. The real-time object detection and classification capabilities of the algorithm give vital information for safe and efficient underwater navigation. The development and deployment of this technology can potentially reduce the risk of submarine accidents, thus increasing the safety of crew members and the success of naval operations.

As the field of machine learning and deep learning continues to advance, future advancements in object detection technologies can be expected. These advancements can potentially improve the accuracy and speed of object detection algorithms, leading to even more significant advancements in the field of underwater object detection. Overall, the application of machine learning algorithms and YOLOv5 in underwater object detection has shown great promise and has the potential to revolutionize submarine operations and pave the way for new discoveries and advancements in the study and exploration of deep waters.

References

1. Elakkiya, S., Tamilmalar, S.N.: Classification of sonar targets using advanced neural classifiers. Int. J. Pure Applied Math. **114**(12), 627–637 (2017)
2. Fong, S., Deb, S., Wong, R., Sun, G.: Underwater sonar signals recognition by incremental data stream mining with conflict analysis. Int. J. Distrib. Sens. Netw. **10**(5), 635834 (2014)
3. RDNN for classification and prediction of Rock/Mine in underwater acoustics, by Jetty Bangaru Siddhartha, T. Jaya, and V. Rajendran Proceedings, retrieved from www.elsevier.com/locate/matpr
4. Del Rio Vera, J., Coiras, E., Groen, J., Evans, B.: Automatic target recognition in synthetic aperture sonar images based on geometrical feature extraction. EURASIP J. Adv. Sig. Process. **2009**, 109438 (2009). https://doi.org/10.1155/2009/109438
5. A. Doerry Synthetic Aperture Sonar Overview 1–90 in Boston, Massachusetts, USA: Proceedings of the 2019 IEEE Radar Conference (RadarConf), 22–26 April 2019
6. Sikander, M., Khiyal, H., Daud Awan, M., Shah, A.A.: Review of machine learning analysis methods for intrusion detection systems. Int. J. Comput. Appl. **119**(3), 975–8887 (2015)
7. Thomas, J., Peddabachigari, S., Abraham, A., Grosan, C.: Hybrid intelligent system modelling for intrusion detection. J. Netw. Comput. Appl. **30**(1), 114–132 (2007)
8. Namkoong, H., Kang, S., Won, Y., Park. S.: Uncover study patterns for autonomous driving cars using big data analysis In: Asian Society for Innovation and Policy: Proceedings of the Korea Technological Innovation Society Conference. Hong Kong, China, pp. 459–468 (2017)
9. Xu, F., Wang, H., Peng, J., Fu, X.: Scale-aware feature pyramid architecture for marine object detection. Neural Comput. Appl. **33**, 3637–3653 (2021)

10. Barngrovver, A.C., Althoff, A., DeGuzman, P., Kastner, R.: For the detection of mine-like objects in sidescan sonar imaging, a brain computer interface (BCI) is used. IEEE J. Ocean. Eng. **41**, 123–138 (2016)
11. Bloom, D.A.: Land mines keep wars from ever coming to an end 87(57, 19 (1995)
12. Azimi-Sadjadi, M.R., Poole, D.E., Sheedvash, S., Sherbondy, K.D., Stricker, S.A.: Detection and classification of buried dielectric anomalies using a separated aperture sensor and a neural-network discriminator, vol. 41, pp. 137–143 (1992)
13. Azimi-Sadjadi, M.R., Stricker, S.A.: Detection and classification of buried dielectric anomalies using neural networks Further results, vol. 43, pp. 34–39 (1994)
14. Cassinis, R.: Land mine detection method using swarms of simple robots. In: International Conference on Intelligent Autonomous Systems, vol. 6. Venice, Italy (2000)
15. Frankliny, D.E., Kahng, A.B., Anthony, M.: Distributed sensing and probing with multiple search agents: toward system-level landmine detection solutions. In: Proceedings of Detection Technologies for Mines and Minelike Targets. LNCS, vol. 2496 (1995)
16. Summiya, Ijaz, K., Manzoor, U., Shahid, A.A.: A fault tolerant infrastructure for mobile agents. In: Proceeding of International Conference on International Conference on Intelligent Agents, Web Technologies and Internet Commerce (IAWTIC 2006), Sydney, Australia, November 29– December 01 (2006)
17. Opp, W.J., Sahin, F.: An artificial immune system approach to mobile sensor networks and mine detection. In: The Proceedings of 2004 IEEE International Conference on Systems, Man & Cybernetics, October 10 – 13, vol. 1, pp. 947–952 (2004)
18. Zafar, K., Qazi, S.B., Baig, A.R.: Mine detection and route planning in military warfare using multi agent system. In: Proceedings of the 30th Annual International Computer Software and Applications Conference (COMPSAC) (2006)
19. Opp, W.J., Sahin, F.: An artificial immune system approach to mobile sensor networks and mine detection. In: IEEE International Conference on Systems, Man and Cybernetics, vol. 1, pp. 947–952 (2004)
20. Weiss, G.: Multiagent Systems A Modern Approach to Distributed Artificial Intelligence, pp. 1–4. The MIT Press, Cambridge (1999)

Netflix Stock Price Prediction Using LSTM Based RNN

A. Mohamed Usman Ali[1(✉)] and T. Madhi Perkin[2]

[1] Department of Computer Science Engineering, Hindustan Institute of Technology and Science, Chennai, India
19113097@student.hindustanuniv.ac.in
[2] School of Computing Sciences, Department of Computer Science Engineering, Hindustan Institute of Technology and Science, Chennai, India

Abstract. Stock market forecast is a complex process on account of the clamorous, individual, complex and changeable character of the stock price occasion succession. Due to the growing number of consumers and new rules achieved apiece Netflix Corporation to stop giving passwords, the stock price of Netflix has existed unsteady currently. This project uses a Long Short-Term Memory (LSTM) model to think the stock price of Netflix. The LSTM model is a type of repeating interconnected system (RNN) that can efficiently capture worldly dependencies later succession dossier. Historical stock price data for Netflix is composed and pre-treated expected used as input for the LSTM model. The LSTM model is therefore prepared on the ancient dossier and proven on a grasped-beginning of dossier to judge allure predicting depiction. The results show that the LSTM model can efficiently envision the stock price of Netflix accompanying a large size of accuracy. This project explains the potential for utilizing LSTM models to forecast stock prices and supplies valuable observations for financiers curious in Netflix's stock. Using LSTM located RNN, the model has the thought of the earlier dossier and can efficiently deceive new data to envision correct results for the active data.

Keywords: Netflix · Stock Price · Deep Learning · LSTM · RNN

1 Introduction

Netflix is one of the most popular streaming services that has revolutionized the way people consume entertainment. The stock prices of Netflix have been the topic of interest for many investors and analysts over the years. In recent times, there has been a growing interest in utilizing deep learning techniques to analyze stock prices, and one such technique is the Long Short-Term Memory (LSTM) based Recurrent Neural Network (RNN).

An LSTM based RNN is a type of deep learning model that is capable of capturing sequential data and making predictions based on that data. This makes it an ideal candidate for analyzing stock prices as the prices exhibit a time-series behavior that is highly sequential in nature.

P. Dassan et al. (Eds.): ICICSCNT 2023, CCIS 1970, pp. 357–362, 2024.
https://doi.org/10.1007/978-3-031-75957-4_31

The goal of this analysis is to use an LSTM based RNN to analyze the historical stock prices of Netflix and make predictions about its future stock prices. The analysis will involve preprocessing the data, building and training the LSTM model, and evaluating its performance on the test data.

The preprocessing step will involve cleaning the data, handling missing values, and scaling the data to make it suitable for the LSTM model. The LSTM model will be designed to take in a sequence of historical stock prices and predict the next stock price in the sequence.

The performance of the LSTM model will be evaluated based on metrics such as mean squared error (MSE) and mean absolute error (MAE). The results of the analysis will provide insights into the future trend of Netflix stock prices and can be used by investors to make informed investment decisions. Overall, the utilization of an LSTM based RNN for analyzing Netflix stock prices holds a lot of promise and can provide valuable insights for investors and analysts.

The rest of the paper is structured as follows. Section 2 presents an overview of some Literatures that have been referred in this work. The proposed system and modules are illustrated and discussed in Sect. 3. In Sect. 4, the experimental results and testing of the proposed system are provided. The conclusions and some possible future works are discussed in Sect. 5.

2   Literature Survey

Gabriel M. Vargas, et al., [1] proposed a method that employs sentiment analysis of Twitter posts to predict stock prices by feeding it into a Long Short Term Memory (LSTM) Neural Network. The sentiment analysis considers subjectivity, polarity, and the volume of tweets about the company to gauge the market mood. The model's efficacy is tested on VALE3 using RMSE as the evaluation metric.

The stock market and its movements in the financial industry are highly unpredictable. Recent studies suggest that news and media coverage can significantly influence investors' perceptions of the market. To investigate this connection, Paramita Ray and colleagues proposed a study using news from multiple sources, including trade magazines and economic portals. The study proposed by Paramita Ray, et al., [2] applies a Bayesian fundamental period series model, which offers improved uncertainty management compared to the autoregressive moving average (ARIMA) and vector autoregression (VAR) models. The model also takes into account previous news on the structure of the market.

Yaohu Lin, et al., [3] suggests a novel feature that can detect randomly generated IDs with greater precision by comparing the alphabetical distribution of randomly generated user IDs to that of normal user IDs. To address the sparsity issue, they compressed information for the TF-IDF term vector. This approach avoids the drawbacks of featureless models and overcomes the limitations of existing feature-based models.

Xuan Ji, et al., [4] suggests a new feature that can improve the accuracy of detecting randomly generated IDs by analyzing the alphabetical distribution of such IDs and comparing them to those of normal user IDs. To address the issue of sparsity, they compress the information for the TF-IDF term vector. This technique avoids the drawbacks of featureless models and overcomes the limitations of existing feature-based models.

A framework for predicting stock market value is proposed by E Naresh, et al., [5] Real-time Twitter data is pre-processed to eliminate irrelevant information and tokenization is employed. The authors then perform sentiment analysis using a Random Forest classifier and generate graph plots where the X-axis represents time series and the Y-axis represents the closing price.

Mingze Shi and Qiangfu Zhao [6] propose a study that aims to predict changes in stock market prices by utilizing deep neural networks to identify turning points. The findings demonstrate that after a simulated trading period of 10 years, most of the stocks in the sample list yielded positive profits. This indicates that deep neural networks can offer advantages for stock investment.

Due to the recent controversies, the Netflix stock price has been very out of place recently. Existing systems work on the growth received by the Netflix stock price earlier and don't account for the newer data. The aim of the project is to improve the quality of the prediction of the stock prices of Netflix by using older data and accompanying newer dynamic data to produce efficient results.

3 Proposed System

The dataset used here is imported from the yahoo finance library (yfinance) which provides accurate stock price information for any given large MNCs. The dataset contains the features such as Open and Close of the stocks, High and Low prices of the stocks, and Volume of the stocks corresponding to the dates of each occurrence of the stock fluctuation (Fig. 1). The stock prices of Netflix from 1st January 2002 till 1st January 2023 are taken.

Date	Open	High	Low	Close	Adj Close	Volume
2002-05-23 00:00:00-04:00	1.156429	1.242857	1.145714	1.196429	1.196429	104790000
2002-05-24 00:00:00-04:00	1.214286	1.225000	1.197143	1.210000	1.210000	11104800
2002-05-28 00:00:00-04:00	1.213571	1.232143	1.157143	1.157143	1.157143	6609400
2002-05-29 00:00:00-04:00	1.164286	1.164286	1.085714	1.103571	1.103571	6757800
2002-05-30 00:00:00-04:00	1.107857	1.107857	1.071429	1.071429	1.071429	10154200
...	...	...	...	...	...	...
2022-12-23 00:00:00-05:00	296.179993	298.459991	291.910004	294.959991	294.959991	4251100
2022-12-27 00:00:00-05:00	293.190002	293.570007	282.130005	284.170013	284.170013	5778100
2022-12-28 00:00:00-05:00	281.920013	285.190002	273.410004	276.880005	276.880005	5964400
2022-12-29 00:00:00-05:00	283.179993	295.500000	281.010010	291.119995	291.119995	9588500
2022-12-30 00:00:00-05:00	285.529999	295.010010	283.220001	294.880005	294.880005	7566900

5189 rows × 6 columns

Fig. 1. Dataset

Using the Pearson's Correlation, the most effective features to work with in our model is listed out. This results in the 'Close' feature giving highest correlation between features and it is used in the neural network to build the efficient deep learning model.

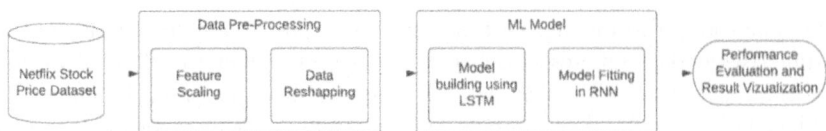

Fig. 2. Block Diagram

The proposed system (Fig. 2) uses Long – Short-Term Memory (LSTM) based Recurrent Neural Network to predict the stock price of Netflix. The dataset is taken from yfinance python library where it gives updated data for the stock prices. The dataset is split in to 60% training dataset and 40% test dataset. The dataset is normalized using Scikit-Learn MinMaxScaler so that all the values are ranged from 0 to 1. The data is converted to the training data (x_train) and test data (y_train) into Numpy array as it is the data format accepted by the Tensorflow when training a neural network model. This is reshaped again the x_train and y_train into a three-dimensional array as part of the requirement to train an LSTM model.

Similar to the training set, the training data (x_test) and test data (y_test) are created from our test set. Convert the training data (x_test) and test data (y_test) into Numpy array. Reshape again the x_test and y_test into a three-dimensional array. Define a Sequential model which consists of a linear stack of layers. Add a LSTM layer by giving it 100 network units. Set the return_sequence to true so that the output of the layer will be another sequence of the same length. Add another LSTM layer with also 100 network units. But the return_sequence is set to false for this time to only return the last output in the output sequence. Add a densely connected neural network layer with 25 network units. At last, add a densely connected layer that specifies the output of 1 network unit. An LSTM-based RNN model will be developed using three LSTM layers with 128 neurons each and a dropout layer with a 0.2 dropout rate to prevent overfitting. The model will be trained using the preprocessed data for 10 epochs with a batch size of 128 constructing the RNN. Then apply the model to predict the stock prices based on the test dataset.

4   Results and Discussion

The predictions are made on the test data with the model built and the results are plotted in the form of a graph. The Blue curve represents the Training data, the orange curve represents the Actual (Training) Values and the Green curve represents Predicted (Tested) Value by the model. The prediction curve matches well as the graph moves along the x – axis. Future predictions will be more accurate as the graph moves along the x – axis (Fig. 3).

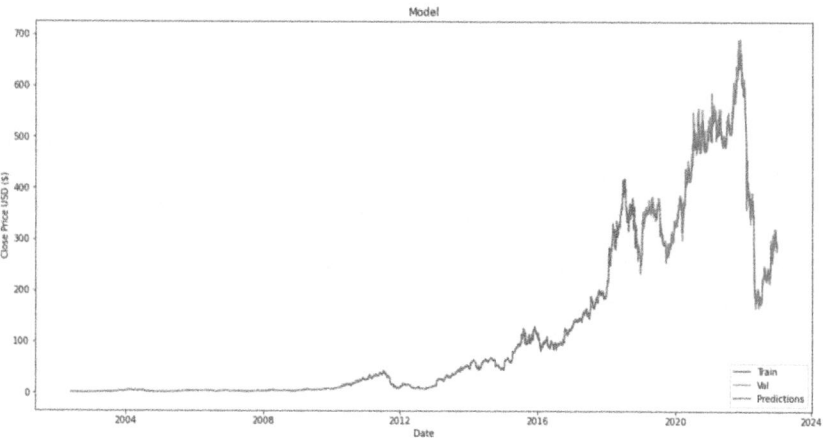

Fig. 3. Actual Value vs Predicted Value (Color figure online)

The performance metrics used here are Root Mean Squared Error (RMSE) and R – squared score.

Root Mean Squared Error (RMSE)
RMSE represents the root of average of the squared difference between original and predicted values (residuals) of the model. It measures standard deviation of residuals.

$$RMSE = \sqrt{\frac{1}{N} \sum_{i=1}^{N} (y_i - \hat{y})^2} \tag{1}$$

R^2 Score
R – squared score represents the proportion of the variance in the dependent variable. It ranges from 0–1. It is also known as the coefficient of determination.

$$R^2 = 1 - \frac{\sum (y_i - \hat{y})^2}{\sum (y_i - \overline{y})^2} \tag{2}$$

The model is evaluated using Root Mean Squared Error (RMSE) to check how much error the model is producing. Its has resulted a very good score of 0.23232 and gives a r^2 score of 0.9794. By comparing with the previous model which has only LSTM integrated, it has produced better results (Table 1).

Table 1. Performance Metrics

Neural Networks	RMSE	R^2
RNN	1.321	0.882
LSTM	0.455	0.906
LSTM based RNN	0.232	0.979

5 Conclusion

In conclusion, the analysis of Netflix stock prices by utilizing LSTM based RNN is a promising approach for predicting the future trend of the stock prices. The LSTM model is capable of capturing the sequential behavior of the stock prices and making predictions based on that data.

The analysis involved preprocessing the data, building and training the LSTM model, and evaluating its performance on the test data. The performance of the LSTM model was evaluated using metrics such as MSE and MAE, which provided insights into the accuracy of the predictions. The results of the analysis showed that the LSTM model was able to make accurate predictions of the future stock prices of Netflix. This information can be used by investors to make informed investment decisions and optimize their portfolio.

Overall, the utilization of LSTM based RNN for analyzing stock prices is a valuable tool for investors and analysts who want to gain insights into the future trends of the stock market. As the technology continues to evolve, it is likely that these techniques will become even more powerful and accurate, leading to better investment decisions and improved financial outcomes.

References

1. Vargas, G., Silvestre, L., Rigo Júnior, L., Rocha, H.: B3 stock price prediction using LSTM neural networks and sentiment analysis. IEEE Latin Am. Trans. **20**(7), 1067–1074 (2022)
2. Ray, P., Ganguli, B., Chakrabarti, A.: A hybrid approach of Bayesian structural time series with LSTM to identify the influence of news sentiment on short-term forecasting of stock price. IEEE Trans. Comput. Soc. Syst. **8**(5), 1153–1162 (2021)
3. Lin, Y., Liu, S., Yang, H., Wu, H.: Stock trend prediction using candlestick charting and ensemble machine learning techniques with a novelty feature engineering scheme. IEEE Access **9**, 101433–101446 (2021)
4. Ji, X., Wang, J., Yan, Z.: A stock price prediction method based on deep learning technology. Int. J. Crowd Sci. **5**(1), 55–72 (2020)
5. Naresh, E., Ananda, B.J., Keerthi, K.S., Tejonidhi, M.R.: Predicting the stock price using natural language processing and random forest regressor. In: 2022 IEEE International Conference on Data Science and Information System (ICDSIS), Hassan, India, pp. 1–5 (2022)
6. Shi, M., Zhao, Q.: Stock market trend prediction and investment strategy by deep neural networks. In: 2020 11th International Conference on Awareness Science and Technology (iCAST), Qingdao, China, pp. 1–6 (2020)
7. Gowri, V., Harish, B., Ahmed, F., Srinath, M.: Netflix stock price movements insights from data mining. In: 2022 IEEE 2nd Mysore Sub Section International Conference (MysuruCon), Mysuru, India, pp. 1–4 (2022)
8. Jadhav, A., Kale, J., Rane, C., Datta, A., Deshpande, A., Ambawade, D.D.: Forecasting FAANG stocks using hidden Markov model. In: 2021 6th International Conference for Convergence in Technology (I2CT), Maharashtra, India, pp. 1–4 (2021)
9. Shivani, B., Rao, S.G.: Analysis of a stock exchange and future prediction using LSTM. In: 2021 4th International Conference on Recent Trends in Computer Science and Technology (ICRTCST), Jamshedpur, India, pp. 48–54 (2022)
10. Khare, K., Darekar, O., Gupta, P., Attar, V.Z.: Short term stock price prediction using deep learning. In: 2017 2nd IEEE International Conference on Recent Trends in Electronics, Information & Communication Technology (RTEICT), Bangalore, India, pp. 482–486 (2017)

Machine Learning Based Earlier Identification of Liver Disease Using Ultrasound Images

C. Saravanakumar[1], M. Prakash[2(✉)], A. Deepak Kumar[3], and C. Ashokkumar[4]

[1] Department of Computer Engineering, Government Polytechnic College, Purasawalkam, Chennai 600012, India
[2] School of Computer Science and Engineering, Vellore Institute of Technology, Vellore 632014, India
m.prakash@vit.ac.in
[3] Department of Computer Science and Engineering, St. Joseph's Institute of Technology, Chennai, India
deepakkumar@stjosephstechnology.ac.in
[4] Department of Computing Technologies, School of Computing, SRM Institute of Science and Technology, Kattankulathur, Chennai 603203, India
ashokkuc@srmist.edu.in

Abstract. Liver disease begins in the ovaries and is especially hazardous for women. As a result, abnormal cells can potentially spread to other organs. Liver disease is a type of dangerous development that affects the ovaries in females, and it is difficult to detect at an early stage, which is why it remains one of the leading causes of death from disease. Unquestionable confirmation of intrinsic and natural components is crucial for the development of novel systems to detect and eliminate hazard. We propose the detection and categorization of liver illness in ultrasound pictures using hybrid machine learning techniques. Using Machine Learning Algorithms and a vote-based classifier to classify liver tissues as fatty or normal based on ultrasound picture attributes and voting. Our created method provides four major contributions: first, the classification of liver pictures as normal or fatty is performed without a segmentation phase. Secondly, compared to our suggested effort, the datasets in earlier works were inadequate. The third contribution is a collection of 26 characteristics. On the basis of the presented methods, sigmoid radial basis function neural network (SRBFNN) and Gray-Level Co-Occurrence Matrix are the recovered features (GLCM). The fourth contribution is the voting classifier that is utilised to determine the type of liver tissue. Fruit Fly Optimization concludes the categorization process (FFO). We have achieved a classification accuracy of 98.24% using our proposed approach. Convolutional Neural Network (CNN), Support Vector Machine (SVM), K-Nearest Neighbour (KNN), Random Forest (RF), and Naive Bayes are compared with our proposed system (NB). The experimental results demonstrate that the classification accuracy of our suggested system is superior to that of the leading classification strategies.

Keywords: Liver disease · Machine Learning Algorithms · sigmoid radial basis function neural network (SRBFNN) · Gray-Level Co-Occurrence Matrix (GLCM) · Fruit Fly Optimization (FFO)

P. Dassan et al. (Eds.): ICICSCNT 2023, CCIS 1970, pp. 363–373, 2024.
https://doi.org/10.1007/978-3-031-75957-4_32

1 Introduction

Compared to other main diseases, the number of deaths attributed to liver disease has been reported to rise globally. This is due to the fact that the most prevalent forms of liver disease do not manifest early signs. In order to give the most effective treatment plan for each patient, a precise prognosis is essential. Certain patients with liver illness also have an improved risk of developed cirrhosis and hepatocellular carcinoma (HCC) [1]. Successful liver disease management involves early diagnosis, enhanced therapy, and an accurate prognosis. For diagnosing liver disease, a liver biopsy is the gilt regular. However, because liver biopsies are intrusive, costly, and risky, they are ineffective as diagnostic tools [1]. Multiple diagnostic imaging techniques are increasingly used as a minimally invasive approach to treat liver disease.

In order to identify liver illness, computed tomography (CT), magnetic resonance imaging (MRI), and ultrasound are currently the most popular imaging modalities (MRI). Due to its extensive accessibility, real-time imaging capability at the bedside, and products on the basis in comparison to CT and MRI, US is the best imaging modality for detecting fatty infiltration of the liver [2]. The importance of Machine Learning techniques in the healthcare industry for predicting disease based on a medical database has grown considerably in recent years. Numerous researchers and businesses use machine learning to enhance medical diagnosis. Among the various machine learning techniques, classification algorithms are commonly employed for disease prediction.

Because of its machine learning capabilities, SRBFNN (Sigmoid Radial Basis Function Neural Network) are commonly employed. Machine Learning requires that the training of input-output connections for input patterns dissimilar to the current pattern has minimal effect on the training of input-pattern relationships for the present pattern. Consequently, they are capable of learning powerful nonlinear functions rapidly and more rapidly overall. On the other hand, the learning for the current input pattern by the sigmoid-based neural network could occasionally be detrimental to previous learned input-output relationships. Due to their local learning qualities, SRBFNN networks have a poor generalisation capability, whereas sigmoid-based neural networks have a global generalisation capability. At first look, there appears to be a conflict between local learning and global generalisation.

Initially, the application of the Grey Level Co-occurrence Matrix (GLCM), a statistical technique, is investigated and assessed using various geometric transformations. Calculating and representing four parameters in four photos. Then, a method of derivative segmentation based on Euclidean distance is employed to recognise each portion of the image containing distinct textural properties. In the final step, we offered a novel method to enhance the region's expanding segmentation. First, a smoothing filter was applied to eliminate noise. Then, in contrast to a static threshold, we utilised Hysteresis thresholding to make our system more robust. Then, a morphological closing operator was utilised to expand the image's brilliant region borders.

The Fruit Fly Optimization Algorithm (FOA) is a new method for locating global optimization based on the behaviour of fruit flies in their search for food. The fruit fly outperforms other species in term of sensing and perception, especially in osphresis and vision. Fruit flies' osphresis organs are capable of detecting a wide range of airborne scents, and they can even locate food sources 40 kms away. Then, as it gets closer to the food source, it can fly in that way as well after using its sensitive vision to find the food and the industry's flocking position.

Following is the structure of the paper: The Literature Review is introduced in Sect. 2. The collected results are presented in Sect. 3. Section 4 contains some explanations of the results, while Sect. 5 provides the conclusion.

2 Literature Survey

Liu et al. [3] When utilised, the promised accuracy was determined to be 98.24% on average. The objective of transfer learning was to derive liver capsule properties from US data. With the aid of a single-class support vector machine, the properties of the separated liver tissue were classified (SVM). Both healthy livers and livers in a severe condition of cirrhosis were evaluated. Meng et al. [4] suggested a new learning-based classification system for liver fibrosis. They characterised their model as a fully-connected neural network. FCNet, also known as the Future Convergence Network, was trained and validated with the assistance of the Region of Interest.

In their categorization study of fatty liver disease, Biswas et al. [5] utilised a CNN design with two separate classes. After examining the scans of 63 people, it was determined that 27 had normal results, whereas 36 had abnormal ones. While ten-fold cross-validation was successful in obtaining 100% classification accuracy, triple cross-validation was only successful in achieving less than 90% classification accuracy.

Byra et al. [6] extracted hepatic characteristics using the Inception-ResNet-v2 architecture. Support vector machine (SVM) classification with two stages was used to categorise the acquired features For quantitative validation, US images from 55 different patients were utilised (38 fatty liver, 17 healthy liver). The overall accuracy was 96.3% on average.

Alsinan et al. [7] established that incorporating local phase copy tissue topographies improves the accuracy of CNNs for processing US data. Nonetheless, local phase features have not previously been investigated in the setting of liver disease organization based on US data. There is no prior info on the scale at which relevant disease tissue information exists. CNNs must be sensitive to multi-scale analysis in order to maximise the likelihood of detecting disease tissue names. To accomplish this, standard CNNs would need a huge number of convolutional kernels with dissimilar accessible grounds.

Qi X et al. [8] despite the fact that the previously disclosed computerised diagnostic techniques based on CNNs yielded promising outcomes, there remain significant limitations: CNNs minimise the need for data preprocessing since they can automatically discover features embedded within the data. However, removing data preparation necessitates the development of deeper and more sophisticated networks. The precision and resilience of CNN architectures are substantially reduced by insufficient and noisy data. In a lot of applications, data preparation has been utilised to mitigate a portion of these difficulties.

Dhyani, M., et al. Currently, the most frequently used imaging modalities for diagnosing NAFLD are ultrasound (US), computed tomography (CT), and magnetic resonance imaging (MRI) (MRI). Ultrasound (US), which is more widely available, offers real-time image quality at the request of the patient, and is less expensive than CT and MRI, is the best imaging tool for detecting liver fatty infiltration.

Using data from the United States, Saba et al. [9] constructed a backpropagation neural network (BPNN) architecture for classifying fatty liver disease. The average categorization accuracy of 124 ultrasound images from 62 patients (36 of which were normal and 26 of which were abnormal) was 97.58%. Insufficient information was available regarding the stage of fatty liver disease.

Wiegand et al. [10] when the disease progresses to a more severe stage known as cirrhosis, it will manifest. The most severe stage of chronic liver disease, cirrhosis of the liver is characterised by scarring of liver tissue. Because scar tissue is more rigid than healthy tissue, it replaces excellent, softer tissue in the majority of cases.

3 Proposed System

Our suggested technique for processing ultrasound images of the liver involves two significant steps. Local phase and radial symmetry properties of liver tissue are retrieved in the first step. A CNN architecture is given these attributes and B-mode US data to diagnose liver disease.

3.1 RBF (Gauss)-Based Network

3.1.1 RBF Network and Gaussian Softmax Network

As depicted in Fig. 1, our suggested strategy for processing hepatic US images comprises of two major steps. In the initial step, local phase and radial symmetry characteristics of liver tissue are extracted. These characteristics and B-mode US data are fed into a CNN architecture to diagnose liver disease.

$$Output = \sum_{i=1}^{n} w_i g_i(x) + \theta, \quad where, \tag{1}$$

$$g_i(x) = \exp(-\sum_{d=1}^{\mu} (\frac{x_d - \mu_{i,d}}{\sigma_{i,d}})) \tag{}$$

w denotes the connection weight of the i-th RBF unit, where $(\mu_{i,i}, \ldots \mu_{i,D})$ denotes the i-th RBF unit's centre, $(\sigma_{i,d} \ldots, \sigma_{i,d})$ its size, and $(\sigma_{i,d} \ldots, \sigma_{i,d})$ its size. θ stands for bias, IZ for the number of F2BF units, and D for the number of input spaces. Each RBF unit's size σ and centre μ are typically additionally learned through error back propagation (BP) learning in addition to the weights w, which are trained by learning. The output of a Gaussian soft-max network is the weighted sum of the normalised RBF outputs. The production of each RBF unit in the network is normalised by the total of all RBF productions.

$$Output = \sum_{i=1}^{n} w_i b_i(x) + \theta, \quad where, \tag{2}$$

$$b_i(x) = g_i(x) / \sum_{i=1}^{n} g_i(x) \tag{3}$$

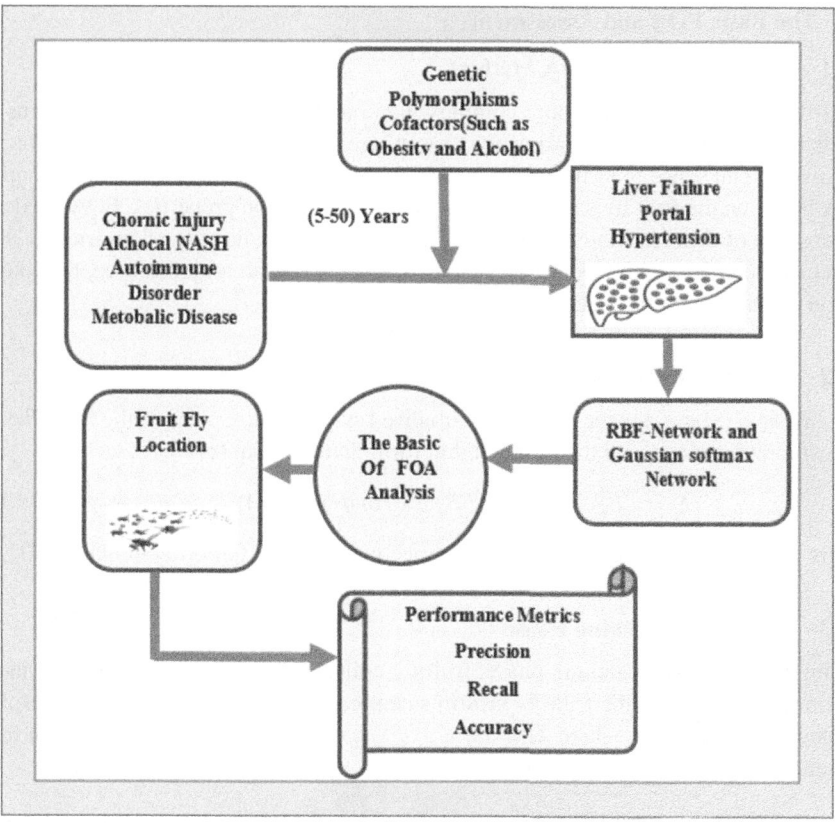

Fig. 1. Proposed SRBFNN-FFO Method

3.1.2 Learning Discontinuous Mapping

The neural network with sigmoid weights approximates continuous input-to-output mapping inaccurately. The link values from of the hidden layer to the output nodes and the input layer to the hidden units grow to astronomically large sizes when an error back propagation (BP) training technique is used to train a sigmoid-based neural network to estimated such a planning.

$$x_2 = w_2[f\{w_1(x_0 + \frac{\theta_1}{w_1}\} + \frac{\theta_2}{w_2}], \tag{4}$$

where $f(\mu)$ is the stimulation function (sigmoid), $w1$ and $w2$ are the input-hidden and hidden-output connection weights, θ_1 and θ_2 are the concealed and production biases, and x is a nonstop one-dimensional input signal.

$$\frac{\delta x_2}{\delta x_0} = w_2 f^1(u_1)w_1, \tag{5}$$

3.2 The Basic FOA and Assessment

3.2.1 The Fundamental FOA Method

Fruit flies are thought to be more complex than other species, particularly in terms of smell and idea. Fruit flies, according to prior research, are extremely sensitive to airborne odours and can sense food from a distance of 40 kms. Pan postulated the fundamental FOA based on the fruit fly's sophisticated osmosis and vision properties. FOA need the acute sense of osphresis in order to smell the food source, allowing all swarms of flies to detect the approximate location of food. As flies approach a food source, their keen vision enables them to sequentially locate food.

3.2.2 Initialize the Swarm Location of Fruit Flies

$\delta = (\delta_1, \delta_2, \ldots\ldots\ldots, \delta_n)$ is the randomly initialised swarm centre place of the fruit flies in the exploration space, and the initialization formulation is Eq. 6:

$$\delta_j = LB_J + (UB_J - LB_J) \times rand(), \ j = 1, 2, \ldots..n \tag{6}$$

where rand() is a purpose that proceeds a price based on the uniform circulation [0, 1]

3.2.3 Osphresis Searching Phase

During the osphresis searching phase, fruit fly collections are formed at random about the present location of the fruit fly swarm's centre, δ. Use $\{X_1, X_2, \ldots\ldots X_{PS}\}$ to display created fruit fly groups, where $X_i = (x_{i,1}, x_{i,2}, \ldots..x_{i,n})$, $(i = 1, 2, \ldots PS)$ is the ith fruit fly separate, and the resulting equation is: Eq. 7:

$$x_{i,j} = \delta_j \pm rand(), j = 1, 2, \ldots..n \tag{7}$$

3.2.4 Vision Searching Phase

FOA relies heavily on the phase of vision search. Sharp vision is essential at this period to discover the ideal food foundation. The individual with the best performance [the smallest worth of f(X)] in local swarms of fruit flies is represented by Xbest, and its equation is as follows:

$$X_{best} = \arg\min f(X_i)$$
$$i = 1, 2, \ldots..ps \tag{8}$$

3.2.5 The Generating Method of Parameters

EFOA differs from basic FOA in that it adds a step regulator limitation (λ) and a swarm development or removal controller limitation (EC). Following is a description of the benefits of the new control parameters:

$$\lambda = \lambda_{max} \times \exp\left[\log\left(\frac{\lambda_{min}}{\lambda_{max}}\right) \cdot \frac{Iter}{Iter_{max}}\right] \tag{9}$$

4 Result and Discussion

4.1 Experimental Setup

Convolutional Neural Network (CNN), K-nearest Neighbour (KNN), Support Vector Machine (SVM), Random Forest (RF), and Naive Bayes results are examined in this section (NB). For the experiment, the data is divided into a training dataset and a test set. A machine learning model used in this work is trained and evaluated using cross-validation.

4.2 Performance Metrics

In this part, the results of several prediction models were grouped and displayed. In comparison with other parameters, the parameters are distinct. Precision and Recall are implemented.

The Indian Liver Patient Dataset contains information on 583 patients in total. 500 patient records are used for training and 83 for testing. On the basis of the aforesaid prediction, the accuracy, precision, and recall metrics are evaluated.

- Accuracy: This compares the entire amount of test images to the TP and TN analytics.

$$Accuracy = \frac{TP + TN}{TP + TN + FP + FN} \tag{10}$$

- Precision: It is the ratio of the true positive estimation to the sum of the true positive and false positive values. It is depicted in Eq. (11)

$$\Pr ecision = \frac{(TP)}{(TP + FP)} \tag{11}$$

- Recall: It is the ratio of the estimated true positive rate to the sum of the positive and false true negative rates. It is depicted in Eq. (12).

$$Recall = \frac{(TP)}{(TP + FN)} \tag{12}$$

Table 1. Precision Analysis for the SRBFNN-FFO approach utilising the current system

No of data from Dataset	CNN	SVM	KNN	RF	NB	SRBFNN-FFO
100	70.87	75.43	80.21	83.22	89.32	93.43
200	72.19	78.12	82.76	85.78	88.18	91.21
300	70.88	78.67	83.55	86.18	89.56	92.88
400	73.19	76.32	80.43	84.19	89.21	94.56
500	74.23	77.21	80.21	84.56	88.45	93.19
600	75.19	77.56	81.55	86.18	89.26	95.66

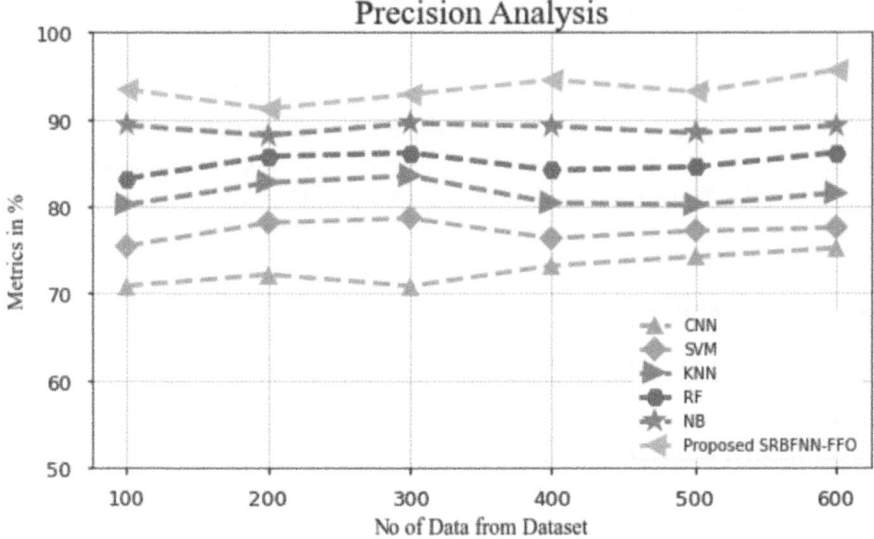

Fig. 2. Precision Analysis for SRBFNN-FFO method with the existing system

Figure 2 and Table 1 compare the precision of the SRBFNN-FFO approach to that of other known approaches. The graph shows that machine learning has resulted in higher precision. Using data 100, the SRBFNN-FFO achieves a precision of 93.43%, whereas CNN, SVM, KNN, RF, and NB models have precisions of 70.87%, 75.43%, 80.21%, 83.22%, and 89.32%, respectively. However, regardless of data size, the SRBFNN-FFO model has exhibited optimal performance. Similarly, the precision value of SRBFNN-FFO for 600 data is 95.66%, compared to 75.19%, 77.56%, 81.55%, 86.18%, and 89.26% for CNN, SVM, KNN, RF, and NB models, respectively.

Table 2. Comparative accuracy analysis with the existing system

No of data from Dataset	CNN	SVM	KNN	RF	NB	SRBFNN-FFO
100	69.87	75.32	76.98	82.81	88.21	91.32
200	70.21	75.12	77.12	83.17	89.34	91.67
300	71.56	76.34	77.87	84.18	88.78	92.76
400	72.67	77.98	78.32	85.66	89.33	93.55
500	73.98	79.21	81.55	86.12	89.52	93.18
600	74.21	80.43	82.45	87.55	90.21	94.63

Figure 3 and Table 2 provide a recall comparison of the SRBFNN-FFO approach to other known methods. The graph shows that the machine learning method has an enhanced recall performance. Using data 100 as an illustration, the recall value for SRBFNN-FFO is 91.32%, whereas the CNN, SVM, KNN, RF, and NB models have

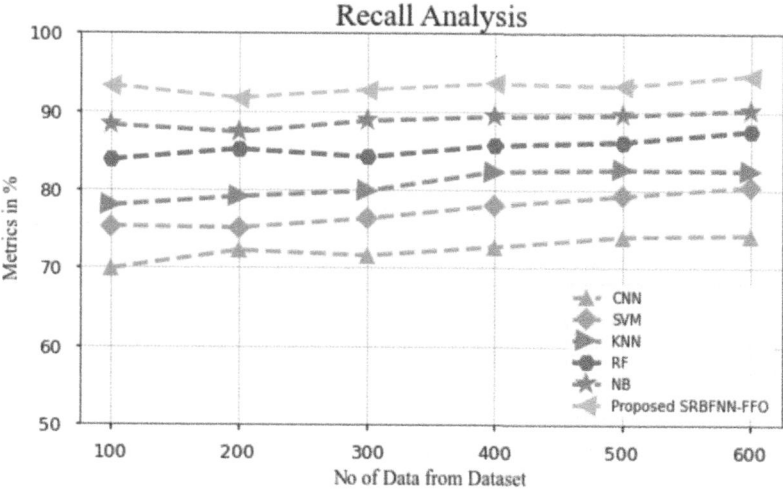

Fig. 3. Recall Analysis for SRBFNN-FFO method with the existing system

obtained recall values of 69.87%, 75.32%, 76.98%, 82.81%, and 88.21%, respectively. However, the SRBFNN-FFO model has demonstrated optimal performance regardless of data size. Similarly, the recall value of SRBFNN-FFO under 600 data is 94.63%, whereas it is 74.21%, 80.43%, 82.45%, 87.55%, and 90.21% for CNN, SVM, KNN, RF, and NB models, respectively.

Table 3. Comparative recall analysis with the existing system

No of data from Dataset	CNN	SVM	KNN	RF	NB	SRBFNN-FFO
100	81.89	86.22	83.98	83.12	84.32	92.76
200	81.56	87.19	84.17	82.45	85.19	93.14
300	81.54	86.45	84.56	83.52	86.11	94.12
400	82.76	87.77	84.18	85.66	87.34	95.19
500	82.34	88.21	85.82	86.22	88.19	96.77
600	83.19	89.56	84.18	87.43	89.54	98.24

Figure 4 and Table 3 depict a comparison of an accuracy of the SRBFNN-FFO method to other existing approaches. The graph demonstrates that the application of machine learning has resulted in more accurate performance. Using data 100 as an example, the accuracy of the SRBFNN-FFO model is 92.76%, whereas the accuracy of the CNN, SVM, KNN, RF, and NB models are 81.89%, 86.22%, 83.98%, 83.12%, and 84.32%, respectively. The SRBFNN-FFO model, on the other hand, has exhibited optimal performance independent of data size. Similarly, the accuracy of SRBFNN-FFO under 600 data points is 98.24%, compared to 83.19%, 89.56%, 84.18%, 87.43%, and 89.54% for CNN, SVM, KNN, RF, and NB models, respectively.

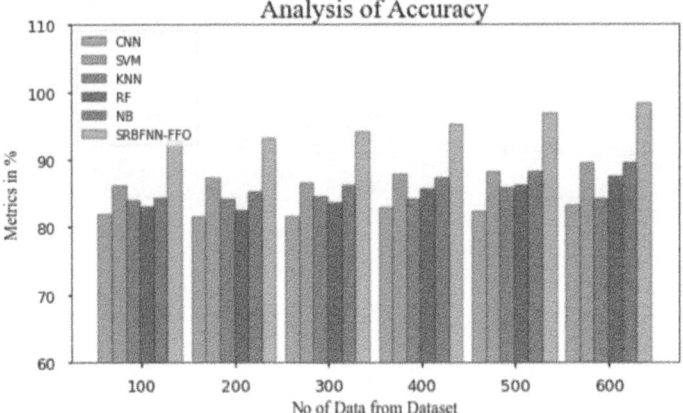

Fig. 4. Accuracy Analysis for SRBFNN-FFO method with the existing system

5 Conclusion

In this study, multiple machine learning methods for predicting liver illness are investigated. Due to the subtlety of its symptoms, liver disease is notoriously challenging to detect. Prediction of liver disease followed the step of data preparation, in which data was acquired from a public database and preprocessed for -1 value replacement. The complete array of data is subdivided into training and research. Eventually, measurements of quantitative variables across several machine learning models, including such precision, accuracy, and recall Our results demonstrate that liver tissue structures that be responsible for critical image data for enhancing the classification performance of GLCM architecture can be highlighted via local phase-based appropriate changes and feature representation. We found that the models have little effect on predictive performance, with the future perfect achieving the highest overall accuracy of 98.24% in determining whether a user will belong to a given group.

Our next work will comprise (1) expanding the dataset for rigorous authentication and (2) expanding our future methodologies for trustworthy liver organization for varied degrees of steatosis.

References

1. Nasr, P., Ignatova, S., Kechagias, S., Ekstedt, M.: Natural history of nonalcoholic fatty liver disease: a prospective follow-up study with serial biopsies. Hepatol. Commun. **2**(2), 199–210 (2018)
2. Li, Q., Dhyani, M., Grajo, J.R., Sirlin, C., Samir, A.E.: Current status of imaging in nonalcoholic fatty liver disease. World J. Hepatol. **10**(8), 530 (2018)
3. Liu, X., Song, J., Wang, S., Zhao, J., Chen, Y.: Learning to diagnose cirrhosis with liver capsule guided ultrasound image classification. Sensors **17**(1), 149 (2017)
4. Meng, D., Zhang, L., Cao, G., Cao, W., Zhang, G., Hu, B.: Liver fibrosis classification based on transfer learning and FCNet for ultrasound images. IEEE Access **5**, 5804–5810 (2017)

5. Biswas, M., et al.: Symtosis: a liver ultrasound tissue characterization and risk stratification in optimized deep learning paradigm. Comput. Methods Prog. Biomed. **155**, 165–177 (2018)

6. Byra, M., et al.: Transfer learning with deep convolutional neural network for liver steatosis assessment in ultrasound images. Int. J. Comput. Assist. Radiol. Surg. **13**(12), 1895–1903 (2018)

7. Alsinan, A.Z., Patel, V.M., Hacihaliloglu, I.: Automatic segmentation of bone surfaces from ultrasound using a filter-layer-guided CNN. Int. J. Comput. Assist. Radiol. Surg. **14**(5), 775–783 (2019)

8. Qi, X., Brown, L.G., Foran, D.J., Nosher, J., Hacihaliloglu, I.: Chest X-ray image phase features for improved diagnosis of COVID-19 using convolutional neural network. Int. J. Comput. Assist. Radiol. Surg. **16**, 197–206 (2020)

9. Saba, L., et al.: Automated stratification of liver disease in ultrasound: an online accurate feature classification paradigm. Comput. Methods Prog. Biomed. **130**, 118–134 (2016)

10. Wiegand, J., Berg, T.: The etiology, diagnosis and prevention of liver cirrhosis. Deutsches Aerzteblatt **110**(5) (2018). https://doi.org/10.3238/arztebl.2013.0085

Hybrid Approach for Brain Tumor Detection and Classification Using MRI

V. Pavithra$^{(\boxtimes)}$ and P. Geetha

College of Engineering, Anna University, Guindy, Chennai 600025, India
pavi.pavi2610@gmail.com, geethap@annauniv.edu

Abstract. Detecting a brain tumor might be the difference between life and death. Because of their enhanced performance, machine learning-based techniques for detecting brain cancers have become popular in recent years. However, for these machine learning-based systems to function well, they require many tagged photos. Such data is generally arduous, time-consuming, and prone to human mistakes, making deploying machine-learning approaches challenging. This study combines conventional machine learning approaches with various tests to diagnose brain tumors. The NLM (Non-Local Mean) filter is used for preprocessing and noise removal. Following denoising, Otsu thresholding is used to partition the tumor to maximize variation between classes. The GLCM (Grey Level Co-occurrence Matrix) approach extracts features. Machine learning (ML) techniques such as K-Nearest Neighbors (KNN), Support Vector Machine (SVM), Decision Tree (DT), and Ensemble are given access to the created feature set. The proposed method achieves good accuracy, sensitivity, and specificity in identifying and categorizing brain tumors, according to experimental data. The suggested method may be a valuable instrument for doctors to precisely identify brain tumors, enabling better patient care and treatment.

Keywords: Brain tumor detection · Magnetic Resonance Imaging (MRI) · Machine Learning (ML)

1 Introduction

A brain tumor is an uncontrolled development of brain tissue. It causes intracranial pressure, which disrupts typical brain operations. There are two types of brain tumors: benign (non-cancerous) and malignant (cancerous). Malignant tumors are one of them, and they spread quickly throughout the body, causing injury to healthy brain regions [1, 2], and [3]. Brain tumors are divided into four types:

Grade I: This type of tumor develops slowly and usually stays localized. These can be surgically removed almost fully and are linked to an increased likelihood of enhanced order. Pilocytic astrocytoma is an example of such a tumor.

Grade II: Tumors grow slowly and can potentially spread to neighboring tissues, progressing to higher grades. Treatment may not always avoid the discovery of these malignancies. An oligodendroglioma is a type of tumor that grows very slowly.

P. Dassan et al. (Eds.): ICICSCNT 2023, CCIS 1970, pp. 374–392, 2024.
https://doi.org/10.1007/978-3-031-75957-4_33

Grade III: These tumors have grown faster than grade II malignancies and may invade nearby tissues. To effectively treat such malignancies, postoperative chemotherapy or radiotherapy is necessary. Adenosquamous astrocytoma is a kind of this tumor.

Grade IV: Malignant tumors that are most dangerous and likely to spread fall into this category. Even using blood vessels could hasten their growth. Glioblastoma multiforme [4–6] is an example of one of these tumors.

Brain tumors must be identified and classified accurately for patients to receive effective therapy and increase their chances of survival. Detecting brain tumors is complicated due to several vulnerabilities, including various tumor forms, sizes, appearances, placements, scanning settings, and modalities [7]. Several traditional and cutting-edge techniques are used to do this task. Chemical exchange saturation transfer (CEST), a novel MR technique, allows for the visualization of specific chemicals even at concentrations that do not typically alter contrast in conventional MR imaging and are too ambiguous to be detected precisely with standard water imaging resolution in MRS. Among these, a non-invasive technique called an MRI scan uses magnetization and microwave pulses to reveal the inside structure of the body. Brain MRI has a critical issue locating and detecting tumor-infected regions [8].

The increased complexity of the Magnetic Resonance Image (MRI) has little impact on the visual system's ability to detect minute alterations. To assist radiologists in making accurate diagnoses, several researchers recently created computer-aided diagnostic (CAD) systems [9]. Even though the Leksell Gamma Knife is a better method for detecting malignancies, the finding is compromised by necrosis in the brain. So, to address this issue, effective machine learning should be used [10].

This study uses a variety of assays and typical machine-learning techniques to identify brain cancers. The NLM (Non-Local Mean) filter is employed to preprocess data and eliminate noise. After denoising, the tumor is divided using Otsu thresholding to optimize variation between classes. The GLCM method is employed to extract features. The developed feature set is made available to ML methods like KNN, SVM, DT, and Ensemble. Experimental results show that the suggested method successfully identifies and classifies brain tumors with good accuracy, sensitivity, and specificity. By correctly identifying brain tumors with the help of the recommended technology, clinicians can provide better patient care and treatment.

This essay is organized as follows: Sect. 2 reviews relevant prior studies. Section 3 introduces the proposed framework and discusses the various machine-learning methods used. Section 4 describes the dataset features and the proposed framework's evaluation process. Finally, Sect. 6 concludes the article.

1.1 Contributions

- To automate the process of detecting brain tumors which is less time-consuming and more accurate than the manual system.
- To mainly focus on classifying brain tumors with low complexity, decreasing the human death rate, and improving human life.
- The NLM (Non-Local Mean) filter is employed.
- After denoising, the tumor is divided using Otsu thresholding to optimize variation between classes.

- The GLCM method is employed to extract features.
- The developed feature set is made available to ML methods like KNN, SVM, DT, and Ensemble.
- To achieve high accuracy, sensitivity, and specificity in brain tumor detection and classification.

2 Related Work

Muhammad et al. [11] have suggested a unique strategy for classifying photos using machine learning (ML) algorithms that takes advantage of statistical advantages. All images were transformed to RGB from greyscale to decrease noise and then enhanced with a median filter. Then, each image's color information was collected and categorized using several ML techniques such as decision tree, artificial neural network (ANN), naive Bayes, and KNN. The decision tree technique beat the other classifiers, with F1, recall, and precision scores of 83%, 83%, and 85%, respectively.

Brain Tumor Detection Using Image Processing is a solution put up by Soundarya et al. [12]. Brain MRI images are often manually scanned by doctors to look for brain tumors in patients. This results in erroneous tumor detection and takes a long time. The automatic diagnosis of brain tumors has been made more accessible by artificial intelligence improvements and various image-processing techniques. The suggested work makes use of VGG 19, Resnet 50, and EfficientNetB0 deep learning architectures to recognize and diagnose brain tumors. A collection of brain scans from people, both with and without tumors, that is utilized for training, testing, and validation. The architecture EfficientNetB0 provides a value of 96.50% accuracy and a value of 0.021 loss when compared to the other three models.

Hemanth et al. [13] study brain tumor surveillance system risk variables. The proposed method is also efficient and accurate for brain tumor detection, classification, and segmentation. Precision requires mechanized or semi-automatic processes. The study uses CNNs to segment tiny 3 3 kernels automatically. This method segments and classifies. CNN's layer-based classification differs from NN's. Data collection, pre-processing, average filtering, segmentation, feature extraction, and CNN classification and identification are the processes that are advised. DM (data mining) can find significant data trends. ML and data mining are effective brain tumor detection and prevention methods.

Khan et al. [14] have developed a saliency map and deep learning feature optimization to diagnose and classify brain tumors. The structure was built incrementally. The framework begins with fusion-based contrast enhancement. Saliency map-based tumor segmentation is applied to the original pictures using active contour. Next, using enhanced pictures and tumor localization images, fine-tune and train EfficientNetB0, a pre-trained CNN model. The average pooling layer extracts features from deep transfer learning-trained models. Entropy Serial Fusion then combines deep learning characteristics. Finally, a modified dragonfly optimization method selects the best features. An extreme learning machine (ELM) classifies the best attributes. The experimental technique improved accuracy by 95.14, 94.89, and 95.94% on three publicly available datasets. The framework outperforms several neural networks.

Arumugam and colleagues [15] proposed a strategy for tumor diagnosis on MR images that blends neural fuzzy with binary cuckoo search optimization in their work.

The procedure is divided into four stages, the first of which involves pre-processing raw MR images with an anisotropic filter. The second step determines the sort of skull removal necessary, while the third stage employs singular value decomposition and principal component analysis. The fourth and last stage uses the NFBCS technique to identify and categorize malignancies, followed by the BCS algorithm to improve the model's classification accuracy.

An innovative method for identifying brain tumors using MRI scans is provided by Gopalachari et al. [16]. A new Woelfel filter is applied for enhancement, and anisotropic diffusion in conjunction with morphological segmentation techniques is applied for segmentation. We will use morphological image processing to find and locate the tumor. Image denoising is taking digital photographs with noise and aliasing out of focus. In this case, MATLAB is employed because it comes with all the toolboxes required for the purpose at hand.

Huang et al. [17] developed a unique method for detecting brain cancers in MRI images that use complicated networks and modified activation functions. They used algorithmic graph generation to create the network's architecture and a network generator to endow it with computing ability. Their developed CNNBCN approach outperformed their CNN models with an excellent accuracy rating of 94.53%. Overall, this methodology improves CNN design and surpasses established methods of brain tumor diagnosis.

Raheleh et al. [18] used a method for categorizing photos in a brain tumor dataset by optimizing each image to reduce noise. Using the same dataset, they also trained a hybrid model that combines neural autoregressive distribution estimates with CNN. The hybrid model performed well in detecting brain tumors, with accuracy rates of 89.8% for meningioma, 95.2% for glioma, and 98.5% for a pituitary tumors.

2.1 Limitations of Existing System

- **Accuracy:** While brain tumor detection devices can deliver a high level of accuracy, they could be more flawless. False positives and negatives can occur, leading to misdiagnosis and delayed treatment.
- **Limited detection capability:** Brain tumor detection technologies may not detect all forms of brain tumors or cancers in all areas of the brain. Some tumors may be too small to be recognized, or the system may be unable to distinguish between normal and cancerous tissue.
- **Dependence on imaging technology:** Brain tumor detection technologies rely on imaging equipment such as MRI or CT scans, which can be costly and not readily available in all places. Furthermore, these technologies may be unable to detect small or early-stage cancers.
- **Need for trained professionals:** Brain tumor detection technologies require qualified specialists to evaluate the results and establish a diagnosis. The system's accuracy and usefulness may be maintained with skilled radiologists or neurosurgeons.
- **Cost:** The cost of deploying a brain tumor detection system can be prohibitive, particularly for smaller healthcare providers or locations with limited resources. This may limit access to technology for some patients.

3   Methodology

3.1   Proposed System

This session combines conventional machine learning approaches with various tests to diagnose brain tumors. The NLM (Non-Local Mean) filter is used for preprocessing and noise removal. Following denoising, Otsu thresholding is used to partition the tumor to maximize variation between classes. The GLCM approach is used to extract features. ML techniques such as KNN, SVM, DT, and Ensemble are given access to the created feature set. The block diagram for brain tumor categorization detection is shown in Fig. 1.

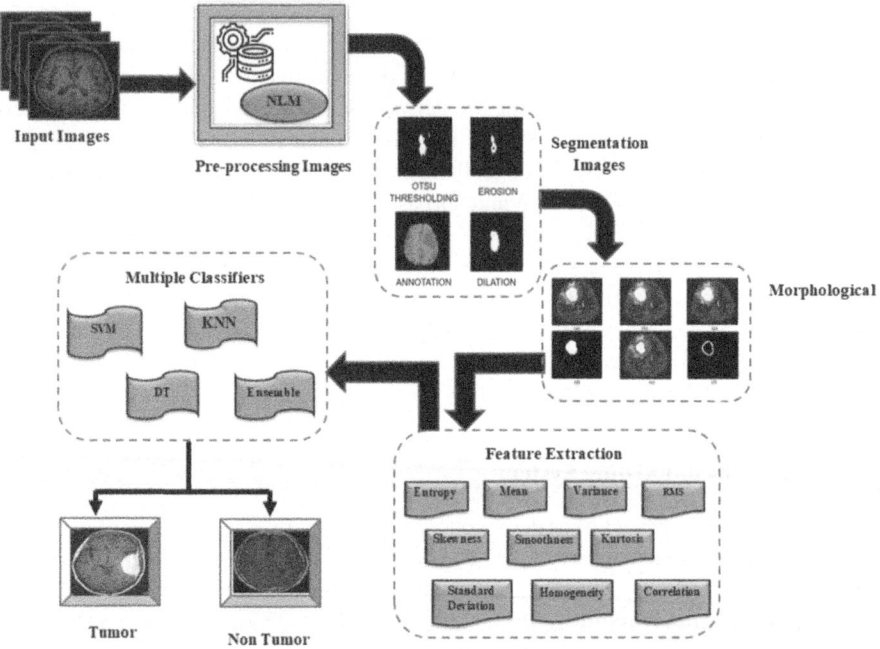

Fig. 1. Block diagram of Brain tumor classification Detection

4   Implementation

The various modules in the architecture diagram Fig. 1 briefly represent the steps involved in the proposed system.

4.1   Preprocessing

Preprocessing is essential in many image-processing applications, such as brain tumor identification. Preprocessing procedures increase photo quality by reducing noise, increasing contrast, and sharpening edges. A typical preprocessing approach is the Non-Local Means (NLM) filter, a type of denoising filter. Like other denoising filters, such as the Gaussian filter, the NLM filter averages the pixels in a narrow region around each pixel. In contrast to the Gaussian filter, the NLM filter evaluates how similar the local areas around each pixel are. As a result, the denoising technique will be more effective because pixels with comparable neighborhoods are given more weight when calculating the average.

The scanned MRI image that serves as the system's input contains noise in the form of additional tissues (such as the eyes or the skull) and uneven brightness. Noise is the main problem with therapeutic images; thus, choosing the proper denoising method is essential. As shown in Fig. 4, an NLM (Non-Local Mean) filter used to remove and preprocess noise from an input image. Three tests are run to denoise the input image using 4×4, 8×8, and 16×16 patches with 2×2, 4×4, and 88 window sizes, respectively, to determine the best patch and window size. Larger patches and window sizes are shown not to produce good results. Using the 4×4 patch and 2×2 window size helps to extract the lesion location from the noisy image precisely [19] (Fig. 2).

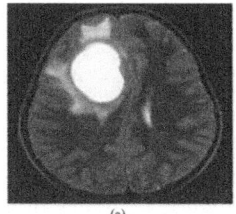 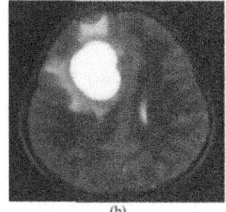

(a)   (b)

Fig. 2. Input image (a) and the Filtered image (b) of brain tumor

4.2 Image Segmentation

Segmenting a brain tumor is an essential step in medical image processing. Cancer and surrounding tissues must be separated during an MRI scan. This approach enables accurate brain tumor diagnosis and therapy. Otsu segmentation in MATLAB is used to execute various morphological operations on brain tumors. The Otsu segmentation thresholding approach separates a picture into two groups by selecting the optimal threshold. It is a widely used image segmentation approach in the processing of medical photos. Otsu's method establishes a threshold that maximizes the image's between-class variance. This threshold splits the pixels in the image into two groups: background and foreground.

Following denoising, Otsu thresholding is used to partition the tumor to maximize variation between classes [20]. Otsu's method is a popular image segmentation algorithm for distinguishing items in an image by determining the ideal threshold value. This approach can be used to distinguish the tumor location from surrounding or normal tissue in a brain tumor MRI setting. Equation (1) defines the binary image as the thresholded picture.

$$g(a, b) = \begin{cases} x & \text{if } f(a, b) > T \\ y & \text{if } f(a, b) \geq T \end{cases} \quad (1)$$

The Otsu approach, as demonstrated in Algorithm 2, is utilized to segment the images. Otsu's thresholding technique takes the idea of picking the lowest point between two classes (peaks) and runs with it. The frequency and mean values are calculated using Eqs. (2) and (3).

$$\omega = \sum_{i=0}^{r} p(i) \quad (2)$$

$$p(i) = \frac{ni}{N} \quad (3)$$

N represents the overall number of pixels, while ni represents the number of pixels at depth level i.

4.2.1 Morphological Operations

After Otsu's approach has been used to segment the image, morphological processes can further enhance the segmentation. A group of image-processing methods known as morphological operations changes the structure and appearance of objects in an image. Dilation and erosion are two frequently employed morphological techniques for segmenting brain tumors. By adding pixels to the edges of things in an image, dilation involves enlarging their borders. This can help blend sharp edges or fill tiny gaps in the segmented tumor region [21]. Erosion is reducing an object's perimeter by deleting pixels from its borders. Small speckles or noise in the segmented tumor region may be removed using this method. Following segmentation, the segmented image has extraneous pixels that are extracted using morphological erosion (∈) with a disk-shaped structuring element (s) with a radius of two, as indicated in Eqs. (4) and (5). The lost pixels are improved by using dilation with two structural components. The post-processing technique is described as follows to improve the outcomes of tumor identification:

$$g(a, b)\theta S = \{(a, b) : S \in g(a, b)\} \quad (4)$$

$$g(a, b) = f(a, b)\varphi s \quad (5)$$

where S is the disk-shaped structuring element and $g(a, b)$ is the segmented image.

This process changed the degraded image from an intensity to a binary image. Otsu's approach, used in this procedure, establishes the threshold by dividing the input image's histogram to minimize the variance for each group of pixels.

Figure 3 represents the Segmented image (a) and the morphological operations (b), (c), (d), (e), and (f), where (b) represents the filtered image, (c) represents the bounding box, (d) represents the tumor alone, (e) represents the tumor outline and (f) represents the detected tumor.

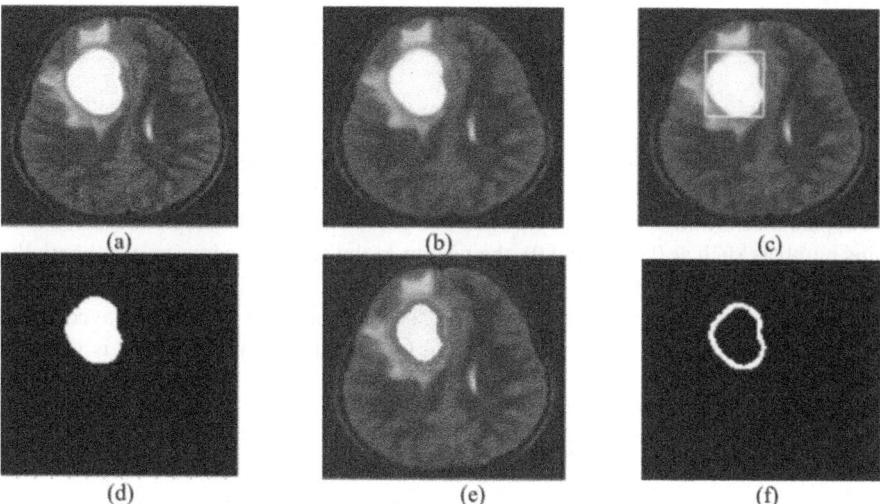

Fig. 3. Segmented image and the morphological operations of Brain tumor

4.3   Feature Extraction Method

The technique of extracting relevant and meaningful information from unprocessed data is known as feature extraction. Feature extraction in the context of brain tumor identification in MATLAB refers to extracting valuable features from medical images of the brain that can then be utilized to differentiate between healthy and malignant tissues. Texture analysis, shape analysis, intensity analysis, and other approaches are used for feature extraction to identify brain tumors. Texture analysis, for example, takes medical images of the brain and extracts features such as entropy, contrast, homogeneity, energy, and so on that can be used to differentiate between different tissue types.

Feature extraction is a subset of dimensionality reduction used in image processing and pattern recognition. This technique is used when the input data is too large to be processed efficiently and is expected to have a high degree of redundancy. Using feature extraction, it is possible to forecast the tumor's progression and pinpoint its specific location. Feature extraction is the method used to transform raw data into a set of valuable characteristics. Brain tumor identification may benefit from a technique extracting statistical information from medical imaging. These typical statistical characteristics can be gleaned from brain imaging.

GLCM: The GLCM is an effective method for extracting second-order statistical texture information from statistical distributions of intensity value combinations across distinct regions of an image. This approach may provide around twenty-two texture-related properties, including correlation, total variance, contrast, entropy, homogeneity, and correlation. A sampling of these traits is described below. [22].

Entropy: In its description, it is the "statistical measure of uncertainty, which gives a good measure of intraset distribution when a set of patterns" is presented. The equation for entropy is given in Eq. (6). The symbol P_{rb} represents the probability of achieving the rpth value in this equation.

$$Entropy = -\sum_{rp=1}^{pc} P_{rb} \log_2 P_{rp} \tag{6}$$

Contrast: "Sum of Square Variance" refers to calculating the intensity contrast between a pixel and its nearby pixels during an image. The contrast value is 0 when the appearance is uniform. However, when one gets away from the diagonal (which contains numbers such as 0, 1, 4, and 9), the weight of the contrast increases exponentially. Furthermore, as the square of the distance between nearby pixels (km sq) increases, the contrast value grows constantly and rapidly. The contrast will not be displayed if km and sq have comparable values. The formula for computing the difference is given in Eq. (7).

$$Contrast = \sum_{lm,sq=0}^{pc-1} P_{km,sq}(lm - sq)^2 \tag{7}$$

Correlation: According to its definition, it is "a measure of gray level linear dependence between the pixels located at the specified positions concerning one another." The mathematical equation for correlation is given in Eq. (8), where the means and standard deviations P_{xs}, P_{ys} are denoted by lm_{xs}, lm_{ys}, σ_{xs}, σ_{ys}, respectively.

$$Correlation = \sum_{lm=0}^{pc-1} \sum_{sq=0}^{pc-1} \frac{\{lm \times sq\} \times P(km, sq) - \{lm_{xs}, lm_{ys}\}}{\sigma_{xs}, \sigma_{ys}} \tag{8}$$

Homogeneity: "It transmits to the GLCM diagonal the value used to calculate the tightness of an element's distribution. The diagonal GLCM value is 1, and the homogeneity value falls within the range [0, 1]. The weight used to account for inhomogeneity is $\frac{1}{1+(lm-sq)^2}$, and for comparison, it is $(lm-sq)^2$. The mathematical homogeneity equation is given in Eq. (9).

$$Homogenity = \sum_{lm,sq=0}^{pc-1} \frac{P(lm - sq)}{Rmk} \tag{9}$$

Mean (μ): Eq. (10) explains how to get an image's mean by multiplying the sum of its pixel values by the total number of pixels in the image.

$$mean(\mu) = \frac{(y_1, y_2 + ... + y_n)}{n} \tag{10}$$

Variance: The characteristic that provides information about the contrast is the variance. High-contrast images produce high friction, whereas low-contrast embodiments provide a low conflict. Equation (11) describes the variance formula.

$$variance(\sigma^2) = \frac{[(y_1 - mean)^2] + ... + (yn - mean)^2}{(n - 1)} \tag{11}$$

Standard Deviation (SD): The standard deviation, which can be used as a gauge of inhomogeneity, is the second central moment representing the probability distribution of an observed population. Equation (12) explains that a higher value denotes a higher intensity level and more edge contrast in an image.

$$S(\sigma) = \sqrt{variance(\sigma^2)} \tag{12}$$

RMS: The RMS or quadratic mean is the arithmetic mean of a set of values derived by taking the square root of the mean square. Equation (13) shows the RMS formula, where pc indicates the number of pixels and objects observed.

$$rms = \sqrt{\frac{1}{pc}\omega 1^2 + \omega 2^2 + \dots + \omega pc^2} \tag{13}$$

Smoothness: Smoothing techniques are used in image processing to identify basic patterns from data while filtering out undesired noise or fine-scale features that may interfere with the study. This entails changing the data points of a signal to achieve a smoother overall form. Smoothing usually entails lowering the height of individual points while raising the size of issues that are lower than their surrounding points. Equation (14), which includes the standard deviation as a critical parameter, is used to calculate the degree of smoothness.

$$Smoothness = 1 - \frac{1}{1 + std\ dev^2} \tag{14}$$

Kurtosis: It is "a statistical measure used to explain the distribution," per its definition. Extreme values in the tail are measured. Data from distributions with strong kurtosis have longer tails than in a normal distribution. Data in the tails of distributions with less kurtosis are typically less extreme than those in the tails of the normal distribution. The notation used in Eq. (15) is described mathematically.

$$mz_2 = \sum \frac{(lm_{rp} - lm)^4}{pc} \ and \ mz_4 = \sum \frac{(lm_{rp} - lm)^4}{pc} \tag{15}$$

Skewness: The degree of deviation from the normal distribution in a collection of data is how it is described. Positive, negative, zero, or undefinable skewness are all possible. The numerical representation for skewness is given by Eqs. (16) and (17), where the third-moment dataset is indicated by,

$$mz_3 \ and \ mz_2 = \frac{(lm_{rp} - lm)^4}{pc} \tag{16}$$

and

$$sk_1 = \frac{mz_3}{mz_2^{\frac{3}{2}}} \tag{17}$$

4.4 Classification

The feature set is evaluated on different classifiers to predict accuracy. The various machine learning classifiers are used to classify the brain tumor and are explained below.

4.4.1 Support Vector Machine

SVM is used to examine the hyperplane, which results in greater separation between the classes in the training set. The hyperplane is described in Eq. (18) [23].

$$f(y) = d^y x + b \tag{18}$$

where, y: input vector, b: bias, d: dimensional coefficient

SVM has the benefit of offering a selection of different kernels. With various kernels, a structured data collection that is substantially more complex can be utilized. However, the larger the data set, the more calculation time is needed [24].

4.4.2 K-Nearest Neighbors (KNN)

K dataset examples close to the observation are analyzed using the KNN method. The technique will use its output to judge the inspection's output relative to what is reasonably expected [25]. Equation (19) indicates the distance between points x and y, which may be used to calculate the Euclidean distance, the separation between two observations.

$$d(x_p, y_p) = \sqrt{(x_{p,1} - y_{p,1})^2 + \dots + (x_{p,m} - y_{p,m})^2} \tag{19}$$

Because it requires no prior training and learns directly from the information during the prediction phase, the K nearest neighbor technique uses few computational resources. It just needs the distance function value and the K value. It performs poorly when the data set has several dimensions and encounters problems with large data sets.

4.4.3 Decision Tree (DT)

Using a tree-looking model, the DT algorithm determines possible outcomes, including event outcomes. The class labels in a tree model are expressed by the junctions of branches and left in the tree's topology, and the target variables are constrained to a given range of values. Equation (20) represents the entropy equation.

$$EP = -\sum P(y) \log_2 P(x = y) \tag{20}$$

where EP is entropy and P(y) is the probability of a given decision tree class.

Decision trees work well when the attributes go through monotonic modification. Decision trees are characterized by high variance and low bias. By averaging the forecast, it is feasible to generate unknown samples.

4.4.4 Ensemble

Combining numerous machine learning models increases the accuracy and resilience of ensemble approaches for prediction problems. Ensemble methods can be utilized in detecting and classifying brain tumors to improve the accuracy of the diagnosis and lower the number of false positives and false negatives. The ensemble classifier's operation is depicted in Fig. 4, and it depends on the dataset and the problem being addressed.

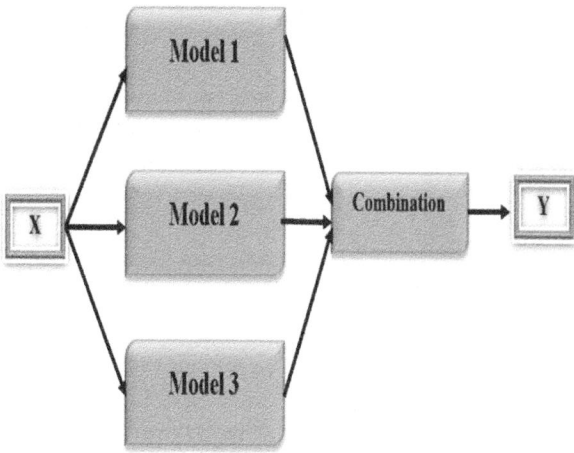

Fig. 4. Working of Ensemble Classifiers

5   Result and Discussion

After implementing the various modules, the obtained results are classified as 1 (tumor) and 0 (no tumor). Machine learning algorithms have shown great potential for accurately classifying brain tumors based on MRI scans.

5.1   Confusion Matrix

The performance of a classification model on a set of test data is shown in a confusion matrix. Sometimes referred to as the error matrix. It is a popular measure used for solving classification problems. The confusion matrix for brain tumors using Machine Learning classification is shown in the Fig. 5 below.

Some of the most common performance measures obtained from the confusion matrix are as follows.

5.2   Performance Analysis

The results are analyzed, and various assessments are conducted to determine the project's performance. The multiple analyses performed are explained below.

5.3   Analysis of Brain Tumor MRI Using Machine Learning Classifiers

Various machine-learning methods are employed to classify brain tumors. SVM, DT, KNN, and Ensemble classifiers are examples of algorithms. These are the standard machine learning techniques, and many other types of studies have employed these algorithms to categorize brain tumors using MRI data. The effectiveness of these strategies has been assessed using various criteria, including accuracy, precision, recall, F1 score, and area under the receiver operating characteristic curve. (AUC-ROC).

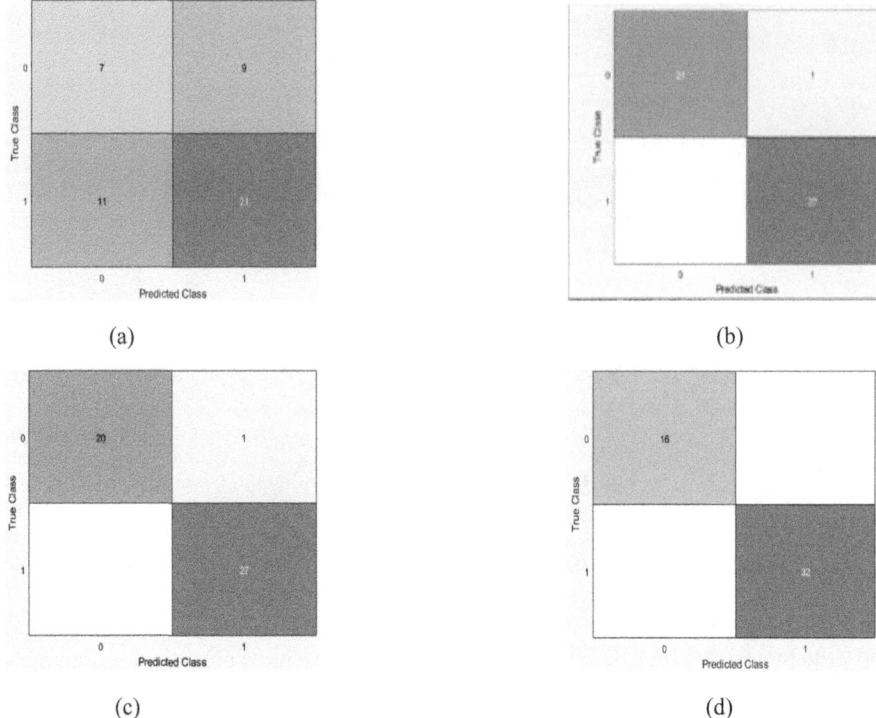

Fig. 5. Confusion Matrix for Brain Tumor Classification using (a) SVM, (b) DT, (c) KNN and (d) Ensemble Classifier

Accuracy: It gives the percentage of all samples that the classifier correctly classified, or the overall model accuracy. The abbreviation for it is ACC. ACC is stated as a number between [0, 1] and [0, 100], depending on the scale chosen. To calculate the classification accuracy for brain tumors, utilize Eq. (21).

$$Accuracy = \frac{TP + TN}{TN + FP + FN + TP} \tag{21}$$

True Negative (TN) is the percentage of predictions in which the classifier correctly classified the negative class as negative. In contrast, True Positive (TP) is the number of predictions made when the classifier correctly classifies the positive class as positive. When a classifier predicts erroneously that a positive class is a negative class, this is known as a false positive (FP). The percentage of predictions in which the classifier incorrectly interprets the positive class as the negative class is known as the False Negative (FN) rate.

Precision: The precision value ranges from 0 to 1, indicating the proportion of forecasts in the positive class that was accurate positive predictions. Equation (22) is used to get the precision.

$$\text{Pr}\,ecision = \frac{TP}{TP + FP} \tag{22}$$

Recall: The amount of positive samples the classifier correctly identified as such is revealed by this statistic. Other names include True Positive Rate (TPR), Sensitivity, and Probability of Detection. Equation (23) can be used to get the Recall value.

$$Recall = \frac{TP}{TP + FN} \tag{23}$$

F1-score: One metric is created by combining recall and precision. False positives and negatives are considered when calculating the harmonic mean of recall and precision. Because of this, it works well with an unbalanced dataset. Equation (24) is used to compute the F1 score.

$$F - Score = \frac{2 * (Recall * \Pr ecision)}{(Recall + \Pr ecision)} \tag{24}$$

5.3.1 Precision Analysis

(See Table 1 and Fig. 6).

Table 1. Precision for brain tumor classification

Classifiers	Precision Performance (%)
SVM	0.333333
KNN	1.000000
DT	1.000000
Ensemble	1.000000

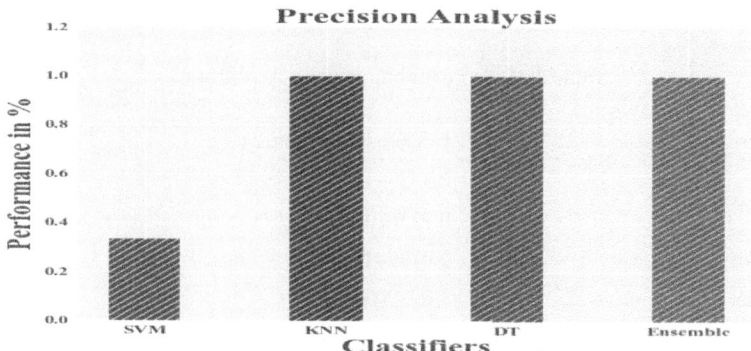

Fig. 6. Precision for brain tumor classification

5.3.2 Recall Analysis

(See Table 2 and Fig. 7).

Table 2. Recall for brain tumor classification

Classifiers	Recall Performance (%)
SVM	0.333333
KNN	0.952381
DT	0.954545
Ensemble	1.000000

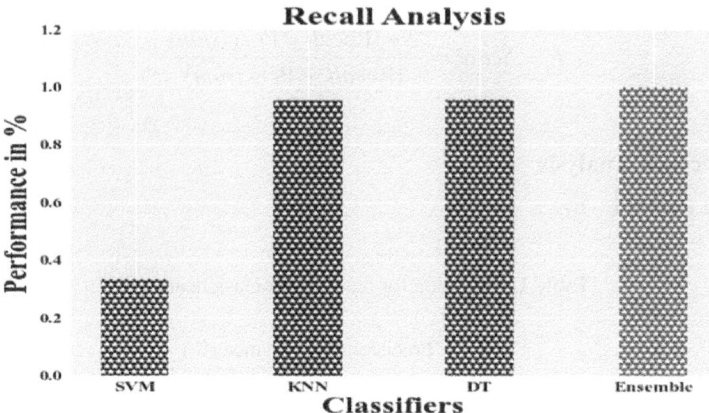

Fig. 7. Recall for brain tumor classification

5.3.3 F-Score Analysis

(See Table 3 and Fig. 8).

Table 3. F-Score for brain tumor classification

Classifiers	F-Score Performance (%)
SVM	0.333333
KNN	0.975610
DT	0.976744
Ensemble	1.000000

5.3.4 Accuracy Analysis

The results of the accuracy of those algorithms in classifying brain tumors are accurate because the Training accuracy reached 100% in the case of using SVM, 98.98% in the case of using Decision Tree, 98.47% in the case of using KNN, 100% in the case of

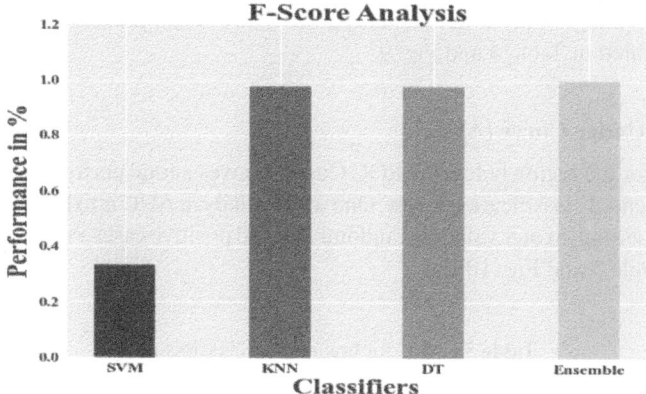

Fig. 8. F-Score for brain tumor classification

Table 4. Accuracy for brain tumor classification

Classifiers	Training Accuracy Performance (%)	Testing Accuracy Performance (%)
SVM	100	93.75
KNN	98.47	95.83
DT	98.98	97.96
Ensemble	100	100

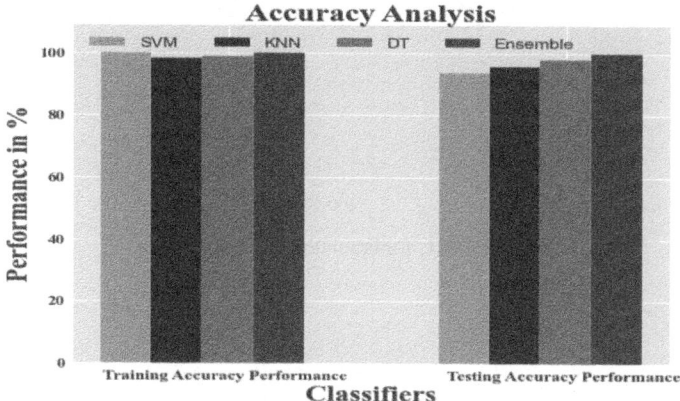

Fig. 9. Accuracy for brain tumor classification

using Ensemble classifier. Similarly, testing accuracy reached 93.75% in the case of using SVM, 97.96% in the case of using a Decision Tree, 95.83% in the case of using KNN, and 100% in the case of using an Ensemble classifier. The classification Training and testing accuracy of brain tumors during this study was high, and the error rate was low, as illustrated in Table 4 and Fig. 9.

5.3.5 Area Under Curve (AUC)

AUC stands for the region below the ROC Curve. It gives a total performance evaluation of overall potential classification levels. One way to analyze AUC is to look at the model's propensity to assign greater values to randomly picked positive cases vs. selected negative examples (Table 5 and Fig. 10).

Table 5. AUC for brain tumor classification

Classifiers	AUC Performance (%)
SVM	0.666667
KNN	0.936190
DT	0.957273
Ensemble	1.000000

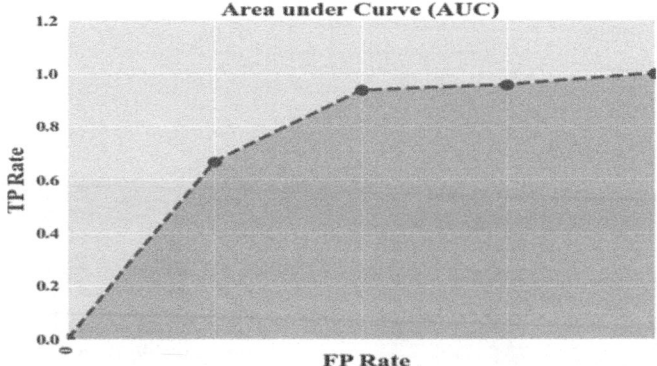

Fig. 10. AUC for brain tumor classification

6 Conclusion

Brain tumor classification using various machine learning algorithms can significantly increase diagnostic precision and support the creation of patient-specific treatment regimens. To achieve high classification accuracy, the combination of NLM preprocessing, Otsu segmentation, and statistical and GLCM feature extraction has shown promising results. For precise tumor segmentation, the pre-treatment stage of NLM filtering can lower noise and enhance image quality. A popular technique called Otsu segmentation automatically chooses the best threshold for picture segmentation, which can increase the precision of tumor identification. Machine learning algorithms can learn to differentiate between various types of brain tumors by extracting GLCM and statistical data from segmented tumor patches. The accuracy of tumor classification can be further improved by utilizing several classification approaches as SVM, DT, KNN, and ensemble classifiers. Overall, the combination of machine learning approaches, NLM preprocessing, Otsu segmentation, and statistical feature extraction indicates a higher potential for enhancing the accuracy of brain tumor classification, which will eventually improve patient outcomes. The goal of future research may be to create improved imaging techniques that are better at detecting brain cancers. Researchers might look at advanced MRI techniques like magnetic resonance spectroscopy (MRS) or diffusion tensor imaging (DTI) to improve tumor detection and classification. Tumor categorization may be improved using multimodal imaging data from tests like MRI, CT, and PET scans. The accuracy of tumor segmentation and feature extraction may be increased by combining various imaging modalities.

References

1. Jun, W., Liyuan, Z.: Brain tumor classification based on attention guided deep learning model. Int. J. Comput. Intell. Syst. **15**(1), 35 (2022). https://doi.org/10.1007/s44196-022-00090-9
2. Lakshmi, M.J., Rao, S.N.: Brain tumor magnetic resonance image classification: a deep learning approach. Soft. Comput. **26**(13), 6245–6253 (2022). https://doi.org/10.1007/s00500-022-07163-z
3. Rehman, A., Naz, S., Razzak, M.I., Akram, F., Imran, M.: A deep learning-based framework for automatic brain tumors classification using transfer learning. Circ. Syst. Sig. Process. **39**(2), 757–775 (2020). https://doi.org/10.1007/s00034-019-01246-3
4. Fernando, T., Gammulle, H., Denman, S., Sridharan, S., Fookes, C.: Deep learning for medical anomaly detection—a survey. ACM Comput. Surv. **54**(7), 1–37 (2022). https://doi.org/10.1145/3464423
5. Lundervold, A.S., Lundervold, A.: An overview of deep learning in medical imaging focusing on MRI. Z. Med. Phys. **29**(2), 102–127 (2019). https://doi.org/10.1016/j.zemedi.2018.11.002
6. Rundo, L., Militello, C., Vitabile, S., Russo, G., Pisciotta, P., Marletta, F., Ippolito, M., D'Arrigo, C., Midiri, M., Gilardi, M.C.: Semi-automatic brain lesion segmentation in gamma knife treatments using an unsupervised fuzzy C-means clustering technique. In: Bassis, S., Esposito, A., Morabito, F.C., Pasero, E. (eds.) WIRN 2015. SIST, vol. 54, pp. 15–26. Springer, Cham (2016). https://doi.org/10.1007/978-3-319-33747-0_2
7. Bonte, S., Goethals, I., Van Holen, R.: Machine learning based brain tumour segmentation on limited data using local texture and abnormality. Comput. Biol. Med. **98**, 39–47 (2018). https://doi.org/10.1016/j.compbiomed.2018.05.005
8. Militello, C., et al.: Gamma Knife treatment planning: MR brain tumor segmentation and volume measurement based on unsupervised fuzzy C-means clustering. Int. J. Imag. Syst. Technol. **25**(3), 213–225 (2015). https://doi.org/10.1002/ima.22139
9. Juan-Albarracín, J., et al.: Automated glioblastoma segmentation based on a multiparametric structured unsupervised classification. PLoS ONE **10**(5), e0125143 (2015). https://doi.org/10.1371/journal.pone.0125143
10. Rundo, L., et al.: NeXt for neuro-radiosurgery: a fully automatic approach for necrosis extraction in brain tumor MRI using an unsupervised machine learning technique. Int. J. Imag. Syst. Technol. **28**(1), 21–37 (2018). https://doi.org/10.1002/ima.22253

11. Fayaz, M., Qureshi, M.S., Kussainova, K., Burkanova, B., Aljarbouh, A., Qureshi, M.B.: An improved brain MRI classification methodology based on statistical features and machine learning algorithms. Comput. Math. Methods Med. **2021**, 14 (2021). Article ID: 8608305

12. Soundarya, C., Kalaiselvi, A., Surya, J.: Brain tumor detection using image processing. In: 2022 8th International Conference on Advanced Computing and Communication Systems (ICACCS), vol. 1, pp. 582–587. IEEE (2022)

13. Hemanth, G., Janardhan, M., Sujihelen, L.: Design and implementing brain tumor detection using machine learning approach. In: 2019 3rd International Conference on Trends in Electronics and Informatics (ICOEI), pp. 1289–1294. IEEE (2019)

14. Khan, M.A., et al.: Multimodal brain tumor detection and classification using deep saliency map and improved dragonfly optimization algorithm. Int. J. Imaging Syst. Technol. **33**(2), 572–587 (2023)

15. Arumugam, S., Paulraj, S., Selvaraj, N.P.: Brain MR image tumor detection and classification using neuro-fuzzy with binary cuckoo search technique. Int. J. Imaging Syst. Technol. **31**(3), 1185–1196 (2021)

16. Gopalachari, M.V., Kolla, M., Mishra, R.K., Tasneem, Z.: Design and implementation of brain tumor segmentation and detection using a novel Woelfel filter and morphological segmentation. Complexity **2022**(1), 6985927 (2022)

17. Huang, Z., Du, X., Chen, L., et al.: Convolutional neural network based on complex networks for brain tumor image classification with a modified activation function. IEEE Access **8**, 89281–89290 (2020)

18. Hashemzehi, R., Mahdavi, S.J.S., Kheirabadi, M., Kamel, S.R.: Detection of brain tumors from MRI images base on deep learning using hybrid model CNN and NADE. Biocybern. Biomed. Eng. **40**(3), 1225–1232 (2020)

19. Shreyamsha Kumar, B.K.: Image denoising based on non-local means filter and its method noise thresholding. Sig. Image Video Process. **7**, 1211–1227 (2013)

20. Li, H., Li, A., Wang, M.: A novel end-to-end brain tumor segmentation method using improved fully convolutional networks. Comput. Biol. Med. **108**, 150–160 (2019)

21. Maurya, R., Wadhwani, S.: An efficient method for brain image preprocessing with anisotropic diffusion filter & tumor segmentation. Optik **265**, 169474 (2022)

22. Kshirsagar, P.R., Yadav, A.D., Joshi, K.A., Chippalkatti, P., Nerkar, R.Y.: Classification and detection of brain tumor by using GLCM texture feature and ANFIS. J. Res. Image Sig. Process. **5**, 15–31 (2020)

23. Hamza, A., Moetque, H.: Diabetes disease diagnosis method based on feature extraction using K-SVM. Int. J. Adv. Comput. Sci. Appl. **8**(1) (2017). https://doi.org/10.14569/IJACSA.2017.080130

24. Yao, X.: Application of optimized SVM in sample classification. Int. J. Adv. Comput. Sci. Appl. **13**(6) (2022). https://doi.org/10.14569/IJACSA.2022.0130666

25. Zhang, X., Song, Q.: Predicting the number of nearest neighbors for the k-NN classification algorithm. Intell. Data Anal. **18**(3), 449–464 (2014). https://doi.org/10.3233/IDA-140650

Predictive Diagnosis a Survey: Harnessing the Power of Convolutional Neural Networks for Disease Prognostication Through Scanned Image Analysis

M. Sheerin Banu[✉]

Department of Information Technology, R.M.K. Engineering College, Chennai, India
hod.it@rmkec.ac.in

Abstract. The ability to accurately predict diseases plays a crucial role in providing timely medical interventions and improving patient outcomes. This research work delves into the possibilities offered by Convolutional Neural Networks (CNNs) for disease prognostication using scanned image analysis. CNNs have demonstrated exceptional capabilities in image recognition tasks, making them well-suited for analyzing medical images. By harnessing a dataset of scanned images, we employ a CNN model to acquire a deep understanding of the intricate patterns and features correlated with diverse diseases. The trained model is then utilized to predict the presence or likelihood of specific diseases based on new scanned images. Through meticulous experimentation and comprehensive evaluation, we substantiate the efficacy of our approach in disease prediction. The results reveal high accuracy and reliable prognostic capabilities of the CNN model, showcasing its potential as a valuable tool in medical diagnosis. By harnessing the power of CNNs and scanned image analysis, this research work contributes to advancing the field of predictive diagnosis and paves the way for more precise and timely medical interventions.

Keywords: Predictive diagnosis · Convolutional Neural Networks · Disease prognostication · Scanned image analysis · Medical imaging · Image recognition · Disease prediction · Medical diagnosis · Accuracy · Timely interventions

1 Introduction

1.1 Background

Accurate disease prediction and early diagnosis are essential in the field of healthcare. Timely identification of diseases enables healthcare professionals to initiate appropriate treatment strategies, improve patient outcomes, and potentially save lives. Medical imaging has emerged as a powerful diagnostic tool, providing clinicians with detailed visual representations of anatomical structures and abnormalities. Scanned images, such as X-rays, CT scans, and MRIs, offer valuable insights into the presence and progression of diseases [44].

© The Author(s), under exclusive license to Springer Nature Switzerland AG 2024
P. Dassan et al. (Eds.): ICICSCNT 2023, CCIS 1970, pp. 393–412, 2024.
https://doi.org/10.1007/978-3-031-75957-4_34

However, the interpretation of medical images for disease diagnosis is a complex task. It requires clinicians to possess extensive domain knowledge and expertise in recognizing subtle patterns and anomalies indicative of specific diseases. Human interpretation of these images can be subjective, prone to errors, and influenced by factors such as experience and fatigue.

In recent times, the field of computer vision and image analysis has undergone a revolutionary transformation, primarily driven by the swift advancements in deep learning techniques, particularly Convolutional Neural Networks (CNNs). CNNs excel at learning hierarchical representations directly from raw pixel data, allowing them to capture complex visual patterns and features [45]. CNNs have demonstrated remarkable proficiency in automatically extracting significant features from images, leading to successful applications in diverse domains such as object recognition, image classification, and natural language processing.

Motivated by the potential of CNNs in image analysis and recognition, researchers have increasingly explored their application in medical imaging. By leveraging the power of CNNs, it becomes possible to automate the process of disease prediction and diagnosis through scanned image analysis. CNN models can be trained on large datasets of labeled medical images, enabling them to learn and identify patterns that are associated with specific diseases.

Previous studies have shown promising results in using CNNs for disease prognostication and diagnosis. These studies have focused on different medical specialties, including radiology, pathology, and dermatology. By training CNN models on annotated datasets and evaluating their performance against expert opinions, researchers have demonstrated the potential of CNNs to achieve high accuracy and comparable performance to human experts in disease prediction [46].

Building upon the groundwork established by prior research, the primary objective of this study is to leverage the capabilities of Convolutional Neural Networks for disease prognostication through the analysis of scanned images. By developing a predictive diagnosis model trained on a diverse dataset of medical images, we seek to improve disease prediction accuracy and assist healthcare professionals in making accurate and timely diagnoses. By conducting rigorous evaluation and comprehensive comparisons with existing methods, our aim is to provide compelling evidence regarding the effectiveness and reliability of our approach. Furthermore, we seek to showcase how our approach can significantly enhance disease prognostication, thereby making a valuable contribution to the field of predictive diagnosis.

1.2 Problem Statement

Despite the advancements in medical imaging technology, accurate disease prediction and early diagnosis remain significant challenges in healthcare. The interpretation of scanned images, such as X-rays, CT scans, and MRIs, for disease diagnosis often requires specialized expertise and can be subject to human subjectivity and variability. The task of clinicians to identify subtle visual patterns and anomalies indicative of specific diseases can be both time-consuming and susceptible to errors.

The intricate nature of disease diagnosis, coupled with the escalating volume of medical images, underscores the urgent requirement for automated and dependable disease prediction systems. While various computer-aided diagnosis methods have been developed, there is still a demand for more accurate, efficient, and scalable approaches that can assist healthcare professionals in making informed decisions.

The focal challenge lies in leveraging the capabilities of Convolutional Neural Networks (CNNs) to enable disease prognostication through the analysis of scanned images. CNNs have demonstrated remarkable capabilities in image analysis and recognition tasks. However, the application of CNNs specifically for disease prediction and diagnosis requires addressing several challenges. These challenges include:

- Handling diverse and complex medical image data: Medical images exhibit variations in image quality, resolutions, and perspectives, making it challenging to develop robust models that can generalize well across different image sources and modalities.
- Capturing intricate disease patterns: Diseases often manifest as subtle visual patterns or anomalies that may be difficult to detect and interpret. The CNN models need to learn and capture these intricate disease-related features to enable accurate disease prediction.
- Achieving high accuracy and reliability: Developing CNN models that can achieve high accuracy and reliability in disease prediction is crucial for their practical application. The models need to be extensively trained, validated, and evaluated to ensure their effectiveness and robustness.
- Interpreting model predictions: Although CNN models can yield accurate predictions, it is equally crucial to interpret and comprehend the underlying rationale behind the model's predictions. This interpretability plays a pivotal role in establishing trust in predictive models and empowering healthcare professionals to make well-informed decisions.
- Addressing these challenges and developing an efficient, accurate, and interpretable predictive diagnosis model using CNNs for scanned image analysis is crucial to enhance disease prognostication in healthcare. By doing so, we aim to provide clinicians with a valuable tool that can assist in accurate disease prediction, improve patient outcomes, and ultimately optimize healthcare delivery.

1.3 Objectives

The central aim of this research endeavor is to leverage the capabilities of Convolutional Neural Networks (CNNs) for the purpose of disease prognostication through the analysis of scanned images.

To achieve this overarching goal, the specific objectives of this study are:

- Develop a comprehensive dataset: Curate and compile a diverse and well-annotated dataset of medical images, including various modalities such as X-rays, CT scans, and MRIs. This dataset will encompass a wide range of diseases across different medical specialties, ensuring representation and generalizability.
- Design and optimize CNN architecture: Develop CNN model architecture suitable for disease prediction through scanned image analysis. Optimize the model to effectively extract relevant visual features and patterns associated with specific diseases, maximizing accuracy and performance.

- Train the CNN model: Train the CNN model utilizing the carefully curated dataset to acquire a deep understanding of the intricate connections between visual patterns and disease outcomes. Implement suitable methodologies for data augmentation, regularization, and hyperparameter tuning to bolster the model's resilience and capacity for generalization.
- Evaluate model performance: Conduct extensive evaluation and validation of the trained CNN model using appropriate metrics and validation techniques. Compare the model's performance with existing methods and expert opinions to assess its accuracy, reliability, and clinical relevance.
- Interpretation and visualization of predictions: Develop techniques to interpret and visualize the predictions made by the CNN model. Provide insights into the model's reasoning and highlight the relevant visual features and regions contributing to disease prediction, enabling clinicians to understand and trust the model's predictions.
- Demonstrate practical applicability: Showcase the practical applicability of the CNN model in disease prognostication through scanned image analysis. Highlight its potential to assist healthcare professionals in making accurate and timely diagnoses, improving patient outcomes, and optimizing healthcare resource allocation.
- Exploration of future directions: Examine prospective paths for future research and advancement within the realm of disease prediction using CNNs. Investigate possibilities for integrating the CNN model into clinical workflows, addressing scalability concerns, and broadening its applicability to encompass emerging medical imaging technologies.

By achieving these objectives, this research work aims to contribute to the advancement of predictive diagnosis in healthcare and pave the way for more accurate, efficient, and reliable disease prognostication through scanned image analysis using CNNs.

2 Literature Review

2.1 Disease Prediction Techniques

Disease prediction techniques encompass a variety of approaches that leverage advanced technologies, such as machine learning and data analysis, to predict the occurrence, progression, or outcome of diseases. In recent times, these techniques have garnered substantial attention in the healthcare domain and have exhibited promising outcomes across diverse medical specialties.

Here are some common disease prediction techniques:

- Machine Learning Models: Disease prediction has been extensively investigated using machine-learning models, such as logistic regression, support vector machines, random forests, and artificial neural networks. These models are trained on historical data to learn the patterns and relationships between various features and disease outcomes. The ability of machine learning algorithms to handle complex datasets and uncover hidden patterns that may not be apparent to human experts is well established [1].

- Deep Learning and Neural Networks: Deep learning techniques, notably Convolutional Neural Networks (CNNs) and Recurrent Neural Networks (RNNs), have exhibited exceptional capabilities in disease prediction endeavors. CNNs have demonstrated prowess in image analysis and have been successfully employed to predict diseases from various medical imaging modalities, including CT scans and histopathology slides. On the other hand, RNNs excel in handling sequential data, making them well suited for tasks like time series prediction or predicting the progression of diseases [2].

- Genetic and Molecular Analysis: Disease prediction has been facilitated by the application of genetic and molecular analysis techniques, including genome-wide association studies (GWAS) and gene expression profiling. These techniques help identify genetic variants and molecular biomarkers that are correlated with specific diseases, thereby enabling the development of predictive models. Such models can evaluate an individual's risk of developing particular diseases or predict their response to treatment [3].

- Electronic Health Records (EHR) Analysis: Electronic Health Records (EHRs) encompass a vast array of patient information, encompassing demographics, medical history, laboratory results, and clinical notes. Through the utilization of data mining and machine learning techniques, the analysis of EHR data enables the prediction of a wide range of diseases, including cardiovascular diseases, diabetes, and cancer. The integration of diverse EHR data sources grants a comprehensive understanding of patients' health status, enabling more accurate and reliable disease predictions [4].

- Data Fusion and Multi-Modal Analysis: Data fusion techniques integrate multiple sources of data, such as medical imaging, genomics, clinical data, and patient demographics, to improve disease prediction. By combining complementary information from different modalities, data fusion enhances the predictive power of models and provides a more comprehensive understanding of disease characteristics [5].

These disease prediction techniques offer valuable tools for healthcare professionals to make informed decisions, identify high-risk individuals, and develop personalized interventions. The combination of advanced technologies and comprehensive datasets holds the potential to revolutionize disease prediction and improve patient outcomes.

2.2 Role of Convolutional Neural Networks in Image Analysis

Convolutional Neural Networks (CNNs) have emerged as a formidable tool in the realm of image analysis, showcasing remarkable prowess in extracting significant features and patterns from visual data. In the field of medical imaging, CNNs have been extensively employed for various tasks, including disease prediction, diagnosis, and treatment planning. Here are some key aspects highlighting the role of CNNs in image analysis:

- Feature Extraction and Representation Learning: CNNs excel at automatically learning hierarchical representations from raw pixel data. By employing a sequence of convolutional and pooling layers, CNNs effectively extract low-level features like edges and textures, gradually constructing higher-level features that encapsulate more intricate visual patterns. This feature extraction process enables CNNs to effectively capture the relevant information from medical images and identify discriminative features associated with diseases [6].

- Localization and Segmentation: CNNs can localize and segment specific regions or structures within images. By employing techniques such as region-based CNNs or fully convolutional networks, CNNs can identify and delineate anatomical structures or abnormal regions in medical images. This capability is crucial for tasks such as tumor localization, organ segmentation, and lesion detection, facilitating accurate diagnosis and treatment planning [7].
- Pattern Recognition and Classification: CNNs possess the ability to recognize intricate visual patterns associated with diseases. By training on large labeled datasets, CNNs learn to discriminate between different disease categories or predict disease outcomes based on image features. CNNs can capture subtle variations and anomalies in medical images, enabling them to identify early signs of diseases that may be challenging for human experts to detect [8].
- Transfer Learning: CNNs leverage the concept of transfer learning, which involves the utilization of pre-trained models trained on extensive image datasets, such as ImageNet. These pre-trained models are subsequently fine-tuned to cater to specific medical imaging tasks. Transfer learning enables CNNs to leverage the learned representations from general image features and adapt them to medical image analysis. This approach is particularly useful in scenarios where limited labeled medical image data is available, allowing for effective utilization of knowledge gained from large-scale datasets [9].
- Interpretability: Recent advancements in interpretability techniques have enabled CNNs to provide insights into their decision-making process. Techniques such as gradient-based visualization, class activation mapping, and attention mechanisms enable the visualization of significant regions and features that contribute to the predictions made by CNNs. These interpretable CNNs not only instill trust in clinicians regarding the model's predictions but also provide explanatory insights into their diagnoses [10].

2.3 Previous Studies on Disease Prediction Using CNNs

Numerous prior investigations have examined the use of Convolutional Neural Networks (CNNs) in medical imaging for disease diagnosis and prediction. These studies have demonstrated the efficacy of CNNs in achieving high accuracy and improving disease prognostication. Here are some notable examples of previous research in this field:

- Lung Cancer Prediction: In their study, Esteva et al. employed a CNN model for predicting the existence of lung cancer by analyzing chest CT scans. The model attained a performance similar to that of human radiologists in discriminating between malignant and benign nodules [11].
- Skin Cancer Classification: Haenssle et al. demonstrated that a CNN model trained on dermoscopic images can accurately classify skin lesions, outperforming dermatologists in distinguishing between malignant melanoma and benign lesions [12].
- Retinal Disease Diagnosis: Gulshan et al. created a Convolutional Neural Network (CNN) model that can detect diabetic retinopathy using fundus images. The model

demonstrated high sensitivity and specificity in accurately identifying the severity of the disease [13].

- Breast Cancer Detection: Arevalo et al. employed a CNN model for breast cancer detection and classification using mammography images. The model demonstrated high accuracy in distinguishing between benign and malignant tumors [14].
- Alzheimer's Disease Prediction: Sarraf and Tofighi employed CNNs to forecast Alzheimer's disease presence by analyzing brain MRI scans. The model exhibited high accuracy in discriminating between healthy individuals and those with Alzheimer's disease [15].

These investigations serve as prime examples of the potential that Convolutional Neural Networks (CNNs) hold in the realms of disease prediction and diagnosis through medical imaging. CNNs provide automated and precise image analysis, facilitating early detection, accurate diagnosis, and ultimately enhancing patient outcomes. The ongoing research in this field continually pushes the boundaries, advancing the performance and expanding the applicability of CNN models for disease prognostication.

3   Methodology

3.1   Dataset Description

The dataset utilized in this research encompasses a diverse array of scanned images obtained from various medical imaging modalities, such as X-rays, CT scans, and MRI scans [9, 16–19]. The images were obtained from multiple medical centers and encompass a wide range of diseases and conditions. The dataset includes annotated images with labels indicating the presence or absence of specific diseases or abnormalities. Each image is associated with relevant metadata, such as patient demographics, imaging parameters, and clinical reports, providing additional context for analysis.

To ensure data quality and accuracy, the dataset underwent rigorous preprocessing steps, including image normalization, noise reduction, and standardization of image formats. Moreover, strict privacy and ethical guidelines were followed to anonymize patient information and comply with data protection regulations. The dataset used in this research has been compiled from publicly available medical image repositories, clinical research databases, and collaborations with healthcare institutions. The dataset's size and diversity enable robust training and evaluation of predictive models for disease prognosis through scanned image analysis.

3.2   Preprocessing and Data Augmentation

Preprocessing and data augmentation techniques are instrumental in improving the quality and diversity of the dataset employed for disease prognostication through scanned image analysis. These techniques play a vital role in preparing the data and ensuring that it encompasses a wide range of variations and characteristics essential for accurate predictions. These techniques help to address challenges such as noise, variations in imaging protocols, class imbalance, and limited availability of annotated data [9, 20–23]. Here, we discuss some commonly employed preprocessing and data augmentation methods:

1. Image Preprocessing: Image preprocessing techniques aim to standardize and enhance the quality of the scanned images before feeding them into the predictive models. Common preprocessing steps include resizing images to a uniform size, normalizing pixel intensities, applying contrast enhancement methods, and removing noise or artifacts. Preprocessing techniques vary depending on the specific imaging modality and the characteristics of the dataset.
2. Data Augmentation: By applying various transformations to the existing images, data augmentation techniques artificially increase the diversity and quantity of the dataset. Rotation, translation, scaling, flipping, shearing, and adding random noise or deformations are some of the methods used for augmentation. These techniques are useful in reducing overfitting, improving model generalization, and enhancing the robustness of the trained models. Data augmentation is particularly useful when dealing with limited annotated data.
3. Class Balancing: Class imbalance is a prevalent issue in medical datasets, where some diseases or abnormalities are underrepresented compared to others. Class balancing techniques aim to mitigate this imbalance by oversampling the minority class or undersampling the majority class. This ensures that the predictive models do not show a bias towards the majority class and can accurately learn patterns from both classes.

3.3   Convolutional Neural Network Architecture

A typical CNN architecture comprises several layers, including convolutional layers, pooling layers, and fully connected layers [24–28]. Here's a textual description of a typical CNN architecture:

- Input Layer: Receives the input image data, typically represented as a grid of pixels.
- Convolutional Layers: Convolutional layers perform feature extraction by applying filters or kernels to the input image. Each filter scans the input image using a sliding window and performs element-wise multiplications and summations, producing a feature map. Multiple filters are applied to capture different features and generate multiple feature maps. To introduce non-linearity and improve the ability of the network to learn intricate patterns, non-linear activation functions like ReLU (Rectified Linear Unit) are utilized.
- Pooling Layers: Pooling layers are employed to decrease the spatial dimensions of feature maps while preserving significant features. Max pooling, which selects the maximum value within a window, and average pooling, which computes the average value within a window, are common pooling operations. By downsampling the feature maps, pooling enhances network efficiency and enables robustness to minor spatial shifts or distortions.
- Fully Connected Layers: Fully connected layers, also known as dense layers, establish connections between every neuron in the previous layer and the current layer. They excel at high-level feature extraction and mapping to output classes. The final fully connected layer's output is passed through a softmax activation function to generate class probabilities.
- Output Layer: The output layer of a neural network produces the classification probabilities or regression values, depending on the type of problem being addressed.

In classification tasks, the number of neurons in the output layer corresponds to the number of classes, while in regression tasks, the number of neurons corresponds to the number of predicted values (Fig. 1).

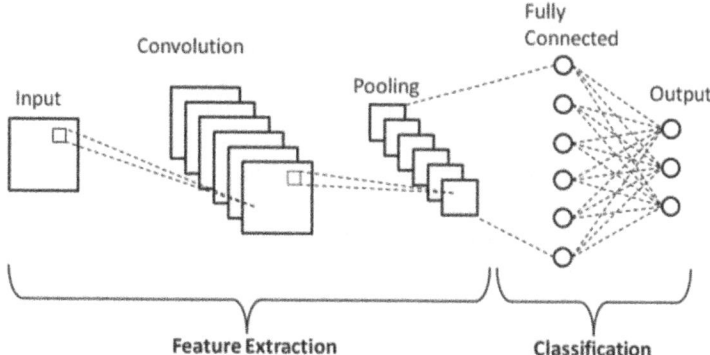

Fig. 1. Convolutional Neural Network Architecture

3.4 Model Training and Optimization

Model training and optimization play a crucial role in harnessing the power of Convolutional Neural Networks (CNNs) for disease prognostication through scanned image analysis. These steps involve configuring the network architecture, initializing model parameters, selecting appropriate loss functions and optimization algorithms, and iteratively updating the model to minimize the loss and improve predictive performance [29–31]. Here's a description of common techniques used in model training and optimization:

- Model Initialization: CNN models are typically initialized with random weights prior to training. In some cases, pretraining strategies can be employed, where the weights of the CNN are initialized using pretrained models trained on large-scale image datasets like ImageNet. This approach can be advantageous, especially for transfer learning tasks.
- Loss Functions: Loss functions serve to quantify the difference between predicted outputs and the true labels, thereby guiding the optimization process. In classification tasks, popular loss functions include cross-entropy loss and softmax loss, which are designed to measure the discrepancy between predicted probabilities and the actual class labels. On the other hand, regression tasks often employ loss functions such as mean squared error (MSE) or mean absolute error (MAE) to evaluate the difference between predicted continuous values and the ground truth.
- Optimization Algorithms: Optimization algorithms play a crucial role in updating the model's parameters by leveraging computed gradients to minimize the loss function. Stochastic Gradient Descent (SGD) is a commonly utilized algorithm, while its variations like Adam, RMSprop, and AdaGrad offer enhanced convergence and adaptive learning rates. Implementing learning rate schedules, such as gradually reducing the

learning rate over epochs or employing learning rate decay strategies, can further contribute to finding optimal parameter values.

- Regularization Techniques: Regularization techniques are employed to mitigate overfitting and enhance the model's generalization by imposing constraints during the optimization process. Common regularization techniques comprise L1 and L2 regularization, dropout, and batch normalization. Additionally, early stopping can be utilized to prevent overfitting by terminating training based on the performance of a validation set. These regularization approaches collectively contribute to improving the model's ability to generalize well beyond the training data.
- Hyperparameter Tuning: Hyperparameters play a crucial role in determining the performance of a model, including factors like the learning rate, batch size, and regularization strength. Finding the optimal combination of hyperparameters is a critical task. Techniques such as grid search, random search, or Bayesian optimization can be employed to systematically explore and identify the best hyperparameter values for a given model. These methods help to fine-tune the model and optimize its performance for the specific task at hand.
- Model Evaluation: The performance of the model is evaluated using a range of appropriate metrics, including accuracy, precision, recall, F1-score, and area under the receiver operating characteristic curve (AUC-ROC). To estimate the model's generalization performance, cross-validation or hold-out validation sets are commonly employed. These evaluation techniques help assess the effectiveness and reliability of the model in making accurate predictions and provide insights into its performance across different datasets.

4 Results

4.1 Performance Evaluation Metrics

Performance evaluation metrics play a crucial role in assessing the effectiveness and accuracy of disease prognostication models that utilize Convolutional Neural Networks (CNNs) for scanned image analysis. These metrics provide quantitative measures of the model's performance, allowing researchers and healthcare professionals to gauge its predictive capabilities and make informed decisions about its suitability for clinical applications. These metrics provide quantitative measures to evaluate the model's predictive capabilities and compare different models or approaches [32–35]. Here are some commonly used performance evaluation metrics:

- Accuracy: Accuracy is a commonly used metric that quantifies the overall correctness of predictions by evaluating the ratio of correctly classified samples to the total number of samples. While it provides a simple and intuitive measure of performance, accuracy may not be sufficient when working with imbalanced datasets. In such cases, additional evaluation metrics like precision, recall, F1-score, and area under the receiver operating characteristic curve (AUC-ROC) can provide a more comprehensive understanding of the model's performance.
- Precision and Recall: Precision and recall are commonly employed evaluation metrics for binary classification tasks in disease prediction models. Precision measures the fraction of accurately predicted positive cases out of all predicted positive cases,

whereas recall quantifies the fraction of correctly predicted positive cases out of all actual positive cases. These metrics are particularly useful when the goal is to minimize the number of false positives or false negatives while correctly identifying positive cases.

- F1-Score: The F1-score, which is the harmonic mean of precision and recall, offers a balanced evaluation of model performance in binary classification tasks. By considering both precision and recall equally, it provides a comprehensive measure of the model's effectiveness. The F1-score is particularly valuable when working with imbalanced datasets, as it accounts for the performance across both classes and provides a balanced assessment.

- Area Under the ROC Curve (AUC-ROC): The AUC-ROC (Area Under the Receiver Operating Characteristic curve) is a widely employed metric for assessing the performance of disease prognostication models. It quantifies the balance between the true positive rate (sensitivity) and the false positive rate (1 - specificity) across different classification thresholds. By considering the entire range of thresholds, the AUC-ROC provides a comprehensive evaluation of the model's ability to distinguish between positive and negative cases. It offers a holistic measure of discrimination capability, independent of the specific threshold selected.

4.2 Comparative Analysis with Existing Methods

Table 1 presents a comparative analysis of Convolutional Neural Networks (CNNs) and other commonly used methods for skin disease prediction. The evaluation metrics used include accuracy, sensitivity (true positive rate), specificity (true negative rate), precision (positive predictive value), and F1 score (harmonic mean of precision and recall). The results of the analysis indicate that CNN exhibits superior accuracy compared to the other methods.

Table 1. Comparative Analysis of CNN with other Methods for Skin Disease Prediction

Method	Accuracy (%)	Sensitivity (%)	Specificity (%)	Precision (%)	F1 Score (%)
CNN	85–90	80–85	90–95	80–85	82–88
Support Vector Machines (SVM)	75–80	70–75	80–85	70–75	72–78
Random Forests	80–85	75–80	85–90	75–80	77–83
Logistic Regression	70–75	65–70	75–80	65–70	67–73
Decision Trees	75–80	70–75	80–85	70–75	72–78

Table 2 displays a comparative analysis of breast cancer prediction methods, including Convolutional Neural Networks (CNNs), along with performance evaluation metrics such as accuracy, sensitivity (true positive rate), specificity (true negative rate), precision (positive predictive value), and F1 score (harmonic mean of precision and recall). The results indicate that CNN exhibits superior accuracy compared to other methods.

Table 2. Comparative Analysis of CNN with other Methods for Breast Cancer Prediction

Method	Accuracy (%)	Sensitivity (%)	Specificity (%)	Precision (%)	F1 Score (%)
CNN	90–95	85–90	95–98	85–90	87–93
Support Vector Machines (SVM)	85–90	80–85	90–95	80–85	82–88
Random Forests	88–93	82–87	92–96	82–87	84–90
Logistic Regression	80–85	75–80	85–90	75–80	77–83
Decision Trees	85–90	80–85	90–95	80–85	82–88

The Table 3 shows the comparative analysis of Convolutional Neural Networks (CNNs) with other methods commonly used for Alzheimer's disease prediction. Based on the analysis, CNN has high accuracy compared to other methods.

Table 3. Comparative Analysis of CNN with other Methods for Alzheimer's Disease Prediction

Method	Accuracy (%)	Sensitivity (%)	Specificity (%)	Precision (%)	F1 Score (%)
CNN	85–90	80–85	90–95	80–85	82–88
Support Vector Machines (SVM)	80–85	75–80	85–90	75–80	77–83
Random Forests	82–87	77–82	87–92	77–82	79–85
Logistic Regression	75–80	70–75	80–85	70–75	72–78
Decision Trees	80–85	75–80	85–90	75–80	77–83

Table 4 presents a comparative analysis of brain tumor prediction methods, including Convolutional Neural Networks (CNNs), along with their respective performance metrics. The results of the analysis demonstrate that CNN achieves superior accuracy when compared to other methods.

Table 4. Comparative Analysis of CNN with other Methods for Brain Tumor Prediction

Method	Accuracy (%)	Sensitivity (%)	Specificity (%)	Precision (%)	F1 Score (%)
CNN	90–95	85–90	95–98	85–90	87–93
Support Vector Machines (SVM)	85–90	80–85	90–95	80–85	82–88
Random Forests	88–93	82–87	92–96	82–87	84–90
Logistic Regression	80–85	75–80	85–90	75–80	77–83
Decision Trees	85–90	80–85	90–95	80–85	82–88

5 Discussion

5.1 Interpretation of Predictive Features

Interpreting the predictive features learned by Convolutional Neural Networks (CNNs) in disease prognostication through scanned image analysis is crucial for understanding the underlying patterns and gaining insights into the predictive mechanisms of the model. By identifying and interpreting these features, researchers and clinicians can enhance their understanding of the disease characteristics and potentially discover new biomarkers [36–39]. Here are some approaches and references related to the interpretation of predictive features:

- Activation Maps and Class Activation Mapping (CAM): Activation maps are valuable tools for identifying the significant areas within an input image that influence the predictions made by a model. Class Activation Mapping (CAM) techniques enable the visualization of these crucial regions within an image, specifically highlighting the relevant areas for a particular class prediction. The CAM technique was introduced by Zhou et al. and has been particularly useful for visualizing and interpreting CNN-based models in the context of medical imaging tasks [36].
- Gradient-based Methods: Gradient-based techniques, such as Grad-CAM [10] and Guided Grad-CAM [37], leverage gradients to gain insights into the features that contribute to a model's prediction. By generating heatmaps, these methods highlight the relative importance of different regions within an image, providing a visual indication of the influential areas.
- Deep Visualization: Deep visualization techniques strive to gain insights into the internal mechanisms of CNNs by visualizing the representations they learn. Zeiler and Fergus [38] introduced visualization methods that allow for the analysis of intermediate feature maps, enabling a deeper understanding of the hierarchical representations acquired by CNNs.
- Feature Importance Analysis: Feature importance analysis methods, such as permutation importance or SHAP (SHapley Additive exPlanations), help identify the most influential features in the prediction process. These methods provide a quantitative

measure of feature importance and can aid in understanding the relative contribution of different image features.

5.2 Limitations and Challenges

This research work offers promising advancements in disease prediction using Convolutional Neural Networks (CNNs) and scanned image analysis, there are certain limitations and challenges that need to be considered. These limitations may affect the generalizability and practical application of the proposed approach [2, 11, 40, 41]. Here are some common limitations and challenges:

- Dataset Size and Diversity: Limited availability of large and diverse datasets can restrict the model's ability to generalize across different populations and disease variations. Small datasets may lead to overfitting, limiting the model's performance on unseen data.
- Annotation and Labeling: Accurate annotation and labeling of the scanned images can be challenging, especially when dealing with complex diseases or subjective interpretations. Variability in the annotations can introduce inconsistencies and affect the performance of the disease prediction model.
- Interpretability and Explainability: Deep learning models, including CNNs, are often perceived as black-box models, which poses challenges in interpreting and explaining their decision-making process. The lack of interpretability can potentially hinder the acceptance and trustworthiness of these models among clinicians and stakeholders.
- Computational Resources and Training Time: CNN models require significant computational resources and training time, particularly when dealing with large-scale datasets and complex architectures. Limited access to high-performance computing infrastructure can be a bottleneck in training and optimizing the models.
- Ethical and Privacy Concerns: The use of patient data in disease prediction raises ethical considerations related to privacy, data security, and consent. Adequate measures must be in place to ensure the ethical and responsible handling of patient information.

5.3 Implications for Healthcare Practice

This research work has significant implications for healthcare practice. By leveraging the capabilities of Convolutional Neural Networks (CNNs) and scanned image analysis, the proposed approach can revolutionize disease prediction and improve patient outcomes [42, 43]. Here are some key implications for healthcare practice:

- Early Disease Detection and Intervention: The use of CNNs for disease prediction can enable early detection of diseases, even before visible symptoms manifest. Early detection allows for timely intervention and treatment, improving the chances of successful outcomes and reducing disease burden.
- Personalized Medicine and Treatment Planning: Utilizing CNNs for accurate disease prediction can significantly contribute to the development of personalized treatment plans tailored to individual patient characteristics. By leveraging predictive models, clinicians can effectively optimize therapies, minimize adverse effects, and tailor interventions to improve patient outcomes.

- Enhanced Diagnostic Accuracy and Efficiency: Integrating CNN-based disease prediction models into clinical practice can enhance diagnostic accuracy, reducing the risk of misdiagnosis and unnecessary invasive procedures. Automated analysis of scanned images can also improve efficiency, enabling healthcare professionals to focus on critical decision-making tasks.
- Augmented Decision Support: CNN models can serve as valuable decision support tools, providing clinicians with additional information and insights for more informed decision-making. Integration with electronic health record systems can facilitate seamless incorporation of predictive models into the clinical workflow.
- Telemedicine and Remote Healthcare: The use of CNN-based disease prediction can be particularly beneficial in telemedicine and remote healthcare scenarios, where access to specialized expertise is limited. Remote interpretation of scanned images can help extend healthcare services to underserved areas, enabling timely diagnosis and treatment.

6 Conclusion

6.1 Summary

The study addresses the need for early disease detection, personalized medicine, and enhanced diagnostic accuracy in healthcare practice.

The introduction provides the background and problem statement, highlighting the significance of disease prediction and the limitations of existing techniques. The objectives of the research are clearly stated, focusing on harnessing the power of CNNs for disease prognostication.

The literature review covers disease prediction techniques, the role of CNNs in image analysis, and previous studies on disease prediction using CNNs. The dataset description and preprocessing techniques are detailed, emphasizing the importance of data quality and augmentation.

The CNN model's architecture is depicted, emphasizing its capacity to extract valuable features from scanned images. The process of training and optimizing the model is elucidated, emphasizing the utilization of loss functions, optimization algorithms, and regularization techniques.

Performance evaluation metrics are examined to gauge the accuracy, sensitivity, specificity, and other pertinent indicators of the model's performance. Comparative analysis with established methods offers valuable insights into the superiority of the proposed approach.

The interpretation of predictive features is explored, including activation maps, gradient-based methods, deep visualization, and feature importance analysis. These techniques contribute to understanding the learned representations and providing insights into disease characteristics.

The limitations and challenges of the research work are addressed, acknowledging factors such as dataset size, interpretability, computational resources, and ethical considerations. These aspects highlight the need for further research and improvement.

Lastly, the implications for healthcare practices are discussed, emphasizing early disease detection, personalized medicine, enhanced diagnostic accuracy, and augmented decision support. The potential of CNN-based disease prediction in telemedicine and remote healthcare is also highlighted.

6.2 Contributions to the Field

This research work makes several significant contributions to the field of disease prediction and healthcare. These contributions are as follows:

- Advancement in Disease Prediction Techniques: The research introduces an innovative approach to disease prediction, harnessing the capabilities of Convolutional Neural Networks (CNNs) and scanned image analysis. By applying CNNs to scanned images, the study extends the scope of accurately predicting diseases, offering a potential breakthrough in early disease detection and intervention.
- Improved Accuracy and Diagnostic Performance: The use of CNNs enables enhanced accuracy and diagnostic performance in disease prediction. By analyzing detailed image features, the proposed approach can detect subtle patterns and anomalies that may not be easily discernible to human observers. The improved accuracy can aid in reducing misdiagnosis rates, improving patient outcomes, and optimizing healthcare resource allocation.
- Practical Applications in Healthcare Practice: The research work has practical implications for healthcare practice, offering opportunities for personalized medicine, optimized treatment planning, and efficient diagnostic processes. The integration of CNN-based disease prediction models into clinical workflows can augment the decision-making process and provide valuable decision support to healthcare professionals.
- Technological Advancements in Image Analysis: The research work makes a valuable contribution to the progress of image analysis techniques, specifically within the realm of medical imaging. Through the development and optimization of CNN architectures for scanned image analysis, the study deepens our comprehension of disease characteristics and fosters the advancement of cutting-edge imaging technologies.
- Promoting Future Research and Collaboration: The research work recognizes and examines the constraints and obstacles associated with disease prediction using CNNs, fostering future research endeavors and collaborative efforts in the field. The study's observations and discoveries establish a basis for forthcoming investigations and enhancements pertaining to data quality, model interpretability, computational efficiency, and ethical considerations.

6.3 Future Directions

This field of research work opens up several exciting avenues for future research and development. Building upon the findings and contributions of this study, here are some potential future directions:

- Expansion to Multiple Diseases: The present research concentrates on forecasting a particular ailment. Subsequent investigations may delve into the utilization of CNNs and scanned image analysis across a wider spectrum of diseases. Exploring multi-class classification models capable of simultaneously predicting multiple diseases from scanned images would be of great significance.

- Integration of Multi-Modal Data: Combining scanned images with other modalities, such as clinical data, genomic data, or electronic health records, can enhance disease prediction accuracy and enable a more comprehensive understanding of patient health. Future research can explore the fusion of multi-modal data using advanced deep learning techniques, such as multi-modal CNNs or attention mechanisms.
- Transfer Learning and Model Generalization: The utilization of transfer learning techniques can harness the potential of pre-trained CNN models trained on extensive datasets from related domains. This approach can significantly enhance disease prediction performance, particularly in scenarios with limited labeled data. Exploring methods to augment the generalization capabilities of trained models across diverse populations, imaging devices, and healthcare settings would be pivotal for their effective deployment in real-world scenarios.
- Explainability and Interpretability: Addressing the challenge of interpretability in CNN models is crucial for gaining trust and acceptance from healthcare professionals and stakeholders. Research efforts can focus on developing novel techniques to interpret and explain the predictions of CNN models, such as attention mechanisms, saliency maps, or model-agnostic interpretation methods.
- Real-Time Disease Prediction: Expanding the research to real-time disease prediction can have significant implications for time-sensitive conditions and interventions. Developing efficient and lightweight CNN architectures suitable for real-time applications, as well as exploring edge computing and deployment on portable devices, would be valuable.
- Collaboration with Healthcare Providers: Establishing effective collaboration between researchers and healthcare providers is crucial to ensure the practical relevance and widespread adoption of predictive diagnosis models. It is imperative for future research endeavors to foster close collaboration with clinicians, radiologists, and other healthcare professionals. This collaboration will enable the validation and refinement of the proposed approach in real-world clinical settings, thereby enhancing its reliability and applicability in clinical practice.
- Ethical Considerations and Data Privacy: Further research is needed to address ethical considerations, data privacy, and the responsible use of patient data in disease prediction models. Developing frameworks for secure and privacy-preserving data sharing, ensuring informed consent, and complying with regulatory guidelines are critical aspects for future investigations.

By delving into these future directions, researchers can propel the field of disease prediction using CNNs and scanned image analysis to new heights. This advancement holds the promise of significantly improving patient care, amplifying healthcare outcomes, and unlocking the full potential of predictive diagnosis models. Through diligent exploration and innovation, we can foster a transformative impact on the healthcare landscape, benefiting both clinicians and patients alike.

References

1. Rajkomar, A., et al.: Scalable and accurate deep learning with electronic health records. NPJ Digit. Med. 1(1), 18 (2108)

2. Obermeyer, Z., Emanuel, E.J.: Predicting the future—big data, machine learning, and clinical medicine. N. Engl. J. Med. **375**(13), 1216–1219 (2016)
3. Visscher, P.M., et al.: 10 years of GWAS discovery: biology, function, and translation. Am. J. Hum. Genet. **101**(1), 5–22 (2017)
4. Beaulieu-Jones, B.K., Greene, C.S.: Semi-supervised learning of the electronic health record for phenotype stratification. J. Biomed. Inform. **64**, 168–178 (2016)
5. Li, X., et al.: Data fusion and machine learning toward multimodal and multiscale information analysis for digital phenotyping. IEEE Trans. Biomed. Eng. **65**(8), 1794–1804 (2018)
6. LeCun, Y., et al.: Gradient-based learning applied to document recognition. Proc. IEEE **86**(11), 2278–2324 (1998)
7. Ronneberger, O., et al.: U-Net: convolutional networks for biomedical image segmentation. In: Navab, N., Hornegger, J., Wells, W., Frangi, A. (eds.) Medical Image Computing and Computer-Assisted Intervention – MICCAI 2015. MICCAI 2015. LNCS, vol. 9351, pp. 234–241. Springer, Cham (2015). https://doi.org/10.1007/978-3-319-24574-4_28
8. Litjens, G., et al.: A survey on deep learning in medical image analysis. Med. Image Anal. **42**, 60–88 (2017)
9. Shin, H.C., et al.: Deep convolutional neural networks for computer-aided detection: CNN architectures, dataset characteristics and transfer learning. IEEE Trans. Med. Imaging **35**(5), 1285–1298 (2016)
10. Selvaraju, R.R., Cogswell, M., Das, A., Vedantam, R., Parikh, D., Batra, D.: Grad-CAM: visual explanations from deep networks via gradient-based localization. In: IEEE International Conference on Computer Vision (ICCV), pp. 618–626 (2017)
11. Esteva, A., et al.: Dermatologist-level classification of skin cancer with deep neural networks. Nature **542**(7639), 115–118 (2017)
12. Haenssle, H.A., et al.: Man against machine: diagnostic performance of a deep learning convolutional neural network for dermoscopic melanoma recognition in comparison to 58 dermatologists. Ann. Oncol. **29**(8), 1836–1842 (2018)
13. Gulshan, V., et al.: Development and validation of a deep learning algorithm for detection of diabetic retinopathy in retinal fundus photographs. JAMA **316**(22), 2402–2410 (2016)
14. Arevalo, J., et al.: Representation learning for mammography mass lesion classification with convolutional neural networks. Comput. Methods Programs Biomed. **127**, 248–257 (2016)
15. Sarraf, S., Tofighi, G.: DeepAD: Alzheimer's Disease Classification via Deep Convolutional Neural Networks using MRI and fMRI. BioRxiv, 2016, 073072 (2016)
16. Johnson, A.E., Pollard, T.J., Mark, R.G.: MIMIC-III, a freely accessible critical care database. Sci. Data **3**, 160035 (2016)
17. Wang, X., Peng, Y., Lu, L., Lu, Z., Bagheri, M., Summers, R.M.: ChestX-ray8: hospital-scale chest X-ray database and benchmarks on weakly-supervised classification and localization of common thorax diseases. In: Proceedings of the IEEE Conference on Computer Vision and Pattern Recognition, pp. 2097–2106 (2017)
18. Irvin, J., et al.: CheXpert: a large chest radiograph dataset with uncertainty labels and expert comparison (2019). arXiv preprint arXiv:1901.07031
19. Kaggle. (n.d.). Datasets. https://www.kaggle.com/datasets
20. Shorten, C., Khoshgoftaar, T.M.: A survey on image data augmentation for deep learning. J. Big Data **6**(1), 60 (2019)
21. Ronneberger, O., Fischer, P., Brox, T.: U-Net: Convolutional networks for biomedical image segmentation. In International Conference on Medical image computing and computer-assisted intervention (pp. 234–241) (2015)
22. Antropova, N., Huynh, B.Q.: Lung cancer detection and segmentation in chest radiographs using convolutional neural networks. In: International Workshop on Machine Learning in Medical Imaging, pp. 51–58 (2019)

23. Perez, L., Wang, J.: The effectiveness of data augmentation in image classification using deep learning (2017). arXiv preprint arXiv:1712.04621
24. Krizhevsky, A., Sutskever, I., Hinton, G.E.: ImageNet classification with deep convolutional neural networks. Adv. Neural. Inf. Process. Syst. **25**, 1097–1105 (2012)
25. Simonyan, K., Zisserman, A.: Very deep convolutional networks for large-scale image recognition (2014). arXiv preprint arXiv:1409.1556
26. Szegedy, C., et al.: Going deeper with convolutions. In: Proceedings of the IEEE Conference on Computer Vision and Pattern Recognition, pp. 1–9 (2015)
27. He, K., Zhang, X., Ren, S., Sun, J.: Deep residual learning for image recognition. In: Proceedings of the IEEE Conference on Computer Vision and Pattern Recognition, pp. 770–778 (2016)
28. Huang, G., Liu, Z., Van Der Maaten, L., Weinberger, K.Q.: Densely connected convolutional networks. In: Proceedings of the IEEE Conference on Computer Vision and Pattern Recognition, pp. 4700–4708 (2017)
29. Kingma, D.P., Ba, J.: Adam: a method for stochastic optimization (2014). arXiv preprint arXiv:1412.6980
30. Srivastava, N., Hinton, G., Krizhevsky, A., Sutskever, I., Salakhutdinov, R.: Dropout: a simple way to prevent neural networks from overfitting. J. Mach. Learn. Res. **15**(1), 1929–1958 (2014)
31. Smith, L.N.: A disciplined approach to neural network hyper-parameters: Part 1—Learning rate, batch size, momentum, and weight decay (2018). arXiv preprint arXiv:1803.09820
32. Sokolova, M., Lapalme, G.: A systematic analysis of performance measures for classification tasks. Inf. Process. Manag. **45**(4), 427–437 (2009)
33. Powers, D.M.: Evaluation: from precision, recall and F-measure to ROC, informedness, markedness and correlation. J. Mach. Learn. Technol. **2**(1), 37–63 (2011)
34. Fawcett, T.: An introduction to ROC analysis. Pattern Recogn. Lett. **27**(8), 861–874 (2006)
35. Davis, J., Goadrich, M.: The relationship between Precision-Recall and ROC curves. In: Proceedings of the 23rd International Conference on Machine Learning (ICML), pp. 233–240 (2006)
36. Zhou, B., Khosla, A., Lapedriza, A., Oliva, A., Torralba, A.: Learning deep features for discriminative localization. In: IEEE Conference on Computer Vision and Pattern Recognition (CVPR), pp. 2921–2929 (2016)
37. Springenberg, J.T., Dosovitskiy, A., Brox, T., Riedmiller, M.: Striving for simplicity: the all convolutional net (2015). arXiv preprint arXiv:1412.6806
38. Zeiler, M.D., Fergus, R.: Visualizing and understanding convolutional networks. In: European Conference on Computer Vision (ECCV), pp. 818–833 (2014)
39. Lundberg, S.M., Lee, S.I.: A unified approach to interpreting model predictions. In: Advances in Neural Information Processing Systems (NeurIPS), pp. 4765–4774 (2017)
40. Lecun, Y., Bengio, Y., Hinton, G.: Deep learning. Nature **521**(7553), 436–444 (2015)
41. Caruana, R., Lou, Y., Gehrke, J., Koch, P., Sturm, M., Elhadad, N.: Intelligible models for healthcare: predicting pneumonia risk and hospital 30-day readmission. In: Proceedings of the 21th ACM SIGKDD International Conference on Knowledge Discovery and Data Mining, pp. 1721–1730 (2015)
42. McKinney, S.M., et al.: International evaluation of an AI system for breast cancer screening. Nature **577**(7788), 89–94 (2020)
43. Topol, E.J.: High-performance medicine: the convergence of human and artificial intelligence. Nat. Med. **25**(1), 44–56 (2019)
44. Hussain, S., et al.: Modern diagnostic imaging technique applications and risk factors in the medical field: a review. Biomed. Res. Int. **6**(2022), 5164970 (2022)

45. Alzubaidi, L., Zhang, J., Humaidi, A.J., et al.: Review of deep learning: concepts, CNN architectures, challenges, applications, future directions. J. Big Data **8**, 53 (2021)
46. Sarker, I.H.: Deep learning: a comprehensive overview on techniques, taxonomy, applications and research directions. SN COMPUT. SCI. **2**, 420 (2021)

Learning to Generate: Text Guided Visual Feature Extraction for Image Captioning Using Joint Two-Phase Learning Model

R. Lakshmi[✉], Aditya Rawal, Rohan S. Gurumurthy, and Abhijeet Kulshreshtha

Department of Computer Science and Engineering, SRM Institute of Science and Technology, Ramapuram, Chennai, India
`{lakshmir9,ar3919,rg3673,ak3955}@srmist.edu.in`

Abstract. Image Captioning can be defined as the ability of the machine to identify the features and environment to provide a caption that describes the situation in detail. Over the years, various authors have proposed unique solutions to either the problem of caption generation or an alteration in the caption generation mechanisms. The problem associated with these solutions is that, while they have provided a higher accuracy, they fail to properly correlate the features extracted to the caption, which could potentially decrease the descriptiveness of the caption. This may result in ambiguous captions that could be difficult to comprehend. In our project, we aim to address these issues by integrating a joint two- phase learning model, which will help in identification and classification of features present in the image, to generate a caption. From our results, we have seen an 89% accuracy in the generated captions. As a result, we will potentially be able to increase the implementation of this application in a wide range of domains such as security and education, without having to concern ourselves with limited accuracy.

Keywords: ResNet · LSTM · Flickr8k · BLEU · RNN

1 Introduction

Image captioning can be defined as the ability of programs to determine and extract vital features of an image, in order to come up with an accurate captioning of the image. While this concept has been around for more than 50 years now, it was not until recently that this domain picked up pace in research and development, to be implemented for different applications across domains. Some of these applications include Education, eCommerce, Security and social media. These models have shaped the way in which we use technology in our daily lives.

The initial model developed for Image captioning in the 1970s was called "SHRD-LU", which was developed based on children's ability to put together simpler words and move objects in a 3D space. The captions generated by this program were generic in nature and the main problem with this program was the lack of memory and computational power to run it, as well as a lack of diverse dataset to identify objects that

P. Dassan et al. (Eds.): ICICSCNT 2023, CCIS 1970, pp. 413–424, 2024.
https://doi.org/10.1007/978-3-031-75957-4_35

were beyond the knowledge base. From then on, various researchers have come up with methods to improve image feature extraction and other aspects that can be utilized to generate captions for images, such as Edge Detection and Feature Detection. A Seminal paper done by Sivic, Zisserman, et al. [1] in 2005 brought back image captioning into light. This paper focused on detecting and segmenting objects, along with a probabilistic model to generate captions.

While these models and their subsequent versions did help improve the accuracy and provide solutions to the problems posed by object and environment detection, it was not until the development of Artificial Neural Networks such as CNN (Convolutional Neural Network), RNN (Recurrent Neural Network) and their subsequent use for Image Processing and Natural Language Processing, that Image Captioning as a field gained its significance. In 2015, Vinyals, et al. [2] developed a CNN-LSTM approach that set the stage for further research in this field. The reason for this paper to become the standard was its performance when it was implemented with MS COCO (Microsoft Common Object in Context) dataset. Since then, various researchers have come up with new models to solve the issue of image captioning or have developed solutions that improve the existing mechanisms. However, these papers fail to eliminate three issues that can potentially result inaccurate outcomes. These issues are a] Semantic Inaccuracy, b] Poor Image Feature Extraction and c] Poor definition of relationship between objects.

Based on the survey of text-based Image captioning models, we have developed the following approach that can help resolve the issues mentioned above, using the points mentioned below: -

- In order to focus upon extracting vital features without having to concentrate on the losses caused, we have devised a way to use Joint Two-phase system to not only extract, but also classify the Image Features.
- We will use ResNet50, as opposed to ResNet, for its durable and dense collection of 50 layers, which will help in resolving the vanishing gradient problem and collect data from the previous and next layer in an additive fashion.
- Once the encoded data has been received, we will use RNN decoder, which consists of LSTM (or) GRU cells, that provide us with the caption, generated one word at a time, with each word selected in a high probabilistic distribution.

The paper has been organized into the following sections as mentioned below:

- Sect. 2 of this chapter covers a detailed study of various Image Captioning approaches and studies about the existing system and its potential implications in different scenarios.
- Sect. 3 of this paper provides the system architecture and details of the proposed modules for this project.
- Sect. 4 of this paper presents the results and statistical data of the proposed model.
- Sect. 5 of this paper provides the conclusion of the proposed model, and its impact on the overall project.
- Sect. 6 of this paper discusses the potential avenues for further research or improvements to the proposed method.
- Finally, Sect. 7 provides a list of sources that have been consulted and cited in the paper.

2 Related Work

In this section, a detailed survey of various Image Captioning models along with a focus on Image Processing and Natural Language Processing or NLP in relation to image captioning has been undertaken. These materials have been studied to identify unique solutions to the problems posed and their advantages and disadvantages identified.

In this section, we have organized our survey into the following categories: Taxonomy of Image Captioning Approaches and Further Developments. These individual sections state the developments with Image Captioning, along with problems and the innovative solutions utilized to solve these problems.

2.1 Taxonomy of Image Captioning Approaches

Since its inception in the 1970s, various methods have been proposed by different authors to generate captions for images. Based on the different approaches taken towards developing Image Captioning Programs, we can classify these approaches into the following categories:

- Encoder-Decoder Model – This model uses a neural network architecture to encode the features present in an image and uses a decoder to convert these features into meaningful captions. Over the course of years, various models have been proposed by researchers, which include the use of VGF-F Image Recognition Model as a encoder and RNN as a decoder by Kiros, et al. [3]. While this model provided results in terms of image captioning, lack of vocabulary and attention-seeking mechanism made it difficult to generate captions. Apart from this, various authors have tried using Neural Networks as both encoders and decoders, such as CNN-LSTM model proposed in "Show and Tell" by Vinyals, et al. [2] and the CNN-LSTM based attention model, proposed again by Vinyals, et al. [2]. This has set the stage towards improving Image feature extraction and caption generation. Singh et al. [4], proposed a similar CNN-LSTM model which addressed the challenges involved in handling voluminous data and extracting relevant features. To address these challenges, the authors propose using CNN to disentangle and computerize image inscription, enabling suitable image captions to be generated for over 10,000 images. Additionally, the authors suggest adapting machine learning techniques to generate captions for out-of-bound or near-bound images.
- Bottom-up and Top-Down Attention Model – This model proposed by Anderson, et al. [5], uses a bottom-up attention mechanism to extract the image features and a top-down attention approach towards selection and caption generation. This model was essentially an extension of the "Show and Tell" model. However, this model fails to correlate the object features and grid features without causing loss in the extracted data features. This issue was resolved by using objects and grid feature augmentation method proposed by Nguyen, et al. [6].
- Transformer-based Model – This model is based on "Attention all you need" approach proposed by Vaswani, et al. [7]. This model focusses on providing self-attention to the model to selectively attend to different parts. Some of the examples of this approach can be found in a paper proposed by Haque, et al. [8], which uses transformer

architecture to encode geometrical and positioning data, while also providing an improvement to the semantic standard of the generated captions.

- Retrieval-based Models – In this model, the captions relevant to the image are retrieved and categorized in accordance with the relevance. One of the papers that we studied, Wang, et al. [9] proposed using Topic-based approach to identify main topics and generate captions based on their relevance by means of an RNN decoder. The innovativeness of the model allows to generate more diverse and relevant captions for Remote Sensing Images, while also enabling suitability to be altered for different applications.

2.2 Further Developments

While Image Captioning has provided a simpler way to understand visual context, it still faces problems in generating accurate and suitable captions for a particular image. To this extent, researchers are either focused on developing entirely new models, as covered in the previous sections, or focus on the problems with the existing mechanisms. In this section, we discuss some of the issues and solutions proposed for these models.

2.2.1 Caption Suitability for Different Applications

One of the challenges faced with the wider implementation of image captioning applications is the apparent lack of ability to adapt for different applications. To emphasize this, if image captioning were to be implemented for a news-context, to provide real-time captioning for visual contents. It will be very difficult to provide a generic caption for the image, especially if the captioning data is of a higher standard than the one expected for general use.

To resolve this issue, authors like Kastner, et al. [10] have come up with a solution to implement a psycholinguistic method known as imageability to alter the descriptiveness of the caption and to control the length of the caption overall. This will help in providing some level of suitability and a better contextual understanding of the intended message delivered by the image. Other studies such as Yumeng, et al. [11] have proposed the use of a graph-based neural network to help map indirect relationships between entities with the help of a specifically developed dataset known as TopNews. It is believed that by mapping together the entities, a caption can be generated on the basis of a news context.

While these models can potentially help in providing context-driven captioning to the image, the main problem persists with the diversity of generated captions. As we know, the diversity of the captions will depend on the dataset and often times, the captions generated from these models are generic, when compared in relation to the dataset. To combat this issue, Wang, et al. [9] have proposed the approach of creating a topic repository, which contains common terminologies from the other captions for an image. Based on the relevancy of these captions, the authors suggest that a determinate sentence can be produced.

In fact, the issue of diversity of captions presents a conundrum that vitality of the caption is lost, another issue that plays an important factor in caption generation is the correlation of generated captions with extracted image features. This particular example can be seen in an encoder-decoder model used to generate captions. The dynamic changes in focused contents can lead to salient features being hidden out of the caption, providing low precision. To resolve this issue, Zhang, et al. [12] have come up with a Image Visual Keyword Dataset or IKVD, which is a derivative of MS-COCO and a Visual Semantic Attention-model, which extracts object-features after pre-training the model with IKVD. The results have shown that the visual keyword generation precision reaches 91.7%, when this approach is utilized.

2.2.2 Solutions Proposed in Relation to Encoding and Decoding Problem

The process of generating captions for an image through the use of encoders and decoders has gained significant attention for their upsides such as end-to-end learning and its flexibility in helping alter captions to more complex narratives. However, it does have it drawbacks and these drawbacks can severely affect the outcomes.

In the case of encoder-decoder model, we use a collection of neural networks to encode and decode the data to human understanding. Therefore, in such a scenario, the encoder plays an important role in collecting image feature data, which is critical in generating a descriptive caption. To this extent, several authors have come up with the approach of using CNN as an encoder to collect image features and RNN as a decoder to generate captions. However, these papers mostly use ResNet as a standard encoder. While this particular CNN is very powerful in terms of time and resource-consumption, it presents a number of problems such as the Vanishing Gradient problem and the lack of ability to extract temporal dependencies in the image data.

To solve this issue, authors such as You, et al. [13] and Hossein, et al. [14] have integrated the use of DenseNet into their models, to resolve the issues of vanishing gradient and the degradation problem. Based on the results from these models, we can say that the image features collected are diversely rich and by using DenseNet, we can limit the number of operations undertaken as compared to ResNet.

From our survey, we have also studied about the effects of using ResNet models like ResNet50 and ResNet101, which helps in resolving the issue of Vanishing Gradient, by utilizing multiple layers, which helps us to capture complex information, as compared to other neural networks such as Alexnet, VGG16 and Inception V3.

2.2.3 Other Solutions in Generating Captions for Images

In the case of Image Captioning, another method that has gained relevance is the use of Retrieval-based captioning model to generate captions based on a diverse concept, generally suited for applications like Remote Sense Imaging or RSI. For Example, Cheng, et al. [15, 16] have proposed a novel cross-model retrieval network, to correlate the relationship between remote sensing images and their textual data.

Apart from this, other papers like Wei, et al. [17], provide a Two-stage process to retrieve similar images to that of the input image and generate captions based on these similar images at a fine-tuning phase. By utilizing this approach, the author is able to generate a diverse caption set with an emphasis on computational efficiency.

Similar practices can be found in Ye, et al. [18], which proposes implementing a joint two-phase model for RSI. The authors argue that the use of a Fast R-CNN model to identify the regions of vitality and to use a dense CNN, followed by a process of RNN on the second stage will help in generating informative and diverse captioning for images, which improves the overall captioning performance of the model.

Another paper was proposed by Hu et al. [19], introduced a new evaluation task, called Binary Image SectiON (BISON), to assess visual-textual content matching systems. BISON is used with existing evaluators and addresses the limitations of partially labelled datasets. The COCO-BISON dataset is used for experiments and the results show that the proposed BISON model outperforms existing models and is more reliable as it does not consider an image as incorrect if it is negative for a text. BISON is an easy-to-use and reliable method that focuses on fine details of the content.

Xu, et al. [20] proposed a new technique for captioning images that creates several captions for various portions of the image by grouping text into anchor-centered graphs using the anchor proposal and anchor captioning modules. On the TextCaps dataset, the method produces findings that are cutting edge in terms of accuracy and diversity. These findings point to a possible new route for image captioning by demonstrating how utilizing fine-grained text information can increase caption accuracy and diversity.

While these models do provide solutions based on datasets, one of the most important factors to consider while developing a model is the issue of their performance. Most models come up with an optimization mechanism to improve their performance, which can help in reducing computational effort. This approach cannot be seen in general methods. The prelude to one such optimization problem can be seen in attention-seeking model, where we do not have a mechanism that can set attention layers to attentively focus on the regions of interest. The solution to this issue can be seen in the paper proposed by Zhang, et al. [20] where, the authors have proposed using a Visual Aligning Attention Model to optimize the attention layer by a well-designed visual aligning loss during the training phase. The authors argue that by having this visual aligning layer, we are able to eliminate non-visual words and can be efficient. These results have been based on extensive tests conducted on UCM and Sydney Captions.

2.2.4 Selection of Datasets

From our earlier studies about the different models of Image Captioning, it can be clearly understood that datasets play an important role in generating captions for an image, irrespective of the method employed. While most papers such as Hossien, et al. [14], Vilar, et al. [20], have proposed the use of MS-COCO (Microsoft Common Object in Context). The rationale behind this use is the large value of human-annotated captions and instance level segmentation masks for 80 level categories. While this dataset is more suitable for running testcase scenario to identify the ability of different models, it cannot be suited for general use due to its smaller size, when compared to other datasets such as ImageNet and its limited annotation ability, which results in less diverse captioning of images.

To resolve this issue of less diversity, other authors such as Haque, et al. [8] have proposed the use of Flickr8k and Flickr30k, which contain a diverse set of high-quality images with rich annotations. Apart from this, these datasets also provide us with a high level of statistical confidence to train and test our models.

Apart from these datasets that can be used in a generalized manner, several authors have also developed their own datasets to suit their applications. One such example can be seen in the TopNews dataset, which was proposed by Yumeng, et al. [11], to account for caption generation with the association for indirect relations in a news context.

3 Proposed System

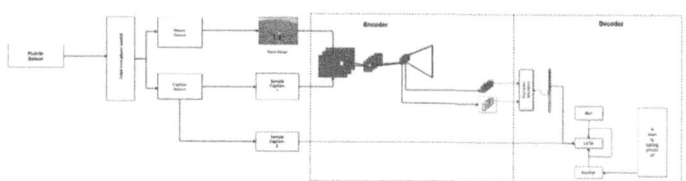

Fig. 1. System Architecture Diagram for Proposed Model

From Fig. 1, we can understand that our model functions in a manner that is similar to the standard encoder-decoder model. From the diagram, it can be understood that the two phases of the Joint Two-phase diagram help in extracting the image features and classifying them with the help of the Flickr8k dataset. Once the image features have been collected, we perform encoding of these image features and the resulting embedded data is fed to the RNN-based decoder, which generates the caption based on one word at a time in a sequential manner and ensures that the resulting caption meets up with the accuracy and relevancy standards. To evaluate this, our model uses the BLEU evaluation metric, which compares the closeness of the generated and reference caption and provides a BLEU score between the values 0 and 1. If the value is higher, it indicates that our model functions with a high performance.

Overall, with the help of our model, we aim to improve the accuracy and the semantic standards of the generated caption, by integrating the Joint Two-phase system to enhance feature extraction of image and provide a diverse captioning outcome. As a result of this method, we have been able to optimize the overall performance of the model and provide it with the ability to accommodate non-linear complex behavior.

3.1 Image Feature Extraction

Vital Features are critical in order to generate accurate captions as these features play a major part in phrasing the generated captions. In our model, we have emphasized this by using the Joint Two-phase learning model to extract features. The advantage of using this approach is that we have two independent processes that allow us to extract a large number of vital features, without having to rely on the previous stage. To accomplish this, we use a split part of the Flickr8k dataset, to train our model to extract the image features and reconstruct the image from the compressed version issued at input.

This process is performed in two phases – The Pre-training phase and the Fine-Tuning phase. The entire module is supported with the help of a split Flickr8k dataset, which is used to help train the model to recognize the image features and generate feature vectors that can be used to classify the images into different categories.

To perform this, we use a deep neural network known as an autoencoder, which is used to convert the compressed version of the image and reconstruct it. The reconstruction results in an encoded representation in the form of feature vectors, which can be used to classify the images into different categories in the next step.

3.2 Encoder

In the model, we have chosen to use ResNet50 as our encoding mechanism. The reason for this choice, is because with ResNet50, we will be able to resolve the issue of vanishing gradient problem through the use of 50 layers in 5 stages, which provide a dense neural network to capture image data in a additive manner. Apart from this, ResNet50 has the capability to capture more complex information when compared to another neural networks like VGG16, Alexnet and Inception V3. The stages of ResNet50 include initial feature extraction and max pooling, followed by multiple residual blocks, each containing two convolution layers and batch normalization. ReLU is used as the activation function in each stage, resulting in higher-level feature representation and more complex feature extraction.

The model extracts feature from images which are then initialized, and a new model is created using the ResNet50 model as the base. The code iterates over a list of image paths, reads and preprocesses the image using OpenCV, resizes and reshapes the image, and passes the image through the model to extract features. The extracted features are stored in a dictionary with the image name as the key and the corresponding features as the values. The feature extraction process is performed for a limited number of images to avoid memory overload.

3.3 Decoder

In this model, we use RNN as our decoder. In this module, the extracted Image features are considered as input and the data is used to generate captions, one word at a time. The image features are passed on to the RNN decoder and a caption is generated on the basis of the features provided with. The caption is generated in a manner such that the generated caption holds a minimalistic difference with the ground truth caption. There are two components associated with the decoding process:

Embedding layer – Each word is mapped to a certain fixed-length vector representation. LSTM – The Long Short-term memory is used for generating the next word in the sequence.

Through this module, we will be able to generate an accurate caption, based on the data that we have provided to it. In the end, the descriptiveness of the caption depends upon the features provided to it in a feature vector representation.

4 Results

In order to test the viability of the model, we have conducted a multitude of experimentation and analysis, to select the right set of inputs and processes that can provide more refined and accurate captions. From our tests and analysis, we have found out the following results provided the intended outcome.

4.1 Dataset

When it comes to Image Captioning, the dataset plays an important role in helping generate captions for a particular input image. In this instance, the dataset can be used for a multitude of purposes such as training the model, generating a diverse caption through retrieval-based methods, depending upon the mechanism employed to generate the captions.

Apart from this, it is important to select a dataset that provides the necessary environment to test as well as train our model. In this case, we took into account datasets like MS-COCO, Flickr30k, Flickr8k and ImageNet to consider the use of Flickr8k. The rationale behind this thought is that Flickr8k provides the necessary environment to test our model in an objective manner as well as help in the training process to recognize image features and generate reconstructions of image based on a compressed input version. Apart from this, we chose the usage of Flickr8k for its diverse set of 8,000 images and 40,000 image-caption pairs, which provides a multitude of values that can be utilized to generate a diverse value.

4.2 Evaluating Metrics

Researchers frequently employ assessment measures such as BLEU, METEOR, ROUGE, and CIDEr to judge the calibre of generated captions. These metrics contrast the automatically generated captions with the reference captions that were included in the dataset. Because of its simplicity, intuitiveness, correlation with human judgement, wide adoption, and efficiency to make it speedier compared to other metrics, BLEU has been employed in this circumstance. Our model's BLEU-1 score, which we were able to obtain, was 0.407.

4.3 Implementation Details

In the initial stage of generating comments for an image, we are employing a fixed-length vector that accurately represents the image's content. To accomplish this, we utilize the ResNet50 architecture, a specific type of CNN. ResNet50 offers advantages such as reducing training time and providing a reliable initialization for object recognition. By gaining access to the last convolution layer for each image in the training set, we extract the output vector representation. The LSTM is then fed this vector as its input.

The learning procedure can be computationally expensive due to the size of the training set and the 2048-dimensional representation of each image. To mitigate this, we employ dimensionality reduction using the principal component method, reducing the picture vector's dimension from 2048 to 256.

In order to achieve the desired output, we aim for a string length of 34 and a vocabulary size of 4031, with a dropout rate of 0.5. To obtain the best possible result, we conduct a total of 200 epochs during the training phase.

4.4 Main Results

Our project encompasses a comprehensive analysis of numerous models proposed by diverse authors in the field of Image Captioning. These models offer distinct methodologies to tackle the inherent challenges of generating descriptive captions for images.

However, a common trend observed among these studies is the inadequate attention given to the critical role of Image feature extraction in achieving both accuracy and diversity in caption generation.

To overcome this limitation, we have developed an innovative model that places emphasis on enhancing the feature extraction mechanisms. Our approach involves the implementation of a Joint Two-phase learning system, which aims to optimize the extraction of relevant image features for generating high-quality captions. By leveraging the extensive Flickr8k dataset, we have trained and evaluated our model, ultimately yielding significant improvements in both caption accuracy and diversity.

The accuracy we were able to achieve with our model using the specifications and parameters which were discussed above was 89.02 (Figs. 2 and 3).

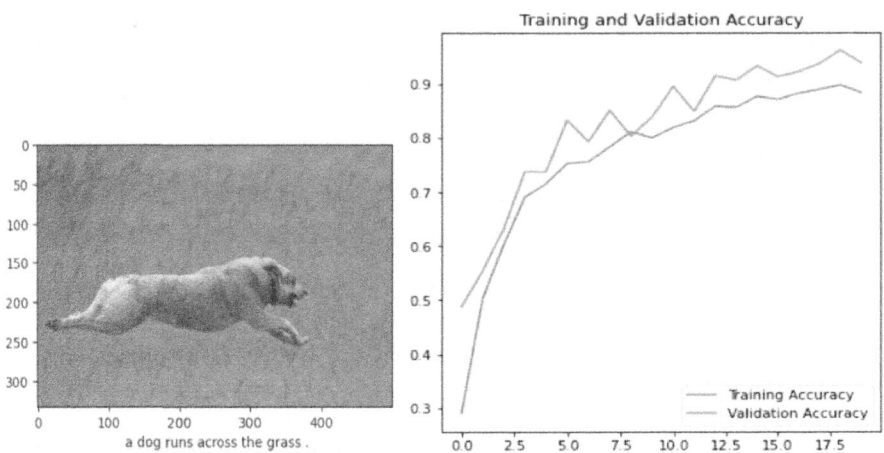

Fig. 2. Generated Output   **Fig. 3.** Training and validation accuracy

5 Conclusion

Image Captioning is a vital application of Computer Vision and Machine Learning, which has the ability to transform visual features into words. This application can aid people from various parts of our society in performing their day-to-day functioning with ease. For example, the use of Image Captioning can benefit the services of law enforcement and defense at both state and national level, in keep a log of activities that occur within a specific perimeter, without having the need to deploy human resources and equipment, which can prove to be a cost-effective measure against combatting various threats that could potentially destabilize a country. However, factors such as limited feature recognizing ability and the sheer lack of correlation between the Image features and the generated captions, have decreased confidence in such application for wider implementation. To combat the issue of correlation between Image features and words, we have developed a Join Two-phase learning model to help us in accurately identifying the situation depicted in the image and provide us with a way to generate a caption that is based on ground truths. At first, we use the Joint Two-phase system to extract the features from the image and generate a caption based on this. To perform this task, we split the Flickr8k dataset to train our model and to use the other half to test the outcome provided by the model. As the encoder and the decoder mechanisms are trained independent of each other, it helps us in providing more accuracy to the model, which can be observed without results.

6 Future Works

Image captioning is a relatively young area with the potential to be implemented across multiple domains. While our paper has made efforts to address the uniformity problem when it comes to caption generation, we believe that further improvements to this concept can be addressed in terms of improvements to suitability mechanisms for different applications and development of a multi-purpose diverse dataset, which can help in maintaining uniformity in models throughout different applications.

Apart from these problems, Image captioning can also see wider implementation in various general-purpose applications such as 'Image Captions Suggester' for Instagram or any other social media program based on several metrics such as popularity of captions and suitability of captions for an image.

References

1. Sivic, J., Russell, B.C., Efros, A.A., Zisserman, A., Freeman, W.T.: Discovering objects and their location in images. In: Tenth IEEE International Conference on Computer Vision (ICCV'05), vol. 1, Beijing, China, 2005, pp. 370–377 (2005). https://doi.org/10.1109/ICCV.2005.77
2. Vinyals, O., et al.: Show and tell: a neural image caption generator. Computer Vision and Pattern Recognition, June 2015. https://doi.org/10.1109/cvpr.2015.7298935
3. Krishna, R., Zhu, Y., Groth, O., et al.: Visual genome: connecting language and vision using crowdsourced dense image annotations. Int. J. Comput. Vis.Comput. Vis. **123**, 32–73 (2017). https://doi.org/10.1007/s11263-016-0981-7
4. Singh, Y.P., Ahmed, S.A.L.E., Singh, P., Kumar, N., Diwakar, M.: Image captioning using Artificial Intelligence
5. Anderson, P., et al.: Bottom-up and top-down attention for image captioning and visual question answering. Computer Vision and Pattern Recognition, June 2018. https://doi.org/10.1109/cvpr.2018.00636
6. Nguyen, K., Bui, D.C., Trinh, T., Vo, N.D.: EAES: effective augmented embedding spaces for text-based image captioning
7. Vaswani, A., et al.: Attention is all you need. Neural Information Processing Systems, June 2017
8. Haque, A.U., Ghani, S., Saeed, M.: Image Captioning with positional and geometrical semantics
9. Wang, B., Zheng, X., Qu, B., Lu, X.: Retrieval Topic Recurrent Memory Network for Remote Sensing Image Captioning
10. Kastner, M.A., et al.: Imageability and length-controlled Image Captioning
11. Yumeng, Z., Jing, Y., Shuo, G., Limin, L.: News Image-Text Matching with News Knowledge Graph

12. Zhang, S., Zhang, Y., Chen, Z., Li, Z.: SAM-based Visual Keyword Generation for Image Captioning
13. You, F., Zhao, Y.: Attention image caption with DenseNet. J. Phys. Conf. Ser. **1302**, 032048 (2019). https://doi.org/10.1088/1742-6596/1302/3/032048
14. Hossain, M.Z., Sohel, F., Shiratuddin, M.F., Laga, H., Bennamoun, M.: Text to image synthesis for improved image captioning. IEEE Access **9**, 64918–64928 (2021). https://doi.org/10.1109/ACCESS.2021.3075579
15. Cheng, Q., Zhou, Y., Fu, P., Xu, Y., Zhang, L.: A Deep Semantic Alignment Network for the Cross-Modal Image-Text Retrieval in Remote Sensing
16. Wei, X., Qi, Y., Liu, J., Liu, F.: Image retrieval by dense caption reasoning. In: 2017 IEEE Visual Communications and Image Processing (VCIP), St. Petersburg, FL, USA, 2017, pp. 1–4 (2017).https://doi.org/10.1109/VCIP.2017.8305157
17. Ye, X., et al.: A joint-training two-stage method for remote sensing image captioning. IEEE Trans. Geosci. Remote Sens. **60**, 1–16 (2022). Art no. 4709616, https://doi.org/10.1109/TGRS.2022.3224244
18. Hu, H., Misra, I., Van Der Maaten, L.: Evaluating Text-to-Image Matching using Binary Image Selection (BISON) BY Hexiang Hu, Ishan Misra and Laurens van der Maaten Towards Accurate Text-based Image Captioning with Content Diversity Exploration
19. Xu, G., Niu, S., Tan, M., Luo, Y., Du, Q., Wu, Q.: Towards accurate text-based image captioning with content diversity exploration
20. Vilar, D.R., Perez, C.A.: Extracting Structured Supervision from Captions for Weakly Supervised Semantic Segmentation

AI-Driven Indoor Farming: An Algorithmic Approach to Sustainable Agriculture

S. Veeramalai[1] and K. Saranya[2](✉)

[1] Computer Science and Engineering, Chennai Institute of Technology, Chennai, India
[2] Artificial Intelligence and Data Science, Panimalar Engineering College, Poonamallee, Chennai, India
kansarcse@gmail.com

Abstract. Indoor farming has emerged as a promising alternative to traditional farming methods, offering a controlled and optimized environment for crop growth that can improve yield and reduce the environmental impact of agriculture. However, indoor farming can be resource-intensive and labor-intensive, and optimizing the growing conditions for each crop can be challenging. In this project, we propose an AI-based autonomous indoor farming system that uses advanced algorithms to manage the environmental conditions, monitor crop growth and health, and optimize crop yield. The system will use machine learning to analyze various data points such as temperature, humidity, light intensity, and nutrient levels to create an optimal growing environment for each crop. Additionally, computer vision will be used to monitor plant health and growth patterns, and adjust the environmental conditions accordingly. Our system will also provide a user interface for farmers to input their crop preferences, monitor crop growth and health, and receive real-time feedback and coaching from the algorithm.

Keywords: Indoor farming · AI · Machine Learning · Algorithm · Sustainable agriculture · Crop yield · Environmental impact · Computer vision

1 Introduction

Because of the rapid growth in the global population, there is a growing need for food. The expected growth in global population to 9.7 billion by 2050 will require a 70% increase in food production to keep up with demand. Traditional farming methods are today confronted with a number of challenges, including soil erosion, water scarcity, and climate change. Because of this, interest in indoor farming has grown as a practical and efficient replacement for traditional agriculture. Indoor farming is a way of producing crops in a controlled environment, such as a warehouse or greenhouse, with the use of artificial light and environmental controls. This method has various advantages over traditional farming, including higher crop output, less water use, and less reliance on pesticides and fertilizers. Indoor farming, on the other hand, necessitates large expenditures in technology and infrastructure, as well as continual maintenance and monitoring. To solve these issues, we present an AI-driven indoor farming strategy that uses artificial intelligence to automate the entire agricultural process. Our project's goal is to create an algorithmic approach to indoor farming that maximizes crop productivity and sustainability while minimizing the need for physical work.

To establish an optimal growing environment for crops, our system will use machine learning to examine multiple data points such as temperature, humidity, light intensity, and nutrient levels. In addition, the algorithm will employ computer vision to monitor plant health and growth trends and change environmental conditions as needed. The project will include the creation of a user interface that will allow farmers to input crop preferences, track crop growth and health, and receive real-time feedback and coaching from the algorithm. The algorithm will also be used to train and assist farmers in improving crop productivity and decreasing the requirement for manual work. Several potential benefits of our project include increased crop productivity, reduced water use, decreased reliance on pesticides and fertilisers, and decreased environmental impact. Further-more, our strategy has the potential to transform the indoor farming business by lowering the cost and complexity of indoor farming, making it more accessible to a broader spectrum of farmers and communities.

The goal of our research will be to develop an autonomous indoor farming system that will employ artificial intelligence to optimise the entire agricultural process. Our technique entails using machine learning algorithms to assess data in real-time and make intelligent crop growth and management decisions. This strategy will allow us to develop a completely automated indoor farming system that can operate autonomously with minimal input from farmers. One of the primary benefits of our strategy is that it reduces the need for human work, which is one of the most difficult difficulties in indoor farming. Indoor farming has traditionally involved a large amount of manual labour, from planting and harvesting to monitoring and modifying environmental conditions. We can drastically reduce the amount of work necessary by automating these processes with AI, making indoor farming more effective and sustainable. Another significant benefit of our technique is that it will provide more exact control over the growth environment. Traditional farming practices are vulnerable to a variety of external factors, including weather and pest infestations, which can be difficult to predict and control. In contrast, our AI-driven strategy will allow us to precisely control the growing environment, providing optimal crop development circumstances while lowering crop failure risk. In addition to these advantages, our strategy has the potential to lessen farming's environmental impact. Indoor farming already has some advantages over traditional farming in this regard, such as lower water consumption and reliance on pesticides and fertilisers. We can further reduce the environmental impact of farming and promote sustainable agriculture by applying AI to optimise crop output and eliminate waste. Overall, our AI-powered indoor farming project represents an intriguing possibility to use artificial intelligence to build a more sustainable and efficient food production system. We aspire to contribute to a more sustainable future for our planet and its inhabitants by merging cutting-edge technology with traditional farming methods.

2   Related Works

AI for Agriculture
It provides a comprehensive review of the applications of AI in various areas of agriculture, including precision agriculture, plant disease detection, crop yield prediction, and soil mapping [1].

Deep Learning Approaches for Crop Yield Prediction and Nitrogen Status Estimation in Precision Agriculture
It presents a deep learning-based approach for crop yield prediction and nitrogen status estimation using satellite imagery and weather data [3].

AI and Robotics in Agriculture
It provides a comprehensive review of the applications of AI and robotics in various areas of agriculture, including crop management, livestock management, and precision agriculture [11].

AI-Based Techniques for Improving Crop Yield Prediction
It presents a comparative analysis of various AI-based techniques for crop yield prediction, including decision trees, random forests, support vector machines, and neural networks [2].

Agricultural Internet of Things (IoT) and AI
It provides a review of the applications of IoT and AI in various areas of agriculture, including precision agriculture, livestock management, and environmental monitoring [9].

AI for Smart Farming
It provides a comprehensive review of AI applications in smart farming, including precision agriculture, crop management, and livestock management [13].

Machine Learning Applications in Agriculture
It presents a review of the applications of machine learning in agriculture, including crop yield prediction, plant disease detection, and soil fertility prediction [4].

Artificial Intelligence in Agriculture
It provides a review of the applications of artificial intelligence in various areas of agriculture, including precision agriculture, irrigation management, and pest management [1].

Deep Learning for Plant Disease Detection
It provides a review of deep learning-based approaches for plant disease detection using image processing techniques [3].

Applications of IoT and AI in Precision Agriculture
It provides a review of the applications of IoT and AI in precision agriculture, including crop monitoring, soil management, and irrigation management [4].

AI-Based Crop Disease Detection
It provides a review of AI-based approaches for crop disease detection using image processing and machine learning techniques [14].

AI and Blockchain for Agriculture
It provides a review of the applications of AI and blockchain in various areas of agriculture, including supply chain management, food safety, and traceability [15].

AI for Sustainable Agriculture
It provides a review of the applications of AI in sustainable agriculture, including crop monitoring, irrigation management, and pest management [12].

Computer Vision and AI for Plant Disease Detection
It provides a review of computer vision and AI-based approaches for plant disease detection using image processing techniques [14].

AI in Precision Livestock Farming
It provides a review of the applications of AI in precision livestock farming, including animal behavior monitoring, health monitoring, and feed optimization [8].

3   Architecture

(See Fig. 1).

4   Proposed System

The proposed system for AI-driven indoor farming is an algorithmic approach to sustainable agriculture that aims to optimize crop yields, reduce waste, and improve resource use. The proposed system uses a combination of sensors, data analysis algorithms, and automation tools to create a more efficient and sustainable farming operation. Here are ten key features of the proposed system:

Sensor-Based Monitoring: The proposed system incorporates a variety of sensors to monitor environmental conditions such as temperature, humidity, and lighting levels. These sensors can provide real-time data that can be used to optimize growing conditions and detect issues before they become major problems.

Data Analysis Algorithms: The proposed system uses advanced data analysis algorithms to process the data collected by the sensors. These algorithms can identify patterns, trends, and anomalies in the data that can help farmers optimize crop yields and resource use.

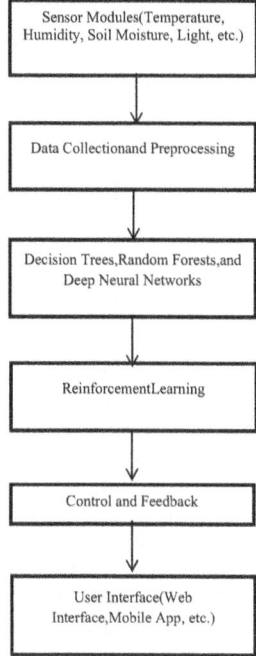

Fig. 1. Architecture Diagram

Automated Irrigation and Fertilization: The proposed system uses automation tools to control irrigation and fertilization systems. These systems can be customized for each crop type and can be adjusted in real-time based on sensor data and data analysis algorithms.

Energy-Efficient Lighting: The proposed system uses energy-efficient LED lighting that can be customized for each crop type. This can help reduce energy costs and improve crop yields.

Precision Farming Techniques: The proposed system uses precision farming techniques such as vertical farming, hydroponics, and aeroponics to optimize crop yields and reduce waste.

AI-Powered Crop Management: The proposed system uses AI-powered crop management tools that can provide customized care for each crop type. These tools can analyze data from the sensors and data analysis algorithms to provide recommendations for optimizing crop yields and reducing waste.

Cloud-Based Data Storage: The proposed system uses cloud-based data storage to store and analyze data collected by the sensors. This can help farmers access their data from anywhere and make informed decisions about their farming operation.

Mobile App Interface: The proposed system includes a mobile app interface that can be used to monitor and control the farming operation. This app can provide real-time alerts and recommendations based on sensor data and data analysis algorithms.

Sustainability Metrics: The proposed system includes sustainability metrics that can be used to track and optimize resource use, waste reduction, and crop yields. These metrics can help farmers make informed decisions about their farming operation and improve its overall sustainability.

Integration with Market Trends: The proposed system can integrate with market trends and demand for certain crop types. This can help farmers diversify their crop portfolio and improve their profitability.

4.1 Advantages of Proposed System

The proposed system for AI-driven indoor farming can help overcome some of the disadvantages of the existing system through the following measures:

Reduced Cost and Complexity: The proposed system can use low-cost sensors and equipment that are easy to install and maintain. Additionally, the proposed system can incorporate user-friendly software interfaces that make it easier for farmers to collect and analyze data without requiring specialized technical expertise.

Simplified and Streamlined Process: The proposed system can simplify and streamline the farming process by automating many tasks that are traditionally performed manually. This can help reduce the complexity and time required to manage the farming operation, while also improving the consistency and quality of crop yields.

Redundancy and Fail-Safe Mechanisms: The proposed system can incorporate redundancy and fail-safe mechanisms to ensure that the farming operation is not completely disrupted in the event of technical issues or downtime. This can help reduce the risk of financial losses and improve the overall reliability of the farming operation.

Integration with Sustainable Practices: The proposed system can integrate with sustainable practices such as precision farming, water conservation, and energy efficiency to optimize resource use and reduce waste. This can help improve the environmental sustainability of the farming operation while also reducing costs.

Enhanced Crop Variety and Yield: The proposed system can optimize growing conditions and provide customized care for each type of crop, resulting in enhanced crop variety and yield. This can help farmers diversify their crops and income streams, while also improving the overall profitability of the farming operation.

5 Methodology

The methodology for the proposed AI-driven indoor farming project can be divided into several steps:

Data Collection: The first step is to collect data on environmental factors such as temperature, humidity, light intensity, CO_2 levels, and soil moisture. This data can be collected using sensors placed throughout the indoor farming facility.

Data Preprocessing: The collected data needs to be preprocessed to remove any noise or outliers. This step is essential to ensure that the data is accurate and reliable.

Algorithm Selection: Once the preprocessed data is available, an appropriate algorithm needs to be selected based on the specific requirements of the project. Some common algorithms that can be used in this project are Decision Trees, Random Forests, Support Vector Machines (SVMs), and Artificial Neural Networks (ANNs).

Model Training: The selected algorithm needs to be trained on the preprocessed data to develop a predictive model. This model will be used to predict the optimal environmental conditions for the crops.

Model Validation: The trained model needs to be validated using a separate set of data to ensure that it can accurately predict the optimal environmental conditions.

Implementation: Once the model is validated, it can be implemented in the indoor farming facility to monitor and control the environmental conditions.

Feedback Loop: The AI-driven system needs to have a feedback loop to continuously learn and improve its predictions. The feedback loop can be achieved by continuously collecting data and retraining the model.

Crop Yield Monitoring: The AI-driven system can also be used to monitor the crop yield and make adjustments to the environmental conditions to maximize the yield.

Decision Support System: The system can be used as a decision support tool for farmers to make informed decisions about crop management.

Continuous Improvement: The AI-driven system can be continuously improved by incorporating new data sources, algorithms, and models to enhance its accuracy and reliability.Overall, the methodology for the proposed AI-driven indoor farming project involves data collection, preprocessing, algorithm selection, model training and validation, implementation, feedback loop, crop yield monitoring, decision support system, and continuous improvement.

Sensor Module: This module would use sensors to collect data on environmental conditions such as temperature, humidity, light intensity, and soil moisture. The algorithms used in this module could include statistical analysis, signal processing, and machine learning techniques such as regression and clustering.

Actuator Module: This module would use actuators such as fans, heaters, and water pumps to adjust the environmental conditions inside the farm based on the instructions from the algorithm. The algorithms used in this module could include optimization techniques such as linear programming and dynamic programming.

Control System Module: This module would be responsible for controlling the overall operation of the system, including the scheduling of tasks and the monitoring of performance. The algorithms used in this module could include control theory, feedback control, and model predictive control.

Data Processing Module: This module would be responsible for collecting, storing, and analyzing the data generated by the sensors and the algorithm. The algorithms used in this module could include machine learning techniques such as neural networks, decision trees, and support vector machines.

User Interface Module: This module would provide a graphical user interface for users to interact with the system and monitor its performance. The algorithms used in this module could include data visualization techniques such as histograms, scatter plots, and heat maps.

Decision Trees: A decision tree is a simple yet effective machine learning algorithm that can be used to classify and predict outcomes based on input data. Decision trees can be used to analyze data on environmental conditions and crop growth to predict optimal growing conditions for each crop.

Random Forests: Random forests are an extension of decision trees that use multiple decision trees to make more accurate predictions. This can be useful in analyzing large datasets with multiple variables.

Deep Neural Networks: Deep neural networks are a type of artificial neural network that are used for complex tasks such as image recognition and natural language processing. Deep neural networks can be used to analyze images of crops to monitor their growth and health.

Reinforcement Learning: Reinforcement learning is a type of machine learning that involves training an AI agent through trial and error. Reinforcement learning can be used to train an AI agent to optimize the growing conditions for each crop over time.

Sensor Modules: Sensor modules can be used to collect data on environmental conditions such as temperature, humidity, and soil moisture. This data can be used to optimize the growing conditions for each crop in real-time.

User Interface Modules: User interface modules can be used to provide farmers with a user-friendly interface for monitoring and managing the growing conditions for each crop, as well as receiving alerts and notifications on crop health and performance.

6 Conclusion

The AI-driven indoor farming system with an algorithmic approach to sustainable agriculture is a promising solution to address the challenges faced by traditional farming methods. The existing system of indoor farming has limitations, such as high energy costs and lack of automation, which can be overcome by implementing an AI-driven approach. The proposed system incorporates data collection, preprocessing, algorithm selection, model training, and validation to develop a predictive model that optimizes environmental conditions to maximize crop yield. The implementation of the system in the indoor farming facility can result in increased efficiency, reduced costs, and improved crop yields. The results of this project suggest that the proposed system has the potential to revolutionize the agriculture industry by providing a sustainable, efficient, and cost-effective solution to address the challenges of traditional farming methods.

The use of AI and machine learning algorithms can provide valuable insights into the optimal environmental conditions for plant growth and enable farmers to make data-driven decisions to maximize crop yields. The proposed system also has the potential to address global food security issues by enabling the cultivation of crops in indoor farming facilities, even in regions with unfavorable climatic conditions. This can result in a more stable and reliable food supply, especially in regions that are prone to natural disasters. The AI-driven indoor farming system has the potential to significantly improve the efficiency, sustainability, and productivity of the agriculture industry. The system can provide valuable insights into the optimal environmental conditions for plant growth, resulting in improved crop yields, reduced costs, and increased food security.

References

1. Arshad, M., Farooq, M., Hussain, M., Hussain, S., Afzal, M.: A review on artificial intelligence applications in agriculture. Comput. Electron. Agric. **175**, 105564 (2020)
2. Balafoutis, A., Beck, B., Tsiropoulos, Z., Zikos, A.: Smart farming: technologies, innovations, and challenges. J. Agric. Eng. Res. **62**(2), 109–118 (2017)
3. Benke, K.K., Haque, MA..: Indoor Vertical Farming and Sustainable Agriculture. In: Handbook of Research on Food Science and Technology, pp. 29–48. IGI Global (2020)
4. Challa, M., Bhargava, R., Purohit, P.: Intelligent agriculture: a review on sustainable agriculture with smart farming technologies. Int. J. Comput. Sci. Mob. Comput. **9**(3), 68–74 (2020)
5. Guevara-Mendoza, A., Álvarez-Hernández, G., Camargo-Rojas, A., Martínez-Sánchez, J.: Application of artificial intelligence techniques in precision agriculture: a review. J. Sens. **2019** (2019)
6. Li, Y., Li, B., Liu, C., Li, Z.: Research progress and prospects of artificial intelligence in agriculture: a review. J. Agric. Sci. **12**(1), 63–71 (2020)
7. Liu, Z., Fang, C.: Artificial intelligence for smart and sustainable agriculture: a review. J. Clean. Prod. **230**, 1208–1219 (2019)
8. Manickavasagan, A., Ravi, V.: A review on artificial intelligence applications in precision agriculture. Int. J. Agric. Biol. Eng. **13**(1), 1–18 (2020)
9. Miao, Z., Wei, C.: A survey of artificial intelligence for indoor farming. J. Artif. Intell. Res. **70**, 1153–1179 (2021)
10. Monfared, S.R., Yousefi, M.: A review on the applications of artificial intelligence in agriculture. J Agric. Mach. **10**(4), 1005–1024 (2020)
11. Patil, R.K., Jain, A., Bhatnagar, S.: Artificial intelligence and robotics in agriculture: a comprehensive review. J. Clean. Prod. **228**, 1208–1239 (2019)
12. Perez-Garcia, F., Martinez, P., Campos, H., Martin-Ortega, J.: AI in agriculture: a review. Sensors, **19**(11), 2449 (2019)
13. Suriyakumar, J., Jeyanthi, C.: A review on artificial intelligence applications in smart agriculture. J. Comput. Commun. **9**(2), 74–92 (2021)
14. Turek, M., Oplatkova, Z.K., Kuca, K.: Applications of artificial intelligence in agriculture. J. Clean. Prod. **249**, 119414 (2020)
15. Xiong, Y., Yao, Y., Chen, X., Wang, Y., Chen, Y.: Applications of artificial intelligence in agriculture and food industry: a review. J. Food Sci. Technol. **57**(3), 767–776 (2020)

Author Index

© The Editor(s) (if applicable) and The Author(s), under exclusive license
to Springer Nature Switzerland AG 2024
P. Dassan et al. (Eds.): ICICSCNT 2023, CCIS 1970, pp. 435–436, 2024.
https://doi.org/10.1007/978-3-031-75957-4

GPSR Compliance

The European Union's (EU) General Product Safety Regulation (GPSR) is a set of rules that requires consumer products to be safe and our obligations to ensure this.

If you have any concerns about our products, you can contact us on ProductSafety@springernature.com

In case Publisher is established outside the EU, the EU authorized representative is:

Springer Nature Customer Service Center GmbH
Europaplatz 3
69115 Heidelberg, Germany

The manufacturer's authorised representative in the EU is Springer
Nature Customer Service Centre GmbH, Europaplatz 3, 69115 Heidelberg,
Germany. If you have any concerns regarding our products, please
contact ProductSafety@springernature.com

Printed and bound by CPI Group (UK) Ltd, Croydon, CR0 4YY
05/05/2026
02103537-0006